Handbook of Indoor Air Quality

Handbook of Indoor Air Quality

Edited by Harlee Knight

SYRAWOOD
PUBLISHING HOUSE

New York

Published by Syrawood Publishing House,
750 Third Avenue, 9th Floor,
New York, NY 10017, USA
www.syrawoodpublishinghouse.com

Handbook of Indoor Air Quality
Edited by Harlee Knight

International Standard Book Number: 978-1-64740-432-1 (Hardback)

Cataloging-in-publication Data

Handbook of indoor air quality / edited by Harlee Knight.
    p. cm.
Includes bibliographical references and index.
ISBN 978-1-64740-432-1
1. Indoor air quality. 2. Air quality. 3. Air--Pollution.
4. Air quality management. I. Knight, Harlee.
TD883.17 .H36 2023
628.53--dc23

TABLE OF CONTENTS

PREFACE

Indoor air quality (IAQ) is refers to the air quality inside and around structures and buildings. It has been shown that IAQ has an impact on the comfort, health and well-being of the building inhabitants. It is measured by collecting samples from building surfaces and air, monitoring the exposure to contaminants, and evaluating airflow within the buildings through computer modeling. IAQ issues are mostly caused due to the indoor pollution sources that discharge particles or gases into the air. The three main causes of indoor air pollution are bioaerosols, combustion and building materials. Diseases such as lung cancer, asthma and chronic obstructive pulmonary disease (COPD) have been linked to poor IAQ. This book is a valuable compilation of topics, ranging from the basic to the most complex advancements in the study of indoor air quality. It is a vital tool for all researching or studying indoor air quality, as it gives incredible insights into emerging concepts.

This book unites the global concepts and researches in an organized manner for a comprehensive understanding of the subject. It is a ripe text for all researchers, students, scientists or anyone else who is interested in acquiring a better knowledge of this dynamic field.

I extend my sincere thanks to the contributors for such eloquent research chapters. Finally, I thank my family for being a source of support and help.

Editor

An Internet of Things-Based Environmental Quality Management System to Supervise the Indoor Laboratory Conditions

Gonçalo Marques and Rui Pitarma *

Unit for Inland Development, Polytechnic Institute of Guarda, Avenida Doutor Francisco Sá Carneiro N° 50, 6300-559 Guarda, Portugal; goncalosantosmarques@gmail.com
* Correspondence: rpitarma@ipg.pt.

Abstract: Indoor air quality (IAQ) is not only a determinant of occupational health but also influences all indoor human behaviours. In most university establishments, laboratories are also used as classrooms. On one hand, indoor environment quality (IEQ) conditions supervision in laboratories is relevant for experimental activities. On the other hand, it is also crucial to provide a healthy and productive workplace for learning activities. The proliferation of cost-effective sensors and microcontrollers along with the Internet of Things (IoT) architectures enhancements, enables the development of automatic solutions to supervise the Laboratory Environmental Conditions (LEC). This paper aims to present a real-time IEQ-laboratory data collection system-based IoT architecture named iAQ Plus (*iAQ+*). The *iAQ+* incorporates an integrated Web management system along with a smartphone application to provide a historical analysis of the LEC. The *iAQ+* collects IAQ index, temperature, relative humidity and barometric pressure. The results obtained are promising, representing a meaningful contribution for IEQ supervision solutions based on IoT. *iAQ+* supports push notifications to alert people in a timely way for enhanced living environments and occupational health, as well as a work mode feature, so the user can configure setpoints for laboratory mode and schoolroom mode. Using the *iAQ+*, it is possible to provide an integrated management of data information of the spatio-temporal variations of LEC parameters which are particularly significant not only for enhanced living environments but also for laboratory experiments.

Keywords: IAQ; enhanced living environments; IEQ; IoT; smart cities; LEC

1. Introduction

Indoor environment quality (IEQ) in buildings incorporates indoor air quality (IAQ), acoustics, thermal comfort, and lighting [1]. It is well known that poor IEQ has a negative effect on occupational health, particularly on children and old people.

Poor IAQ is a significant problem which affects particularly the underprivileged people in the world, as they remain most exposed, presenting itself as a critical problem for global health such as tobacco use or the problem of sexually transmitted diseases [2]. The Environmental Protection Agency (EPA) is responsible for indoor and outdoor air quality supervision in the United States. The EPA assessment recognises that IAQ pollutants concentration can be up to 100 times greater when compared with the outdoor pollutants concentration levels and established poor air quality, which are in the top five environmental risks to global health [3].

Indoor living environments include several types of spaces and workplaces such as offices, hospitals, public service centres, schools, libraries, leisure spaces and also the cabins of vehicles [4]. In particular, schools are an essential place to monitor. Typically, the large number of occupants, the time spent indoors, and the higher density of occupants justifies the need to develop automatic supervision systems to provide a healthful and productive workplace for students, teachers and school staff [5].

Buildings are responsible for 40% of the global energy consumption and 30% of the CO_2 emissions. Thermal comfort assurance is related to a substantial percentage of the referred energy consumption. The adoption of personalised conditioning systems is apparently a reliable approach to improve user acceptability with the environment thermal conditions since thermal comfort is a complicated subject with several interrelated aspects that need to be understood [6].

Currently, IEQ in buildings is based on random sampling. However, these procedures are only providing information relating to a specific sampling and are devoid of details of spatio-temporal variations, which are particularly relevant in laboratory tests such as thermography experiments.

The fundamental concept of the Internet of Things (IoT) is the pervasive presence of several types of devices with communication and collaboration skills between them to reach a mutual purpose [7]. The IoT will not only improve everyday life behaviours and activities in different areas such as smart homes, assisted living environments and smart-health, but also in present innovative data and computational resources to develop novel software solutions [8]. One of the fields where the IoT plays a significant role is IAQ monitoring [9]. The rapid increasing of Information and Communication Technologies (ICTs) and IoT proliferation offer significant opportunities in the development of healthcare information systems. However, challenges still exist in achieving safety, security and privacy for healthcare applications [10,11].

Ambient Assisted Living (AAL) is a research area which studies the conceptualisation of an ecosystem of diverse kinds of sensors, computers, mobile gadgets, wireless networks and software development for enhanced personal healthcare supervision and telemedicine solutions [12]. At 2050, a significant proliferation of diseases will lead to high healthcare costs, lack of caregivers, dependency and notable social impact, because 20% of the world population will be aged 60 or over [13]. Most people (87%) prefer to stay in their houses and pay a significant charge for nursing care [14]. However, there are numerous difficulties in AAL solutions conceptualisation and development such as data structure, user-interface, human–computer communication, ergonomics, usability and accessibility [15]. Furthermore, the AAL solutions adopted by the aged population is influenced by social and ethical obstacles. Therefore, privacy and confidentiality are fundamental to AAL solutions' development. Actually, technology should not substitute human care but must be seen as a useful supplement. IoT and AAL research can be conducted side by side in order to present multiple advantages to various healthcare activities such as not only the identification, authentication and tracking of objects and patients but also in automated data acquisition, consulting and storage [16].

The "smart city" approach has recently been proposed as an important method to include current urban production circumstances in a mutual framework especially to focus the ICTs influence in the last 20 years on city performance [17]. The smart city concept is associated with a recent approach to decrease the obstacles caused by the growth of urban population and rapid urbanisation [18]. The interoperability of heterogeneous technologies and devices is a significant challenge in smart cities; IoT can offer the interoperability needed to create unified urban-scale ICT [19].

IoT incorporates notable features to develop innovative real-life solutions and services for smart cities [20], particularly for environmental quality supervision. The smart home is an indispensable element in smart cities [21]. In the future, smart homes should incorporate IoT wearable gadgets managed by smartphone applications and powered by miniaturised built-in sensors [22]. Smart homes follow the AAL paradigm, which enables access to health-care services for patients and medical staff as well as other AAL applications [23]. The use of the paradigm of cognitive dynamic systems associated with IoT architectures in intelligent houses is proposed by Reference [24]. According to this study, the smart home can benefit from several building blocks of the cognitive dynamic system (CDS) along with the incorporation of IoT.

This paper aims to present an environmental quality solution based on IoT to supervise Laboratory Environmental Conditions (LEC) named *iAQ+*. This low-cost wireless solution for

IEQ supervision incorporates mobile computing technologies for data consulting, easy installation, significant notifications for enhanced living conditions, and laboratory activities.

2. Related Work

Several studies have shown the IAQ importance in schoolrooms. The IAQ of public primary Portuguese schools was analysed in 73 primary classrooms in Porto [25]. This study indicates that IAQ deficiencies can persist in classrooms with pollutant sources and defective ventilation. Therefore, pollutant source regulation procedures are the most effective methods for the prevention of unfavourable health impacts on children in scholar institutions. The effectiveness of manual airing strategies on the IAQ of naturally ventilated Italian schools was presented by Reference [26]. A study conducted in New York State primary and secondary schools organised by 501 teachers concluded that many classroom characteristics are probably associated with bad IAQ and over 40% of the teachers described a minimum of one health symptom associated with the building construction [27]. A review of the association between IAQ and its consequences on respiratory wellbeing in Malaysian students was proposed by Reference [28]. Despite the relatively small-scale epidemiologic evidence, Malaysian research proposes effective and relatively consistent evidence among IAQ and children's respiratory health. Another study that characterises the levels of several indoor air contaminants at scholar institutions in Hong Kong, correlates the calculated concentrations with proper standards, and recommend methods to decrease the exposure of students to unwanted pollutants was proposed by Reference [29].

On the one hand, in most university establishments, laboratories have a large number of polluting sources. On the other side, these spaces are very often used as classrooms. Therefore, university laboratories need to be monitored for two purposes: as a classroom (IAQ parameters, including thermal comfort) and/or to ensure different conditions for sampling and performing experiments with reliable quality and data. Most people consider thermal comfort with higher importance when compared with visual and acoustic comfort and IAQ [30]. The comfort temperature might be as low as 17 °C and as high as 30 °C. Thermal comfort is influenced by several factors such as air temperature, radiant temperature, air velocity, humidity, clothing insulation and metabolic heat. The first four factors can be measured and the last two are personalised factors [31]. Although, for laboratory experiments, the recommendation is 23 °C (± 5 °C) and <70 % RH for temperature and relative humidity respectively. A study on the thermal comfort in a Portuguese school was presented by Reference [32].

Multiple studies on environmental quality supervision are accessible in the literature. This section presents several prominent low-cost solutions which include not only open-source but also mobile computing technologies.

A wireless sensor network (WSN) low-cost solution for proper greenhouse pepper cultivation which incorporates proper supervision methods such as remote administration for drip irrigation and equipment control was proposed by Reference [33].

A real-time WSN architecture for environmental supervision which provides acoustic levels, temperature, relative humidity and particulate matter concentration data collection for smart cities was proposed by Reference [34].

A WSN approach for temperature distribution supervision in large-scale indoor environments was proposed by Reference [35]. This methodology aims to enhance the quality of the data transmitted by wireless signals, classify the temperature distribution model and improve the allocation of supplied air measure flow levels to various supply air terminals which fulfill the area taking into account the temperature distribution model.

Numerous IoT solutions for IAQ supervision which merge open-source technologies not only for processing and data transmission and sensors for data collection but also to provide data consulting from distinct areas at the same time using Web and smartphone applications in real time are proposed by References [36–43].

The *iAQ+* solution proposes a valuable instrument for enhanced living environments in smart cities. The advantages for well-being, comfort and productivity of healthy IAQ levels can be enhanced by reducing the pollutant concentration when the ventilation is still unchanged [44]. Consequently, the authors propose a wireless system which incorporates an ESP8266 module which performs the IEEE 802.11 b/g/n networking protocol. The ESP8266 support built-in Wi-Fi technology, therefore, this module is used not only as a processing unit but also for data transmission. For data consulting, this solution uses a smartphone application developed using SWIFT language for the iOS operating system (Figure 1).

Taking into account the IAQ influence on health, the development of a low-cost, open-source supervision system is a trending idea. Therefore, several monitoring solutions have been created [45–52]. Regarding the quality and important contribution of the referred solutions, a review of these studies is presented in Table 1.

Table 1. A summary comparison of real-time IAQ monitoring studies.

Authors	MCU	Sensors	Architecture	Low Cost	Open-Source	Connectivity	Data Access	Easy Installation	Notifications
P. Srivatsa and A. Pandhare [45]	Raspberry Pi	CO_2	WSN/IoT	√	√	Wi-Fi	Web	×	×
F. Salamone et al. [46]	Arduino UNO	CO_2	WSN	√	√	ZigBee	×	×	×
S. Bhattacharya et al. [47]	Waspmote	CO, CO_2, PM, Temperature, Relative Humidity	WSN	×	√	ZigBee	Desktop	×	×
F. Salamone et al. [48]	Arduino UNO	Temperature, Relative Humidity, CO_2, Ligth, Air velocity	IoT	√	√	ZigBee / BLE	Mobile	×	×
Wang, S.K et al. [49]	Arduino	Temperature, Relative Humidity, CO_2	WSN	√	√	ZigBee	Desktop	×	×
Jen-Hao Liu et al. [50]	TI MSP430	CO, Temperature, Relative Humidity	WSN	√	√	ZigBee	×	×	×
J.Kang and K. Hwang [51]	TI MSP430	CO, Temperature, Relative Humidity, VOC, PM	IoT	√	×	ZigBee	×	×	×
Benammar M. et al [52]	Raspberry Pi	CO_2, CO, SO_2, NO_2, O_3, Cl_2, CO, Temperature, Relative Humidity	IoT/WSN	√	×	ZigBee/Ethernet	×	×	×

MCU: microcontroller; √: apply; ×: not apply.

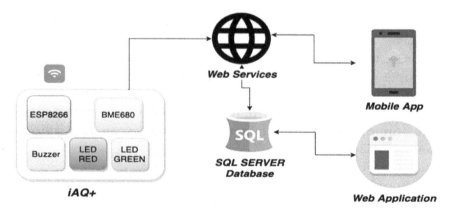

Figure 1. *iAQ+* architecture.

Proposed iAQ+ Management System

Considering the importance of developing a cost-effective system, only the Bosch BME680 sensor was chosen. This sensor was selected since it is an integrated environmental sensor produced particularly for portable applications and wearables where dimension and low power consumption are essential requirements. This sensor provides temperature, relative humidity, barometric pressure and qualitative air quality, the main parameters that should be supervised for both school and laboratory scenarios. The proposed system is composed of a hardware prototype for environmental data acquisition and a Web portal and mobile application compatible for data access (Figure 1). The *iAQ+* is based on open-source technologies and is a totality Wi-Fi solution with various benefits when related to current solutions, such as its modularity, scalability, low-cost, real-time notifications and easy installation. The data collected is stored in a SQL Server database management system and a smartphone application or a Web portal can be used for data consulting. The *iAQ+* incorporates the ESP8266 microcontroller with built-in Wi-Fi communication technology and has been tested in the context of infrared thermography experiments.

Taking into consideration the SQL Server native .NET framework integration feature, this database management system was chosen. Furthermore, the SQL Server incorporates several advantages such as data-recovery features, professional management utilities and built-in data compression and data encryption features. The SQL Server supports easy administration and audit functionalities for enhanced data management.

3. Laboratory Environmental Conditions

LEC supervision is critical from a quality perspective to provide a consistent and regulated state of control for equipment and samples for proper lab activities. LEC is influenced by the building envelope, thermal loads, obstructions and buildings climatisation system, which can produce air flow patterns of hot and cold spots. In some cases, these hot and cold spots can produce environmental conditions variations during experiment activities which lead to unwanted effects on samples as well as manipulation of the testing equipment's samples. Therefore, the LEC must be monitored in real-time not only to assure proper regulation and maintenance of indoor conditions but also to correlate the collected samples with these conditions.

On the one hand, the laboratory activities must be monitored and stored to ensure that they are stable when the test is conducted and at the data collection moment as they influence the quality of the results. LEC has a significant influence on test results and on the precision and reliability of testing records that remain concerned with environmental parameters. On the other hand, monitored results must be saved in the laboratory information management system that should provide the following requisites: 1) must incorporate authentication and access control for data safety; 2) must incorporate data recovery and data adulteration protection methods; 3) must be manipulated in an environment that satisfies the provider or laboratory specs or, in the situation of non-computerized methods, provides conditions which preserve the precision of manual reporting and transcription; 4) must be managed to guarantee the information integrity; and 5) must implement reporting system crashes support to allow prompt corrective operations.

In general, temperature, humidity and barometric pressure are assumed as the main conditions to be monitored for enhanced laboratory environments. The recommendation for LEC is 23 °C ($\pm$5 °C) and <70 % RH for temperature and relative humidity respectively. For temperature data collection, the measurements must be made with calibrated sensors that must be placed far from the equipment under analysis to anticipate every heating consequence which can lead to inaccurate ambient samples. Regarding the humidity measurements, the data collected must be done at the same altitude from the ground as the equipment under test and, preferably, in a similar place if possible. The barometric levels can be consulted from local airports. However, for critical laboratory experiences, barometer sensors are required and must be placed in the laboratory to detect indoor building environmental limitations and barometric fluctuations.

IAQ has a significant influence on LEC [53–56]. On the one hand, IAQ should be supervised in order to provide a healthy and productive workplace for the researchers. On the other hand, IAQ should be monitored to minimise the impact the laboratory experiments' samples. For instance,

IQ is extremely significant in the clinical embryology laboratory activities and is recognised as a significant parameter which influences in vitro fertilisation (IVF) success levels [57].

4. Materials and Methods

Taking into account, not only the necessity to keep a healthful and productive workplace for the students, teachers and the school staff, but also to provide a consistent and regulated state of control for laboratory activities, the *iAQ+* solution has been created by the authors. The *iAQ+* solution provides a low-cost and reliable method for IAQ supervision which incorporates easy configuration and easy installation features. This system is a low-cost, reliable system that can be easily configured and installed by the average user without supporting the cost of an installation done by certified professionals.

The main objective is to provide proper supervision of the LEC, such as air temperature, humidity and barometric pressure and a qualitative air quality index. Therefore, the authors selected a cost-effective BOSCH BME680 sensor, a four-in-one multi-functional microelectromechanical system (MEMS) environmental sensor which integrates a Volatile Organic Compounds (VOC) sensor, temperature sensor, humidity sensor and barometer. The DFRobot Gravity BME680 environmental sensor was used as it provides a Gravity I2C connector, is plug & play and is easy to connect.

The *iAQ+* is based on a microcontroller with built-in Wi-Fi compatibility, a FireBeetle ESP8266 (*DFRobot*). Figure 2 presents the hardware prototype developed by the authors. In this section, the hardware and software development materials and methods will be discussed.

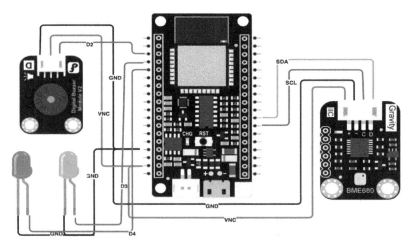

Figure 2. *iAQ+* prototype.

A brief introduction of each component used is shown below.

- **FireBeetle ESP8266** is 32-bit Tensilica L106 microcontroller module which supports IEEE802.11 b/g/n WiFi (2.4 GHz~2.5 GHz). This module support one 10-bit analogue input, 10 digital inputs which incorporate multiple interfaces such as SPI, I2C, IR, and I2S. The clock speed is 80MHz and can reach a maximum 160MHz; in addition, it includes a 50KB SRAM and 16MB flash memory. It supports a low-power-consumption mode of 46uA and the operating voltage is 3.3 V.

- **DFRobot Gravity BME680** is an I2C environmental VOC sensor, temperature sensor, humidity sensor and barometer. It supports an input voltage of 3.3–5.0 V; the operating current consumption is 5 mA without air quality sensing and 25 mA with air quality features. This sensor module size is 30 × 22 mm / 1.18 × 0.87 inches. The temperature range is from -40 °C to +85 °C with a precision of ±1.0 °C (0–65 °C). The humidity range is from 0 to 100% r.H with a precision of a ±3% r.H. (20–80% r.H., 25 °C). The atmospheric pressure measurement range is from 300 to 1100 hPa with a precision of ±0.6 hPa (300–1100 hPa, 0–65 °C).

- **DFRobot Buzzer Module** is a buzzer module that supports an input voltage of 3.3–5.0 V.

- **5V Green LED**—a 5 V green LED is used to notify the end-user of a good IEQ conditions.
- **5V Red LED**—a 5 V red LED is used to notify the end-user of poor IEQ conditions.

The BME680 sensor calculates the sum of VOCs in the surrounding air to provide qualitative air quality data. This sensor incorporates a background auto-calibration feature in order to provide reliable IAQ qualitative data. This process regards the recent measurement records to guarantee that IAQ index ~25 matches to typical good air and IAQ index ~250 states for typical polluted air. The sensor output resistance value varies according to VOCs concentrations, the higher the concentration of reducing VOCs, the lower the resistance and vice versa. The IAQ qualitative range is from 0 to 500 (the larger, the worse). Table 2 illustrates the IAQ qualitative index meaning of the selected sensor.

Table 2. BME680 qualitative IAQ index meaning.

IAQ index	Air Quality
0–50	Good air quality
51–100	Normal air quality
101–150	Little poor air quality
151–300	Poor air quality
201–300	Bad air quality
301–500	Very bad air quality

The mobile application, designated as *iAQ+Mobile,* was created using SWIFT programming language in XCode IDE (Integrated Development Environment), and the minimum requirement is the iOS 12. The *iAQ+Mobile* incorporates significant highlights as it allows not only real-time data access of the last temperature, humidity, air pressure and qualitative IAQ index information but also to receive real-time warnings to notify the occupants when the IEQ has severe deficiencies (Figure 3). The smartphone application also allows viewing the collected data in chart form.

Figure 3. *iAQ+Mobile.*

The *iAQ+* was developed by the authors to be a centralised supervision solution. Therefore, the Web application, *iAQ+Web,* was built in ASP.NET C# using the Visual Studio IDE. Using the *iAQ+Web*, the build manager can consult the IEQ data in real time. This Web portal saves the IEQ conditions history for future analysis (Figure 4).

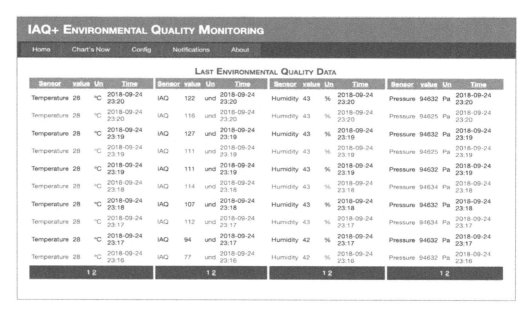

Figure 4. *iAQ+Web*.

The *iAQ+Web* allows historical data export for enhanced reporting and analysis. The Web portal supports hourly, daily and monthly charts and tables of the monitored data for enhanced building management and audits.

To support the two main activities in the university laboratories, the laboratory experiments and teaching activities, the *iAQ+* software (Web and smartphone) supports two functional modes, the laboratory mode and the schoolroom mode. The *iAQ+* software allows the user to configure setpoints for both modes and the laboratory manager can change the status mode using the Web or the smartphone app. The smartphone app is an agile and effective way to change the status. However, the software allows the end-user to schedule the working modes (Figure 5).

Figure 5. *iAQ+Mobile* Settings.

5. Results and Discussion

In Portugal, the majority of indoor environments have natural ventilation. The nature of dwelling, construction, heating and ventilation assumes an important influence on the air permeability variations. It is expected that over 66% of business/services constructions which support natural ventilation mechanisms are notably airtight, and the other 33% have a tendency to be leakier.

For testing purposes, one laboratory of a Portuguese university were on-site supervised using one *iAQ+* prototype. Figure 6a presents the experiment conducted by the authors of LEC supervision of thermography activities. As in most laboratories, the supervised space incorporates natural ventilation and does not have dedicated ventilation slots on the facades. The indoor air is reheated and recirculated through a couple of standard air–water fan-coils of the heating system, and the air exchange is performed through infiltrations and by opening windows manually. The outdoor air is employed to afford ventilation, to reduce the temperature or when the occupants detect the severe or disturbing odour; therefore, the IAQ is frequently deficient.

(a)   (b)

Figure 6. Laboratory layout: (**a**) experiments research area (e.g. thermography experiments supported by *iAQ+*); (**b**) classroom learning space (teaching support).

The *iAQ+* was powered by the power grid using a 230V-5V AC-DC 2A power supply. The LEC exposure data were collected in thermography experiments which showed that indoor conditions can be different from the recommended values and can affect the integrity of the collected data.

The tests conducted show the system capability not only to keep a healthful and productive workplace for the students, teachers and the school staff but also to provide a consistent and regulated state of control for laboratory activities. Figure 6b represents the laboratory space used as a schoolroom.

Figure 7 presents a sample of the graphics of the results achieved in the tests conducted. It should be noted that the graphs displayed the results obtained in the monitored rooms with induced simulations using tobacco smoke.

On one hand, monitoring environmental conditions and maintaining laboratory temperature and humidity requirements is extremely important for high-quality experimental activities. On the other hand, supervising IEQ is significant in creating a healthy and productive workplace for learning and teaching. Using the smartphone application, the user can carry all the monitored data in his pocket.

This solution not only supports alerts and setpoints configuration for real-time notifications via e-mail, SMS and push notifications but also provides an integrated dashboard for the monitored real-time collected data. Quite apart from this, the *iAQ+* incorporates a built-in visual and audio alarm, two LED's and a buzzer respectively. When the configurated setpoints are met, the correspondent

LED is triggered, green to inform the good IEQ status and red to notify the poor IEQ status along with the buzzer activation. In the last case, the buzzer will ring for 30 seconds with the aim to notify the occupants to act in real-time to enhance the indoor conditions.

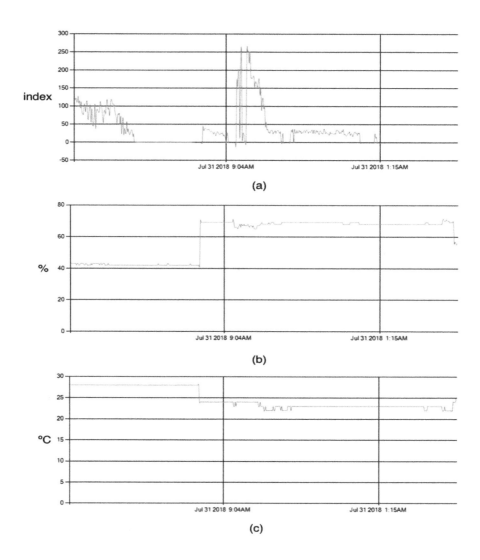

Figure 7. Results of IAQ index, air temperature and relative humidity obtained in the experiments conducted in a real laboratory environment: (**a**) IAQ index; (**b**) relative humidity (%); (**c**) air temperature (°C).

The data collected by the system is analysed before being inserted into the database. If the data exceeds the parameterised limits, the user is notified and an email or SMS is triggered (Figure 8).

The SMS notification is performed using Twilio, a cloud communication platform, as a service. This service makes it possible to programmatically send an SMS and other communication functions using its Web service APIs. The push notifications are performed using the Firebase Cloud Messaging that is a cross-platform solution for messages and notifications for Android, iOS and Web applications.

The real-time alerts promote behaviour changes. In fact, these messages alert the user to act in real-time to perform actions to increase IAQ. On the other hand, with this real-time feature, the building manager can understand when the recurrent unhealthy cases are detected and implement new adjustments to prevent them. Consecutively, *iAQ+* provides the requirements to act in real-time for enhanced living environments and laboratory conditions. The alert indicators architecture is shown in Figure 9.

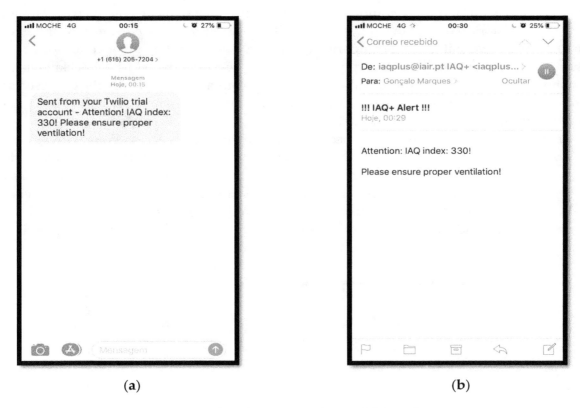

Figure 8. Examples of SMS (**a**) and E-mail (**b**) notifications.

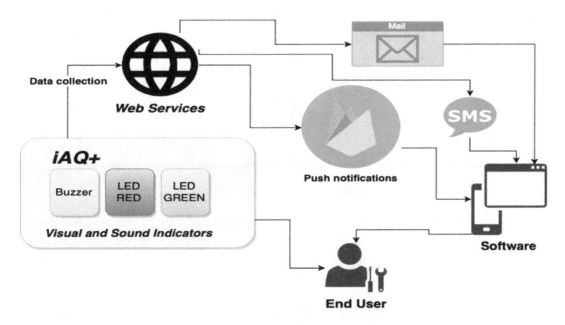

Figure 9. Alerts indicators architecture.

Firstly, *iAQ+* can help in the maintenance of LEC by providing an integrated management system that supports daily records of historical behaviours and variations as well as remote notifications when the configurated setpoints are met. Secondly, this system can not only be used by the building manager to detect unhealthy situations in real-time but also as a decision-making tool to address behavioural changes to promote IEQ for enhanced productive environments.

From a quality point of view, supervising the laboratory environment is crucial to achieving a continuous LEC in the laboratory activities. Irregular maintenance of LEC can lead to unwanted effects to samples as well as to laboratory equipment's ability to produce stable results. The *iAQ+* allows the user to store the historical data of the LEC; therefore, it is possible to check the integrity of the collected data and perform data analysis and comparison taking into account the indoor conditions.

Through charts visualisation of IEQ conditions, the application provides a better acknowledgement of the supervised data when compared to the numeric table format. Consequently, the proposed solution is a significant decision-making tool to plan interventions in order to promote a healthful and productive living environment but also for IEQ analysis.

The *iAQ+* incorporates a wireless connection interface for Internet access along with an easy Wi-Fi network configuration. The *iAQ+* should be connected to a Wi-Fi hotspot for data transmission and to store the network credentials on the flash memory after successful connection. At system start-up, the *iAQ+* searches for a stored Wi-Fi network to connect to. However, if there no saved Wi-Fi networks available, the system will turn to hotspot mode and create a Wi-Fi network with an SSID "*iAQ+*". At this stage, the user must connect to the referred hotspot in order to configure the credentials of the Wi-Fi network to which the system will be connected (Figure 10).

Figure 10. Wi-Fi configuration.

In this way, the *iAQ+* can be easily installed by the end-user which not only follows the original paradigm of IoT solutions but also contributes to the low-cost aim of the presented solution.

Compared to several similar solutions proposed and described in Table 1, the *iAQ+* supports push notifications to alert people in a timely way for enhanced living environments and occupational health by providing ventilation, deactivation of pollutant equipment and by activating air purification systems. When the parameters meet the setpoints for the correspondent working mode, the user is alerted to ensure a productive environment for teaching or correct laboratory environmental parameters. The *iAQ+* offers a work mode feature; the user can configure setpoints for laboratory mode and schoolroom mode. Using the *iAQ+Mobile*, the user can easily change the work mode in real-time.

On one hand, compared to other systems proposed by References [46,47,49,50] based on WSN, the *iAQ+* provides several advantages in scalability and installation in indoor living environments as it is only necessary to configure the Wi-Fi Internet connection and it is not required to configure the sensor nodes and coordinators. On the other hand, compared to the other systems proposed by References [46,50–52] which do not use mobile computing, the *iAQ+* provides smartphone and Web compatibility for data access. The solution proposed by Reference [48] supports mobile access.

However, the proposed mobile app allows only to access the last collected data and does not support historical data access. The *iAQ+Mobile* supports graphical presentation for an easy overview of the IEQ data by the end-user.

The IoT design facilitates the scalability of the proposed solution offering flexibility and expandability as the area can be monitored using only one *iAQ+* unity, but other unities can be added if needed.

Enhancements to the system hardware and software are scheduled in order to incorporate other IEQ parameters such as noise supervision and/or experiment variables. So, as future work improvements and adaptions for specific laboratory tests, such as thermographic experiments applied to wood and trees are planned. The proposed solution has multiple benefits in installation and configuration as the *iAQ+* incorporates wireless communication technology for data transmission, but also provides compatibility with both the standard domestic homes and smart homes.

Apache Kafka is an open-source stream-processing software platform for distributed high-throughput systems. This platform provides low-latency, built-in partitioning, replication and inherent fault-tolerance for real-time data handling. Apache Spark is a framework for large-scale data processing which can be used with Kafka to stream the data to solve the response time and system throughput problems. The SQL Server response time and system throughput tests were not performed. However, these evaluations are planned by the authors as long as the study of the incorporation of open-source platforms such as Apache Kafka and Apache Spark to face response time and system throughput difficulties.

It is imperative to enhance the IEQ conditions, and the authors consider that the first step is to implement real-time supervision to identify its variation and to plan interventions for enhanced living environments.

Managerial Implication

Regarding managerial implications, the results provide insights for enhanced living environments and laboratory experiments. In most university establishments, laboratories are also used as classrooms. Therefore, *iAQ+* is relevant to provide a healthful and productive living environment but also to support experimental activities. LEC has a direct impact on test results and on the accuracy and consistency of test data that are affected by environmental conditions. Thus, the proposed solution offers an integrated management system for historical data analysis in order to guarantee the integrity of test results. The two functional modes of the *iAQ+* solution, the laboratory mode and the schoolroom mode allow the user to configure setpoints for both modes. This will allow switching the functional mode in an agile way, contributing to increasing the quality of the laboratory experiments and teaching activities. The *iAQ+* easy installation feature follows the original paradigm of IoT solutions and contributes to the low cost of the proposed solution. Most importantly, the real-time alerts promote behaviour changes to support the occupants to act promptly in order to provide a healthy and productive workplace for learning and a regulated environment for laboratory activities. The proposed solution provides a mobile application and a Web portal for IAQ data analytics which facilitates data access and analysis.

6. Conclusions

This paper presented an IoT real-time supervision system architecture named *iAQ+* composed by a hardware prototype for data acquisition along with Web and mobile compatibility for data access. The proposed system provides temperature, relative humidity, barometric pressure and qualitative air quality supervision for both school and laboratory scenarios.

With the proliferation of IoT technologies, there is great potential to create automatic IEQ supervision solutions for enhanced living environments and occupational health.

The results achieved are promising, indicating an important contribution to LEC supervision solutions based on IoT. Using *iAQ+*, the monitored data can be particularly valuable to analyse and store the laboratory activities' conditions to ensure that they are stable in the course of the experiments conducted, as they influence the quality of the results.

Compared to existing systems, the *iAQ+* supports push notifications to alert people in a timely way for enhanced living environments and occupational health; in addition, by supporting a work mode feature, the user can configure setpoints for laboratory mode and schoolroom mode. This system offers flexibility and expandability as the user can start with only one *iAQ+* unity and add more unities if needed. However, quality assurance (QA) and quality control (QC) testing of the proposed study and the SQL Server response time and system throughput analysis was not done by the authors. In the future, the QA/QC should be implemented for enhanced product quality traceability and these response time and systems throughput evaluations will be conducted. The authors have also planned software and hardware improvements to adapt the system to specific laboratory tests, such as thermographic experiments applied to wood and trees. In spite of the influence of indoor environments in daily human activities, systems like this will contribute to ensuring a productive environment for teaching and proper LEC.

Author Contributions: G.M. and R.P. designed the study, developed the methodology, performed the analysis, and wrote the manuscript.

Acknowledgments: The financial support from the Research Unit for Inland Development of the Polytechnic Institute of Guarda is acknowledged.

References

1.   Vilcekova, S.; Meciarova, L.; Burdova, E.K.; Katunska, J.; Kosicanova, D.; Doroudiani, S. Indoor environmental quality of classrooms and occupants' comfort in a special education school in Slovak Republic. *Build. Environ.* **2017**, *120*, 29–40. [CrossRef]

2.   Bruce, N.; Perez-Padilla, R.; Albalak, R. Indoor air pollution in developing countries: A major environmental and public health challenge. *Bull. World Health Organ.* **2000**, *78*, 1078–1092. [PubMed]

3.   Seguel, J.M.; Merrill, R.; Seguel, D.; Campagna, A.C. Indoor Air Quality. *Am. J. Lifestyle Med.* **2016**, *11*, 284–295. [CrossRef] [PubMed]

4.   De Gennaro, G.; Dambruoso, P.R.; Loiotile, A.D.; Di Gilio, A.; Giungato, P.; Tutino, M.; Marzocca, A.; Mazzone, A.; Palmisani, J.; Porcelli, F. Indoor air quality in schools. *Environ. Chem. Lett.* **2014**, *12*, 467–482. [CrossRef]

5.   Madureira, J.; Paciência, I.; Rufo, J.; Ramos, E.; Barros, H.; Teixeira, J.P.; de Oliveira Fernandes, E. Indoor air quality in schools and its relationship with children's respiratory symptoms. *Atmos. Environ.* **2015**, *118*, 145–156. [CrossRef]

6.   Rupp, R.F.; Vásquez, N.G.; Lamberts, R. A review of human thermal comfort in the built environment. *Energy Build.* **2015**, *105*, 178–205. [CrossRef]

7.   Giusto, D.; Iera, A.; Morabito, G.; Atzori, L. (Eds.) *The Internet of Things: 20th Tyrrhenian Workshop on Digital Communications*; Springer: New York, NY, USA, 2010; ISBN 978-1-4419-1673-0.

8.   Gubbi, J.; Buyya, R.; Marusic, S.; Palaniswami, M. Internet of Things (IoT): A vision, architectural elements, and future directions. *Future Gener. Comput. Syst.* **2013**, *29*, 1645–1660. [CrossRef]

9.   Ibaseta, D.; Molleda, J.; Díez, F.; Granda, J.C. Indoor Air Quality Monitoring Sensor for the Web of Things. *Proceedings* **2018**, *2*, 1466. [CrossRef]

10. Yin, Y.; Zeng, Y.; Chen, X.; Fan, Y. The internet of things in healthcare: An overview. *J. Ind. Inf. Integr.* **2016**, *1*, 3–13. [CrossRef]

11. Bhatt, Y.; Bhatt, C. Internet of Things in HealthCare. In *Internet of Things and Big Data Technologies for Next Generation Healthcare*; Bhatt, C., Dey, N., Ashour, A.S., Eds.; Springer International Publishing: Cham, Switzerland, 2017; Volume 23, pp. 13–33. ISBN 978-3-319-49735-8.

12. Marques, G.; Pitarma, R. An indoor monitoring system for ambient assisted living based on internet of things architecture. *Int. J. Environ. Res. Public Health* **2016**, *13*, 1152. [CrossRef]

13. United Nations. *World Population Ageing, 1950–2050*; Department of Economic and Social Affairs, Population Division, United Nations: New York, NY, USA, 2002; pp. 11–13. ISBN 92-1-051092-5.

14. Centers for Disease Control and Prevention. The state of aging and health in America 2007. N. A. on an Aging Society. 2007. Available online: http://https://www.cdc.gov/aging/pdf/saha_2007.pdf (accessed on 1 December 2018).

15. Koleva, P.; Tonchev, K.; Balabanov, G.; Manolova, A.; Poulkov, V. Challenges in designing and implementation of an effective Ambient Assisted Living system. In Proceedings of the 2015 12th International Conference on Telecommunication in Modern Satellite, Cable and Broadcasting Services (TELSIKS), Niš, Serbia, 14–17 October 2015; pp. 305–308.

16. Atzori, L.; Iera, A.; Morabito, G. The Internet of Things: A survey. *Comput. Netw.* **2010**, *54*, 2787–2805. [CrossRef]

17. Caragliu, A.; Del Bo, C.; Nijkamp, P. Smart Cities in Europe. *J. Urban Technol.* **2011**, *18*, 65–82. [CrossRef]

18. Chourabi, H.; Nam, T.; Walker, S.; Gil-Garcia, J.R.; Mellouli, S.; Nahon, K.; Pardo, T.A.; Scholl, H.J. Understanding Smart Cities: An Integrative Framework. In Proceedings of the 2012 45th Hawaii International Conference on System Sciences, Maui, HI, USA, 4–7 January 2012; pp. 2289–2297.

19. Zanella, A.; Bui, N.; Castellani, A.; Vangelista, L.; Zorzi, M. Internet of Things for Smart Cities. *IEEE Internet Things J.* **2014**, *1*, 22–32. [CrossRef]

20. Hernández-Muñoz, J.M.; Vercher, J.B.; Muñoz, L.; Galache, J.A.; Presser, M.; Hernández Gómez, L.A.; Pettersson, J. Smart Cities at the Forefront of the Future Internet. In *The Future Internet*; Domingue, J., Galis, A., Gavras, A., Zahariadis, T., Lambert, D., Cleary, F., Daras, P., Krco, S., Müller, H., Li, M.-S., et al., Eds.; Springer Berlin Heidelberg: Berlin/Heidelberg, Germany, 2011; Volume 6656, pp. 447–462, ISBN 978-3-642-20897-3.

21. Skouby, K.E.; Lynggaard, P. Smart home and smart city solutions enabled by 5G, IoT, AAI and CoT services. In Proceedings of the 2014 International Conference on Contemporary Computing and Informatics (IC3I), Mysore, India, 14–17 December 2014; pp. 874–878.

22. Alaa, M.; Zaidan, A.A.; Zaidan, B.B.; Talal, M.; Kiah, M.L.M. A review of smart home applications based on Internet of Things. *J. Netw. Comput. Appl.* **2017**, *97*, 48–65. [CrossRef]

23. Theoharidou, M.; Tsalis, N.; Gritzalis, D. Smart Home Solutions: Privacy Issues. In *Handbook of Smart Homes, Health Care and Well-Being*; Van Hoof, J., Demiris, G., Wouters, E.J.M., Eds.; Springer International Publishing: Cham, Switzerland, 2017; pp. 67–81. ISBN 978-3-319-01582-8.

24. Feng, S.; Setoodeh, P.; Haykin, S. Smart Home: Cognitive Interactive People-Centric Internet of Things. *IEEE Commun. Mag.* **2017**, *55*, 34–39. [CrossRef]

25. Madureira, J.; Paciência, I.; Pereira, C.; Teixeira, J.P.; Fernandes, E.d.O. Indoor air quality in Portuguese schools: Levels and sources of pollutants. *Indoor Air* **2016**, *26*, 526–537. [CrossRef] [PubMed]

26. Stabile, L.; Dell'Isola, M.; Russi, A.; Massimo, A.; Buonanno, G. The effect of natural ventilation strategy on indoor air quality in schools. *Sci. Total Environ.* **2017**, *595*, 894–902. [CrossRef] [PubMed]

27. Kielb, C.; Lin, S.; Muscatiello, N.; Hord, W.; Rogers-Harrington, J.; Healy, J. Building-related health symptoms and classroom indoor air quality: A survey of school teachers in New York State. *Indoor Air* **2015**, *25*, 371–380. [CrossRef] [PubMed]

28. Choo, C.P.; Jalaludin, J. An overview of indoor air quality and its impact on respiratory health among Malaysian school-aged children. *Rev. Environ. Health* **2015**, *30*, 9–18. [CrossRef]

29. Lee, S.; Chang, M. Indoor and outdoor air quality investigation at schools in Hong Kong. *Chemosphere* **2000**, *41*, 109–113. [CrossRef]

30. Yang, L.; Yan, H.; Lam, J.C. Thermal comfort and building energy consumption implications—A review. *Appl. Energy* **2014**, *115*, 164–173. [CrossRef]

31. Havenith, G.; Holmér, I.; Parsons, K. Personal factors in thermal comfort assessment: Clothing properties and metabolic heat production. *Energy Build.* **2002**, *34*, 581–591. [CrossRef]

32. Pereira, L.D.; Cardoso, E.; da Silva, M.G. Indoor air quality audit and evaluation on thermal comfort in a school in Portugal. *Indoor Built Environ.* **2015**, *24*, 256–268. [CrossRef]

33. Srbinovska, M.; Gavrovski, C.; Dimcev, V.; Krkoleva, A.; Borozan, V. Environmental parameters monitoring in precision agriculture using wireless sensor networks. *J. Clean. Prod.* **2015**, *88*, 297–307. [CrossRef]

34. Sanchez-Rosario, F.; Sanchez-Rodriguez, D.; Alonso-Hernandez, J.B.; Travieso-Gonzalez, C.M.; Alonso-Gonzalez, I.; Ley-Bosch, C.; Ramirez-Casanas, C.; Quintana-Suarez, M.A. A low consumption

real time environmental monitoring system for smart cities based on ZigBee wireless sensor network. In Proceedings of the 2015 International Wireless Communications and Mobile Computing Conference (IWCMC), Dubrovnik, Croatia, 24–28 August 2015; pp. 702–707.

35.   Zhou, P.; Huang, G.; Zhang, L.; Tsang, K.-F. Wireless sensor network based monitoring system for a large-scale indoor space: Data process and supply air allocation optimization. *Energy Build.* **2015**, *103*, 365–374. [CrossRef]

36.   Marques, G.; Pitarma, R. IAQ Evaluation Using an IoT CO2 Monitoring System for Enhanced Living Environments. In *Trends and Advances in Information Systems and Technologies*; Rocha, Á., Adeli, H., Reis, L.P., Costanzo, S., Eds.; Springer International Publishing: Cham, Switzerland, 2018; Volume 746, pp. 1169–1177. ISBN 978-3-319-77711-5.

37.   Pitarma, R.; Marques, G.; Ferreira, B.R. Monitoring Indoor Air Quality for Enhanced Occupational Health. *J. Med Syst.* **2017**, *41*, 23. [CrossRef]

38.   Pitarma, R.; Marques, G.; Caetano, F. Monitoring Indoor Air Quality to Improve Occupational Health. In *New Advances in Information Systems and Technologies*; Rocha, Á., Correia, A.M., Adeli, H., Reis, L.P., Mendonça Teixeira, M., Eds.; Springer International Publishing: Cham, Switzerland, 2016; Volume 445, pp. 13–21. ISBN 978-3-319-31306-1.

39.   Marques, G.; Pitarma, R. Smartwatch-Based Application for Enhanced Healthy Lifestyle in Indoor Environments. In *Computational Intelligence in Information Systems*; Omar, S., Haji Suhaili, W.S., Phon-Amnuaisuk, S., Eds.; Springer International Publishing: Cham, Switzerland, 2019; Volume 888, pp. 168–177. ISBN 978-3-030-03301-9.

40.   Marques, G.; Pitarma, R. Using IoT and Social Networks for Enhanced Healthy Practices in Buildings. In *Information Systems and Technologies to Support Learning*; Rocha, Á., Serrhini, M., Eds.; Springer International Publishing: Cham, Switzerland, 2019; Volume 111, pp. 424–432. ISBN 978-3-030-03576-1.

41.   Marques, G.; Pitarma, R. Monitoring Health Factors in Indoor Living Environments Using Internet of Things. In *Recent Advances in Information Systems and Technologies*; Rocha, Á., Correia, A.M., Adeli, H., Reis, L.P., Costanzo, S., Eds.; Springer International Publishing: Cham, Switzerland, 2017; Volume 570, pp. 785–794. ISBN 978-3-319-56537-8.

42.   Marques, G.; Roque Ferreira, C.; Pitarma, R. A System Based on the Internet of Things for Real-Time Particle Monitoring in Buildings. *Int. J. Environ. Res. Public Health* **2018**, *15*, 821. [CrossRef]

43.   Akkaya, K.; Guvenc, I.; Aygun, R.; Pala, N.; Kadri, A. IoT-based occupancy monitoring techniques for energy-efficient smart buildings. In Proceedings of the 2015 IEEE Wireless Communications and Networking Conference Workshops (WCNCW), New Orleans, LA, USA, 9–12 March 2015; pp. 58–63.

44.   Wargocki, P.; Wyon, D.P.; Sundell, J.; Clausen, G.; Fanger, P.O. The Effects of Outdoor Air Supply Rate in an Office on Perceived Air Quality, Sick Building Syndrome (SBS) Symptoms and Productivity. *Indoor Air* **2000**, *10*, 222–236. [CrossRef]

45.   Srivatsa, P.; Pandhare, A. Indoor Air Quality: IoT Solution. In Proceedings of the National Conference NCPCI, 19 March 2016; Volume 2016, p. 19.

46.   Salamone, F.; Belussi, L.; Danza, L.; Galanos, T.; Ghellere, M.; Meroni, I. Design and Development of a Nearable Wireless System to Control Indoor Air Quality and Indoor Lighting Quality. *Sensors* **2017**, *17*, 1021. [CrossRef]

47.   Bhattacharya, S.; Sridevi, S.; Pitchiah, R. Indoor air quality monitoring using wireless sensor network. In Proceedings of the 2012 Sixth International Conference on Sensing Technology (ICST), Kolkata, India, 18–21 December 2012; pp. 422–427.

48.   Salamone, F.; Belussi, L.; Danza, L.; Ghellere, M.; Meroni, I. Design and Development of nEMoS, an All-in-One, Low-Cost, Web-Connected and 3D-Printed Device for Environmental Analysis. *Sensors* **2015**, *15*, 13012–13027. [CrossRef]

49.   Wang, S.K.; Chew, S.P.; Jusoh, M.T.; Khairunissa, A.; Leong, K.Y.; Azid, A.A. WSN based indoor air quality monitoring in classrooms. *AIP Conf. Proc.* **2017**, *1808*, 020063.

50.   Liu, J.; Chen, Y.; Lin, T.; Lai, D.; Wen, T.; Sun, C.; Juang, J.; Jiang, J.-A. Developed urban air quality monitoring system based on wireless sensor networks. In Proceedings of the 2011 Fifth International Conference on Sensing Technology, Palmerston North, New Zealand, 28 November–1 December 2011; pp. 549–554.

51.   Kang, J.; Hwang, K.-I. A Comprehensive Real-Time Indoor Air-Quality Level Indicator. *Sustainability* **2016**, *8*, 881. [CrossRef]

52. Benammar, M.; Abdaoui, A.; Ahmad, S.; Touati, F.; Kadri, A. A Modular IoT Platform for Real-Time Indoor Air Quality Monitoring. *Sensors* **2018**, *18*, 581. [CrossRef] [PubMed]

53. Klein, R.C.; King, C.; Kosior, A. Laboratory air quality and room ventilation rates: An update. *J. Chem. Health Saf.* **2011**, *18*, 21–24. [CrossRef]

54. Stuart, R.; Sweet, E.; Batchelder, A. Assessing general ventilation effectiveness in the laboratory. *J. Chem. Health Saf.* **2015**, *22*, 2–7. [CrossRef]

55. Yau, Y.H.; Chew, B.T.; Saifullah, A.Z.A. Studies on the indoor air quality of Pharmaceutical Laboratories in Malaysia. *Int. J. Sustain. Built Environ.* **2012**, *1*, 110–124. [CrossRef]

56. Ugranli, T.; Gungormus, E.; Sofuoglu, A.; Sofuoglu, S.C. Indoor Air Quality in Chemical Laboratories. In *Comprehensive Analytical Chemistry*; Elsevier: Amsterdam, The Netherlands, 2016; Volume 73, pp. 859–878. ISBN 978-0-444-63605-8.

57. Morbeck, D.E. Air quality in the assisted reproduction laboratory: A mini-review. *J. Assist. Reprod. Genet.* **2015**, *32*, 1019–1024. [CrossRef]

Sniffin' Sticks and Olfactometer-Based Odor Thresholds for n-Butanol: Correspondence and Validity for Indoor Air Scenarios

Marlene Pacharra [1,2,*], **Stefan Kleinbeck** [2], **Michael Schäper** [2], **Christine I. Hucke** [2] and **Christoph van Thriel** [2,*]

[1] MSH Medical School Hamburg, University of Applied Sciences and Medical University, Am Kaiserkai 1, D-20457 Hamburg, Germany

[2] Leibniz Research Centre for Working Environment and Human Factors at TU Dortmund University, Ardeystr. 67, D-44139 Dortmund, Germany; kleinbeck@ifado.de (S.K.); schaeper@ifado.de (M.S.); hucke@ifado.de (C.I.H.)

[*] Correspondence: marlene.pacharra@medicalschool-hamburg.de (M.P.); thriel@ifado.de (C.v.T.);

Abstract: Threshold assessments for the reference odorant n-butanol are an integral part of various research, clinical, and environmental sensory testing procedures. However, the practical significance of a high or low threshold for n-butanol beyond a particular testing environment and procedure are often unclear. Therefore, this study aimed to determine between-method correlations and to investigate the association between the n-butanol threshold and perceptual/behavioral odor effects in natural breathing scenarios in 35 healthy adults. The thresholds for n-butanol derived from the Sniffin' Sticks test and determined by the ascending limit dynamic dilution olfactometry procedure were significantly correlated ($|r| = 0.47$). However, only the thresholds determined by olfactometry were significantly correlated to the odor detection of n-butanol in an exposure lab. Moreover, participants with a higher sensitivity for n-butanol in the olfactometer-based assessment rated ammonia, during a 75 min exposure, to be more unpleasant and showed better performance in a simultaneous 3-back task than participants with lower sensitivity. The results of this study suggest that beyond the strict parameters of a certain psychophysical procedure, the threshold for n-butanol can be a meaningful indicator of odor detection and effects in some cases.

Keywords: odor threshold; olfactometry; Sniffin' Sticks; chemosensory perception; validity assessment

1. Introduction

In clinical, research, and environmental assessment practice, odor sensitivity is currently determined almost exclusively with n-butanol (CAS: 71-36-3) as a reference odorant. As a consequence, parts of the clinical diagnosis of anosmia, the selection of panel members for sensory emission testing, and participation in olfactory research experiments can depend on an individual's threshold for n-butanol [1,2]. Moreover, n-butanol is one of the more abundant and relevant volatile organic compounds (VOCs) in indoor air environments. The German Environment Agency (UBA) mentioned in their indoor air guidance value document for 1-butanol (synonymical to n-butanol) that this VOC was found in 75–90% of indoor air samples in various databases and surveys [3]. Based on the developmental toxicity of 1-butanol, a health hazard guide value (RW II) of 2 mg/m^3 and a precautionary guide value (RW I) of 0.7 mg/m^3 were derived. The UBA report also stated that the RW I is above the odor threshold and that the olfactory perceptions need additional considerations. Regardless of the relevance of n-butanol as an indoor air pollutant, empirical evidence is lacking as to whether sensitivity to n-butanol is an adequate marker for sensitivity to other odorants as well as for n-butanol itself outside of a given lab environment and testing procedure [4,5].

Odor delivery methods and psychophysical testing procedures used to derive the odor threshold for n-butanol vary widely between areas of application. This may give rise to a between-method variability in thresholds. While the Sniffin' Sticks test [6] is very common in research and clinical practice, dynamic dilution olfactometry is the most common method in environmental practice (see DIN EN 13725 [7]). The single staircase, 3-alternative forced choice procedure used in the Sniffin' Sticks test adapts every subsequent step to the individual's previous performance [6]. As this technique is difficult to implement when testing several participants simultaneously, dynamic olfactometry, as used during environmental odor evaluation procedures [8], relies on an ascending limit procedure [2].

While a recent report indicated a non-significant correlation between n-butanol thresholds determined with the Sniffin' Sticks test and ascending limits olfactometry ($r = 0.27$) [4], another study comparing sniff bottles and olfactometry methods for n-butanol and ammonia (CAS: 7664-41-7) reported adequate between-method correlations (e.g., $r = 0.78$) [9]. With regard to the real-life impact of n-butanol thresholds, there is some indication that a lower Sniffin' Sticks threshold for n-butanol is associated with lower pleasantness ratings for different odors presented in glass jars [10]. However, necessary parts of olfactometry and the Sniffin' Sticks tests are (a) prompted sniffing at a clearly identifiable odor source and/or (b) artificial breathing rhythms. Thus, the association between the odor thresholds derived from these methods and the odor detection and evaluation of environmental odors presented more naturally in the ambient air is so far unclear.

Given the practical importance of thresholds for n-butanol in clinical, research, and environmental assessment practice, the aims of the current study were threefold. Firstly, the between-method correlation (concurrent validity) was assessed for n-butanol thresholds determined with the very common Sniffin' Sticks test [6] and the established ascending limit dynamic dilution olfactometry procedure [2]. Secondly, the correspondence of these established threshold tests with the odor detection of n-butanol in indoor air scenarios was tested using an exposure lab. Thirdly, the association of these thresholds with odor effects caused by ammonia in an exposure lab was investigated. As the odors are presented in the ambient air, the exposure lab should more closely mimic the situation in the real world. Thus, the results of the here presented exposure lab experiments should be helpful in determining the ecological validity of the Sniffin' Sticks and olfactometry-based n-butanol thresholds.

To this end, a novel ascending limits procedure presenting a stair-wise increasing concentration of n-butanol under normal breathing conditions in an exposure lab was conducted, and its results correlated with the results of the established methods (Sniffin' Sticks and olfactometry). Moreover, the transferability of the results to the malodorous compound ammonia and its odor effects was tested; it was investigated whether the n-butanol thresholds derived using Sniffin' Sticks or olfactometry are associated with the perceptual and behavioral odor effects of the malodorous compound ammonia in a well-controlled natural breathing scenario simulated by means of an exposure lab experiment [11,12]. To compare the results of individuals more and less sensitive to n-butanol during ammonia exposure and, in this way, to mimic the potential behavior of different selected panelists in real-world scenarios, subgrouping of the sample was performed using cut-off values from a large normative sample (Sniffin' Sticks) [1] or the DIN EN 13725 norm (80 ppb) [7].

2. Experiments

2.1. Participants

Thirty-nine non-smoking participants were recruited for this experiment. Exclusion criteria included pregnancy, asthma, and acute or chronic upper airway diseases. Four participants were excluded from the data analysis to avoid unclear or biased odor thresholds; three participants had increased false alarm rates during the olfactometer threshold test (>mean + 2 SD) (cf. [13]), and one participant indicated that he could not detect an odor at all in the exposure lab threshold test.

Thus, the final sample comprised 35 participants. For descriptive details, see Table 1. To evaluate if the number of subjects was sufficient, a power analysis (G-Power; [14]) was conducted. The expected correlations should be in the range of the test-retest reliabilities of the established olfactory detection threshold tests (e.g., for Sniffin' Sticks, between 0.43 and 0.85 [15]; 0.61 [6]; 0.92 [16]). Thus, for the comparison of different methods, we expected a correlation (Pearson r) of about 0.60 (see also [9], $r = 0.78$ correlation between sniff bottles and olfactometry). With 35 subjects, a statistical power of $1 - \beta = 0.97763$ could be achieved [17].

Table 1. Descriptive statistics for the total sample.

Subject Characteristics	Total Sample
Men/Women (n)	12/23
Age (mean (SD))	23.8 (3.1)
CSS-SHR (mean (SEM))	31.8 (1.3)
Negative affectivity (mean (SEM))	14.0 (0.7)
FEV1 (mean (min-max))	96.4% (84.8–111.1%)

Note: SD = standard deviation, SEM = standard error of the mean, CSS-SHR = Chemical Sensitivity Scale for Sensory Hyperreactivity, FEV1 = forced expiratory volume in 1 s.

2.2. Procedure

The ethics committee of the Leibniz Research Centre for Working Environment and Human Factors (IfADo) approved the study protocol (approval date: 23 March 2016), and written informed consent was obtained from all participants. The participants received no feedback about their test performance in any of the performed tests at any point during the study. They were instructed not to talk to the other participants about their odor perceptions during any of the tests or during the ammonia exposure. The study procedure is depicted in Figure 1.

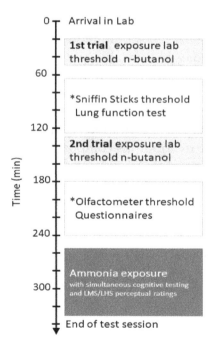

Figure 1. Study procedure. * blocks were switched randomly for half of the participants. LMS = labeled magnitude scale, LHS = labeled hedonic scale.

After arrival in the lab and giving informed consent, groups of 3–4 participants were administered the first trial of the n-butanol threshold procedure in the exposure lab. After completion, a 15 min break followed. Participants were assigned according to an a priori computed randomization scheme to one of two groups, which differed in the order the following detection tests were presented (see Figure 1): half of the participants (Group 1) first completed the olfactometer threshold assessment in groups of two participants and answered the Chemical Sensitivity Scale for Sensory Hyperreactivity [18] and the trait version of the Positive and Negative Affect Schedule [19]. The other half of the participants (Group 2) were first administered, individually, the Sniffin' Sticks threshold test and a lung function test (VitaloGraph, Hamburg, Germany).

In accordance with the GOLD guidelines [20] a forced expiratory volume in 1 s (FEV1) value $\leq 80\%$ in the lung function test was used as an indicator of asthma and chronic obstructive pulmonary disease. Accordingly, subjects with lower FEV1 values would have been excluded from the experimental exposure to ammonia. As only non-smoking, young, and healthy volunteers were enrolled, none of the participants had a FEV1 value below 80% (see Table 1) [20]. Then, all participants completed the second trial of the threshold procedure in the exposure lab. After a 15 min break, participants completed either the Sniffin' Sticks and the lung function test or the olfactometer test and questionnaires, depending on which tests they had already been administered by this point.

After a 15 min break, all participants underwent the 75 min ammonia exposure in the exposure lab. During ammonia exposure, cognitive testing, namely the n-back task [21] and flanker task [22], and perceptual ratings (via labeled magnitude scale, LMS; [23]) were conducted. The LMS is characterized by a quasi-logarithmic spacing of verbal labels and mimics the ratio-like properties of magnitude estimation scaling [23]. Furthermore, for hedonic scaling, the labeled hedonic scale was used (LHS; [24]) that is based on the LMS. The scale values for LHS and LMS in the computerized version used in this study ranged from 0 to 1000.

2.3. Materials

2.3.1. Sniffin' Sticks-Based Threshold for n-Butanol

The Sniffin' Sticks (Burghart, Wedel, Germany) subtest for the assessment of the n-butanol threshold was used [1,6]. Here, the threshold value is defined as the average Sniffin' Stick number (lower numbers indication higher concentrations) of the last four reversals in a single-staircase, 3-alternative forced choice procedure.

Following the newest available norms of the test (see [25]), the cut-off score for individuals more and less sensitive to n-butanol using this test was 9 (median normative sample for age 21–30). Within this age range, there are only negligible differences between males and females, 8.75 for males and 9 for females (age 16–35; [1]), or more recently, 8.5 vs. 8.75 (age 21–30; [26]). As only non-smoking, healthy volunteers participated in the study, a cut-off value of 9 for males and females seemed to be appropriate.

2.3.2. Olfactometer-Based Threshold for n-Butanol

A dynamic dilution olfactometer TO 8 (ECOMA GmbH, Kiel, Germany) was used that complies with DIN EN 13725 [2]. N-butanol was injected into 25 L Tedlar®-bags filled with nitrogen. The mixture was homogenized by heating and rotating the bag.

The standard procedure of the ascending method of limits with a 2-fold geometric dilution series was applied as in previous studies [13,27,28]. In short, the threshold measurement consisted of three trials in which increasing concentration steps of n-butanol were presented, interspersed with blank samples.

Participants had to press a button whenever they thought they detected an odor. The lower of two subsequent correctly identified concentration steps represented the estimate of reliable olfactory detection in that trial. The detection threshold was defined as the geometric mean of the three trial estimates [13,27,28]. As in previous studies [13,28], the detection thresholds were subjected to log-transformations before data analysis.

According to DIN EN 13725 [7], a panel member for environmental odor testing should have an n-butanol threshold between 20 and 80 ppb [2]. There are no established, published thresholds that differentiate between males and females for the here used olfactometry test for n-butanol. Thus, the cut-off value for individuals more and less sensitive to n-butanol using the ascending limits olfactometry test was set to 80 ppb in this study.

2.3.3. Exposure Lab-Based Threshold for n-Butanol

The threshold assessment took place in a 28 m^3 exposure lab with four PC workstations. This environmental chamber has been used in previous experimental exposure studies, i.e., [29]. The assessment followed the same general procedure of the ascending method of limits as used for the olfactometer-based assessment [2]. Due to the higher time and operating costs of the exposure lab compared to the olfactometer, the assessment in the exposure lab consisted of only two instead of three trials.

In each trial, subjects were exposed over 30 min to an ascending concentration series of n-butanol (2-fold geometric series: 20, 40, 80, 160 and 320 ppb; see Supplement Figure S1). Every 5 min, subjects were prompted on a computer screen to indicate whether they detected an odor or not ("Odor? Yes/No"). Due to the technical restrictions in the lab, it was not feasible to insert randomly blank samples into the series. Thus, the first correctly identified concentration step represented the estimate of reliable olfactory detection in that trial. The detection threshold was defined as the geometric mean of the two trial estimates. Just as the olfactometer-based thresholds [13,28], the detection thresholds derived from the exposure lab procedure were subjected to log-transformations before data analysis.

2.3.4. Experimental Ammonia Exposure

The procedure as described in previous studies [11,12] was applied. In short, subjects were exposed to an ascending concentration of ammonia (CAS: 7664-41-7) over 75 min. The maximum concentration after 75 min was 10 ppm (see Supplement Figure S4) corresponding to 50% of the German maximum workplace concentration (MAK value) [30]. This concentration is clearly above previously published odor thresholds but still well below the lateralization thresholds [28]. To estimate the odor effects of ammonia during the exposure, chemosensory perceptions were rated via the LMS [23] and the LHS [24]. Further, cognitive performance was assessed using a 3-back working memory and response inhibition task (see Supplementary Figure S3).

2.3.5. Air Monitoring in the Exposure Lab

The 28 m^3 laboratory was supplied with conditioned air by a climate control unit in a neighboring room (temperature, 24.4 °C; humidity, 46.0%). A predefined amount of n-butanol or ammonia (experimentally determined by volumetric analysis) was mixed into the inlet airstream of the climate control system. The conditioned air was dispersed throughout the laboratory by a branched pipe system, which was located on the floor. The outlet system at the ceiling of the laboratory was actively controlled through four outlets by an exhaust air ventilator; it maintained the laboratory at a negative pressure of 20–30 Pa. The air exchange rate was approximately 300 m^3/h.

Air samples were taken from the airflow of the inlet pipe and from the inside of the exposure laboratory quasi-continuously (every 80 s) during all exposure sessions. Photo acoustic IR spectroscopy was used to analyze the air samples (INNOVA, 1412i Photo Acoustic Field Gas-Monitor, LumaSense, Ballerup, Denmark). An overview of measured concentration values for n-butanol and ammonia is given in the Supplement (Supplementary Figures S2 and S4).

2.4. Statistical Analysis

The statistical analyses were performed in IBM SPSS Statistics 24. The level of significance for all statistical tests was set to 0.05. We checked for outliers by using the more liberal definition of extreme outliers ("outer fences": Q3 + 3 × IQR) [31], and according to this criterion, all participants could be included in the analysis.

Based on the two thresholds, participants were classified into a 2 × 2 table below or above the respective cut-off values. Pearson's chi-square and exact tests were used to analyze the association of the grouping results. Moreover, the group differences for the Sniffin' Sticks scores and the olfactometer-based threshold were analyzed by Mann–Whitney U tests.

A Pearson correlation was computed between the Sniffin' Sticks-based and olfactometer-based threshold for n-butanol to compare the methods.

Next, the two established thresholds were correlated with the exposure lab-based threshold using further Pearson correlations. All correlations were adjusted (Bonferroni method) for the total number of computed multiple comparisons. Bonferroni-adjusted p-values are shown in addition to the non-adjusted correlations for these analyses.

The experimental data from the ammonia exposure were analyzed using full-factorial analyses of variance (ANOVAs), with time as the repeated measures factor and group as the between-subjects factor. Models were calculated taking into account, on the one hand, the grouping factor Sniffin' Sticks threshold (cut-off value: 9, see Sniffin' Sticks norms) and taking into account, on the other hand, the grouping factor olfactometer-based threshold (cut-off: 80 ppb, see DIN EN norm 13725 [7]). If the assumption of sphericity was violated, Greenhouse–Geisser-corrected degrees of freedom were used. Significant interaction effects were further analyzed using Bonferroni-adjusted post hoc tests.

3. Results

3.1. Results of the Psychometric Threshold Assessments

Table 2 presents the descriptive statistics of the three olfactory measures of n-butanol sensitivity for the total sample and after applying the respective cut-offs. Unsurprisingly, when a cut-off was applied based on one of the thresholds, Mann-Whitney U tests indicated a significant difference between resultant groups in this threshold. Moreover, participants more and less sensitive in the Sniffin' Sticks tests also differed significantly in their olfactometry-based threshold. Participants did not differ in relevant psychological variables for odor effects [29] such as negative affectivity and self-reported chemical sensitivity (see supplement Table S1).

Table 2. Description of total sample and classified subgroups.

Subject Characteristics	Total Sample	Sniffin' Sticks Threshold		Olfactometer Threshold	
		<9	≥9	>80 ppb	≤80 ppb
Men/Women (n)	12/23	7/14	5/9	6/10	6/13
Sniffin' Sticks T No. pen (median (IQR))	8.0 (6.5–9.8)	6.8 (6.3–8.0)	9.8 * (9.3–10.8)	8.0 (6.5–9.1)	8.3 (7.3–10.8)
Olfactometer T ppb (median (IQR))	80 (50–160)	101 (64–160)	45.2 * (32–127)	160 (127–228)	50.4 * (32–80)
Exposure lab T ppb (median (IQR))	80 (40–113)	80 (57–113)	68.3 (40–113)	136.6 (48–226)	80 (40–113)

Note. IQR = inter-quartile range, T = threshold, * $p \leq 0.05$ subgroup comparison using Mann-Whitney U tests.

3.2. Results of the between-Method Correlations for n-Butanol Thresholds

The correlation between the Sniffin' Sticks- and olfactometer-based threshold (see Figure 2) was significant ($r = -0.47$; $p = 0.004$, Bonferroni-adjusted $p = 0.012$).

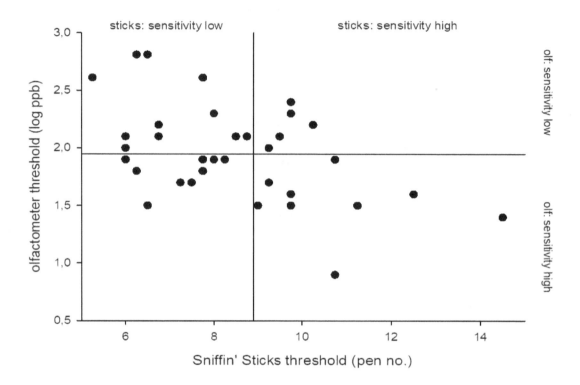

Figure 2. Scatter plot depicting the association between n-butanol Sniffin' Sticks- and olfactometer-derived thresholds. Note that in the Sniffin' Sticks test, a higher pen number corresponds to a higher n-butanol dilution and thus a lower threshold (higher sensitivity). Vertical and horizontal lines depict the respective cut-off values for high vs. low sensitivity groups (for details see text).

When applying the cut-off values for the Sniffin' Sticks- (≥ 9) and olfactometer-based (≤ 1.9 log ppb) thresholds, nine participants (25.7%; lower right quadrant) were classified as individuals with high olfactory sensitivity in both standardized n-butanol threshold assessments (cf. Figure 2). Three of these participants were males (three out of 12; 25%) and the other six were females (six out of 23; 26%). However, the statistical analysis of the 2 × 2 contingency table yielded a non-significant Pearson chi-square value of 0.94 ($p = 0.49$). Thus, there was no significant overlap of the two olfactory sensitivity classification approaches.

Both thresholds for n-butanol (olfactometer and Sniffin' Sticks) were correlated with the exposure lab-based threshold for n-butanol (Figure 3). Due to repeated computation of correlations with the same participants, a Bonferroni adjustment of *p*-values was conducted, resulting in a significant correlation between the olfactometer-based threshold and the exposure-lab based threshold ($r = 0.41$, $p = 0.015$, Bonferroni-adjusted $p = 0.045$, see Figure 3a). However, the Bonferroni-adjusted correlation between the Sniffin' Sticks-based threshold and the exposure lab-based threshold was non-significant ($r = -0.34$, $p = 0.048$, Bonferroni-adjusted $p = 0.144$, see Figure 3b).

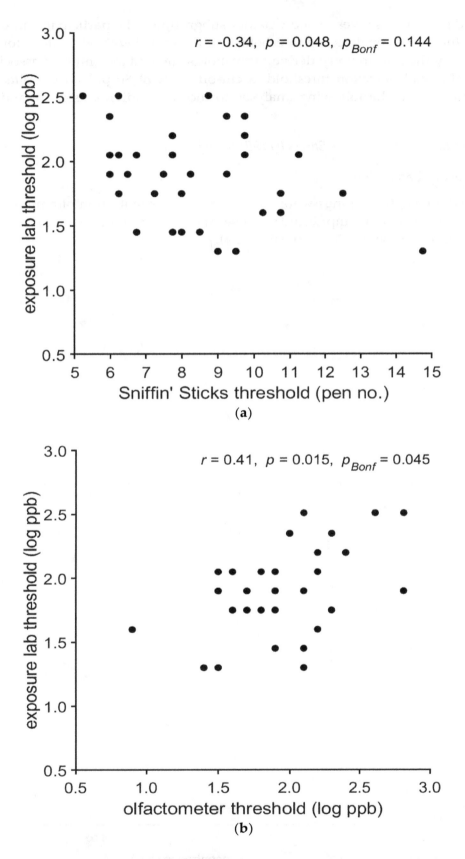

Figure 3. Scatter plots depicting the associations between the n-butanol thresholds derived with the exposure lab and the investigated methods, (**a**) Sniffin' Sticks and (**b**) olfactometer. Note that in the Sniffin' Sticks test, a higher pen number corresponds to a higher n-butanol dilution and thus a lower threshold (higher sensitivity).

In a second step, it was investigated whether subgrouping the participants into more and less sensitive individuals based on detection thresholds was associated with odor effects for the compound ammonia. As only the olfactometry-derived thresholds showed a significant association with the exposure lab n-butanol detection threshold, a cut-off score of 80 ppb in the olfactometer-based assessment was used in the following analyses to indicate individuals more and less sensitive to n-butanol.

3.3. Results of the Modulation of Odor Effects by n-Butanol Thresholds

3.3.1. Chemosensory Perceptions

As expected, perceptual ratings were affected by the concentration of ammonia; participants perceived ammonia to be more unpleasant, intense, and pungent with increasing concentration (all main effects of concentration, $p < 0.001$; see Figure 4).

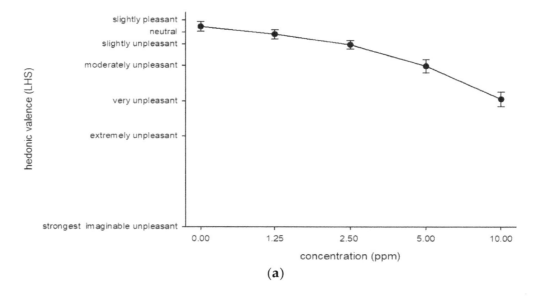

(a)

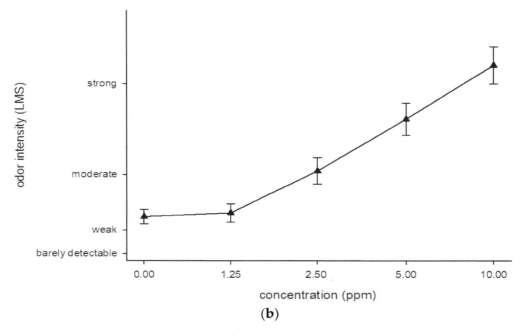

(b)

Figure 4. *Cont.*

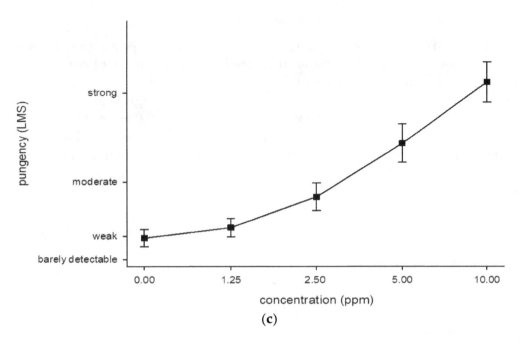

(c)

Figure 4. Impact of different concentrations of ammonia on perceived (**a**) hedonic value, (**b**) odor intensity and (**c**) pungency (mean ± SEM).

A significant main effect of the olfactometer-based threshold on pleasantness ratings emerged, F(1,33) = 4.2, p = 0.049. Participants with a lower olfactometer-based threshold (higher sensitivity) rated the exposure to be more unpleasant (mean = 426, SEM = 12; scale range: 0–1000) than participants with a higher olfactometer-based threshold (mean = 463, SEM = 13) (see Figure 5).

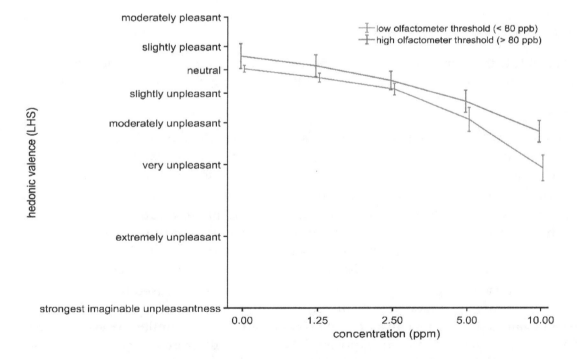

Figure 5. Effect of olfactory sensitivity assessed via the olfactometer-based threshold on pleasantness ratings during ammonia exposure (mean ± SEM).

Figure 5 indicates that the difference between the two groups increased with increasing ammonia concentration. However, the interaction of the sensitivity group and concentrations was not significant.

3.3.2. Odor Effects on Behavioral Task Performance

Participants improved their performance in the 3-back and response inhibition tasks over the course of the test session as indicated by an increase in the percentage of correct responses and a decrease in reaction times (all main effects of concentration, $p \leq 0.05$).

With regard to the olfactometer-based threshold for n-butanol, significant main effects on reaction times, $F(1,33) = 19.7$, $p < 0.001$, and error rates, $F(1,33) = 5.4$, $p = 0.026$, in the 3-back task emerged. Participants with a lower olfactometer-based threshold (higher sensitivity) had shorter reaction times and a higher percentage of correct responses in the 3-back task compared to participants with a higher olfactometer-based threshold (see Figure 6).

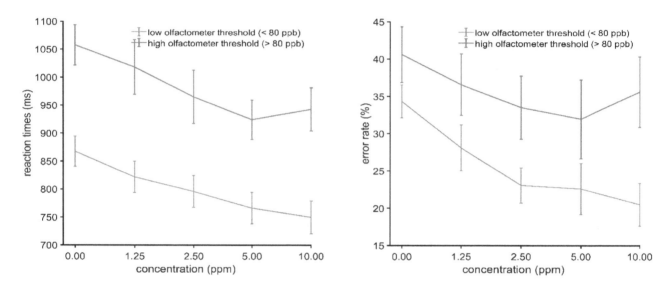

Figure 6. Effect of olfactometer-based threshold on 3-back reaction times (**left**) and error rates (**right**) during ammonia exposure (mean ± SEM).

Comparable to the rating data (see Figure 5), there was no interaction with the increasing ammonia concentration, indicating no additional impact of the increasing chemosensory perceptions.

4. Discussion

Given the importance of n-butanol odor thresholds in many research, clinical, and environmental testing contexts, information on the practical significance of this particular odor threshold beyond the particular testing environment and procedure is scarce. This study sought to remedy that.

In contrast to a previous report [4], a medium-sized, significant correlation between the thresholds derived from the Sniffin' Sticks test and the ascending limit dynamic dilution olfactometry procedure could be shown. This indicates that the determined sensitivity to n-butanol is associated between these two established methods of threshold assessment [2,6] and supports the good between-method correlations (concurrent validity) previously reported for other threshold assessment methods [9,28].

Beyond established threshold procedures, a novel exposure lab-based threshold assessment for n-butanol was proposed that more closely mimics odor detection during natural breathing. Measured concentration values showed that an ascending concentration series similar to the olfactometer-based method [2] could be generated in an exposure lab. After Bonferroni correction, only a significant medium-sized correlation between n-butanol thresholds derived using olfactometry and this novel method emerged. This indicates that olfactometry-derived thresholds can be meaningful indicators of odor detection in a more realistic context. While the Sniffin' Sticks threshold test requires artificial breathing (e.g., sniffing), the olfactometry and exposure lab scenarios have in common that they allow a more natural breathing pattern.

Moreover, the results showed that a lower olfactometer-based threshold for n-butanol is associated with lower pleasantness ratings for ammonia during an exposure lab scenario. This further highlights the external validity of n-butanol thresholds with regard to perceptual effects during natural breathing of another odor and irritant (ammonia). Additionally, it is in line with a previous experimental finding [10], showing that the threshold for n-butanol is associated with lower pleasantness ratings for a range of odors presented in glass jars.

An interesting, secondary finding in this study constitutes the better cognitive working memory performance in those with lower olfactometer-based thresholds irrespective of the ambient ammonia concentration. With regard to the Sniffin' Sticks threshold for n-butanol, Hedner, et al. [32] reported that the threshold is unrelated to cognitive factors such as executive functioning, semantic memory, and episodic memory. However, whether this is also the case for olfactometer-based thresholds is so far unclear. As the two "high odor sensitivity" groups showed only a weak overlap (25.7%), factors unrelated to olfaction but relevant for cognitive task performance (e.g., education and IQ) might have caused this general performance difference.

In recent studies using gas chromatography-olfactometry, a coupling of gas chromatography analysis and human olfaction by panelists was employed to identify single VOCs in mixtures [33]. For n-butanol, a linear relationship was found between the modified detection frequency (frequency of detection × evaluation of intensity) of panelists and concentration of n-butanol as measured by gas chromatography (MS) in adhesives [34].

When humans inhale, ambient air is analyzed when reaching the olfactory epithelium. There, trace components of the air interact with receptor cells [35]. Thresholds and atmospheric lifetime are related in such a way that highly reactive odorants (short-lived molecules) are detected more sensitively [35]. N-butanol, belonging to the family of alcohols, therefore, has a relatively low odor threshold.

All threshold assessments in this study indicated that the median olfactory threshold for n-butanol in the experimental sample was higher than what would be expected from norm values [1] or permissible for panel members during sensory emission testing according to DIN EN 13725 [7] (compare Table 2). This would suggest an overall lower than average sensitivity to n-butanol in the sample. This could be due to (1) a sampling error associated with the low sample size, (2) undetected nasal obstruction in the participants, or (3) olfactory adaptation due to multiple assessments of the odor threshold for n-butanol on the same day.

Despite these possible confounding factors, the results showed that the threshold for n-butanol can be a meaningful indicator of odor detection and odor effects in natural breathing scenarios. This could be seen as a first step in providing much needed confidence in these thresholds [4,5] that are used daily in so many research and other application areas.

5. Limitations of the Study

Before coming to the conclusions, some limitations should be mentioned that need to be addressed in further studies. First, the sample size was sufficient to detect the association between n-butanol odor thresholds and the odor effects of another compound, but the sample was highly selective, and therefore, the transferability to the general population is somewhat limited. Here, a larger sample including older subjects, subjects with mild diseases of the upper respiratory tract (e.g., allergic rhinitis), and subjects reporting an increased odor sensitivity should be investigated. Second, the new method of the exposure lab-based threshold assessment should be tested with other odorants and compared to other threshold assessment procedures like squeezing and sniffing bottles [36–38] or the triangle bag method [39]. Third, odorants other and more pleasant than ammonia should be used to include the highly relevant dimension of pleasantness [40] into this branch of odor research.

6. Conclusions

The results presented here provide further empirical evidence that the olfactory sensitivity of an individual may be an important predictor of odor perceptions in near to realistic scenarios of the human odor experience. The reference compound n-butanol seems to be an adequate choice as shown by the good cross-method correlations. Nevertheless, the role of suprathreshold olfactory functioning such as odor discrimination or identification has not been conclusively studied in this context. Moreover, other reference compounds for panelist selection are currently under discussion (DIN EN 13725:2019) [41]. With respect to the impact of environmental odors on cognitive task performance, our results showed that "high odor sensitivity" was not associated with worse performance in a challenging working memory task. The results were opposite to a distractive effect of malodors as proposed previously [42].

Supplementary Materials: Figure S1: Schematic overview of the experimental procedure during the exposure lab-based threshold assessment. Figure S2: Measured concentration values for n-butanol during the exposure lab-based threshold assessment. Figure S3: Schematic overview of the experimental procedure during the ammonia exposure (cf. [11,12]). Figure S4: Measured concentration values of ammonia during the experimental exposure. Table S1: Descriptive statistics for the total sample and subgroups.

Author Contributions: Conceptualization, M.P. and C.v.T.; methodology, M.P. and M.S.; validation, S.K. and M.S.; formal analysis, M.P. and S.K.; investigation, M.P. and C.v.T.; resources, C.v.T.; data curation, M.S.; writing—original draft preparation, M.P. and S.K.; writing—review and editing, M.P., S.K., M.S., C.I.H. and C.v.T.; visualization, M.P., C.I.H. and S.K.; supervision, C.v.T. All authors have read and agreed to the published version of the manuscript.

Acknowledgments: The authors would like to thank Meinolf Blaszkewicz, Nicola Koschmieder, Eva Strzelec, Michael Porta, and Beate Aust for technical assistance. The publication of this article was funded by the Open Access Fund of the Leibniz Association.

References

1.   Hummel, T.; Kobal, G.; Gudziol, H.; Mackay-Sim, A. Normative data for the "Sniffin' Sticks" including tests of odor identification, odor discrimination, and olfactory thresholds: An upgrade based on a group of more than 3000 subjects. *Eur. Arch. Otorhinolaryngol.* **2007**, *264*, 237–243. [CrossRef]

2.   Mannebeck, D. Olfactometers according to EN 13725. In *Springer Handbook of Odor*; Buettner, A., Ed.; Springer International Publishing: Heidelberg, Germany, 2017; pp. 545–552.

3.   Bekanntmachung des Umweltbundesamtes. Richtwerte für 1-Butanol in der Innenraumluft. *Bundesgesundh. Gesundh. Gesundh.* **2014**, *57*, 733–743. [CrossRef] [PubMed]

4.   Barczak, R.; Sowka, I.; Nych, A.; Sketowicz, M.; Zwozdiak, P. Application of the standard sniffin' sticks method to the determination odor inspectors' olfactory sensitivity in Poland. *Chem. Eng. Trans.* **2010**, *23*, 13–18. [CrossRef]

5.   Feilberg, A.; Hansen, M.J.; Pontoppidan, O.; Oxbol, A.; Jonassen, K. Relevance of n-butanol as a reference gas for odorants and complex odors. *Water Sci. Technol.* **2018**, *77*, 1751–1756. [CrossRef] [PubMed]

6.   Hummel, T.; Sekinger, B.; Wolf, S.R.; Pauli, E.; Kobal, G. 'Sniffin' sticks': Olfactory performance assessed by the combined testing of odor identification, odor discrimination and olfactory threshold. *Chem. Senses* **1997**, *22*, 39–52. [CrossRef] [PubMed]

7.   DIN EN 13725. *Air Quality—Determination of Odour Concentration by Dynamic Olfactometry*; German Version Beuth: Berlin, Germany, 2003. [CrossRef]

8.   Sucker, K.; Both, R.; Bischoff, M.; Guski, R.; Winneke, G. Odor frequency and odor annoyance. Part I: Assessment of frequency, intensity and hedonic tone of environmental odors in the field. *Int. Arch. Occup. Environ. Health* **2008**, *81*, 671–682. [CrossRef]

9.   Hayes, J.E.; Jinks, A.L.; Stevenson, R.J. A comparison of sniff bottle staircase and olfactometer-based threshold tests. *Behav. Res. Methods* **2013**, *45*, 178–182. [CrossRef]

10.   Kärnekull, S.C.; Jonsson, F.U.; Larsson, M.; Olofsson, J.K. Affected by smells? Environmental chemical responsivity predicts odor perception. *Chem. Senses* **2011**, *36*, 641–648. [CrossRef]

11.   Pacharra, M.; Schäper, M.; Kleinbeck, S.; Blaszkewicz, M.; van Thriel, C. Olfactory acuity and automatic associations to odor words modulate adverse effects of ammonia. *Chem. Percept* **2016**, *9*, 27–36. [CrossRef]

12.   Pacharra, M.; Kleinbeck, S.; Schäper, M.; Blaszkewicz, M.; van Thriel, C. Multidimensional assessment of self-reported chemical intolerance and its impact on chemosensory effects during ammonia exposure. *Int. Arch. Occup. Environ. Health* **2016**, *89*, 947–959. [CrossRef]

13. Pacharra, M.; Schäper, M.; Kleinbeck, S.; Blaszkewicz, M.; Wolf, O.T.; van Thriel, C. Stress lowers the detection threshold for foul-smelling 2-mercaptoethanol. *Stress* **2015**, *19*, 1–10. [CrossRef] [PubMed]

14. Faul, F.; Erdfelder, E.; Buchner, A.; Lang, A.G. Statistical power analyses using G*Power 3.1: Tests for correlation and regression analyses. *Behav. Res. Methods* **2009**, *41*, 1149–1160. [CrossRef] [PubMed]

15. Albrecht, J.; Anzinger, A.; Kopietz, R.; Schopf, V.; Kleemann, A.M.; Pollatos, O.; Wiesmann, M. Test-retest reliability of the olfactory detection threshold test of the Sniffin' sticks. *Chem. Senses* **2008**, *33*, 461–467. [CrossRef] [PubMed]

16. Haehner, A.; Mayer, A.M.; Landis, B.N.; Pournaras, I.; Lill, K.; Gudziol, V.; Hummel, T. High test-retest reliability of the extended version of the "Sniffin' Sticks" test. *Chem. Senses* **2009**, *34*, 705–711. [CrossRef] [PubMed]

17. Hemmerich, W. StatistikGuru: Poweranalyse für Korrelationen. Available online: https://statistikguru.de/rechner/poweranalyse-korrelation.html (accessed on 25 April 2020).

18. Nordin, S.; Millqvist, E.; Löwhagen, O.; Bende, M. A short Chemical Sensitivity Scale for assessment of airway sensory hyperreactivity. *Int. Arch. Occup. Environ. Health* **2004**, *77*, 249–254. [CrossRef]

19. Watson, D.; Clark, L.A.; Tellegen, A. Development and validation of brief measures of positive and negative affect: The PANAS scales. *J. Pers. Soc. Psychol.* **1988**, *54*, 1063–1070. [CrossRef]

20. GOLD Guidelines. Diagnosis of Disease of Chronic Airway Limitation: Asthma, COPD and Asthma-COPD Overlap Syndrome (ACOS). Available online: http://goldcopd.org/asthma-copd-asthma-copd-overlap-syndrome (accessed on 26 March 2020).

21. Kirchner, W.K. Age differences in short-term retention of rapidly changing information. *J. Exp. Psychol.* **1958**, *55*, 352–358. [CrossRef]

22. Carbonnell, L.; Falkenstein, M. Does the error negativity reflect the degree of response conflict? *Brain Res.* **2006**, *1095*, 124–130. [CrossRef]

23. Green, B.A.; Dalton, P.; Cowart, B.; Shaffer, G.; Rankin, K.; Higgins, J. Evaluating the 'Labeled Magnitude Scale' for measuring sensations of taste and smell. *Chem. Senses* **1996**, *21*, 323–334. [CrossRef]

24. Lim, J.; Wood, A.; Green, B.G. Derivation and evaluation of a labeled hedonic scale. *Chem. Senses* **2009**, *34*, 739–751. [CrossRef]

25. Normative Data for the Sniffin' Sticks (University Hospital Carl Gustav Carus Dresden). Available online: https://www.uniklinikum-dresden.de/de/das-klinikum/kliniken-polikliniken-institute/hno/forschung/interdisziplinaeres-zentrum-fuer-riechen-und-schmecken/downloads/SDI_Normwerte_2015.pdf (accessed on 6 May 2020).

26. Oleszkiewicz, A.; Schriever, V.A.; Croy, I.; Hahner, A.; Hummel, T. Updated Sniffin' Sticks normative data based on an extended sample of 9139 subjects. *Eur. Arch. Otorhinolaryngol.* **2019**, *276*, 719–728. [CrossRef] [PubMed]

27. Kleinbeck, S.; Schäper, M.; Juran, S.A.; Kiesswetter, E.; Blaszkewicz, M.; Golka, K.; Zimmermann, A.; Brüning, T.; van Thriel, C. Odor Thresholds and breathing changes of human volunteers as consequences of sulphur dioxide exposure considering individual factors. *Saf. Health Work* **2011**, *2*, 355–364. [CrossRef] [PubMed]

28. Smeets, M.A.; Bulsing, P.J.; van Rooden, S.; Steinmann, R.; de Ru, J.A.; Ogink, N.W.; van Thriel, C.; Dalton, P.H. Odor and irritation thresholds for ammonia: A comparison between static and dynamic olfactometry. *Chem. Senses* **2007**, *32*, 11–20. [CrossRef] [PubMed]

29. Pacharra, M.; Kleinbeck, S.; Schäper, M.; Juran, S.A.; Hey, K.; Blaszkewicz, M.; Lehmann, M.L.; Golka, K.; van Thriel, C. Interindividual differences in chemosensory perception: Toward a better understanding of perceptual ratings during chemical exposures. *J. Toxicol. Environ. Health A* **2016**, *79*, 1026–1040. [CrossRef]

30. The Deutsche Forschungsgemeinschaft (DFG). *List of MAK and BAT Values*; WILEY-VCH Verlag GmbH: Weinheim, Germany, 2019. [CrossRef]

31. NIST/SEMATECH e-Handbook of Statistical Methods. Available online: https://www.itl.nist.gov/div898/handbook/prc/section1/prc16.htm (accessed on 6 May 2020).

32. Hedner, M.; Larsson, M.; Arnold, N.; Zucco, G.M.; Hummel, T. Cognitive factors in odor detection, odor discrimination, and odor identification tasks. *J. Clin. Exp. Neuropsychol.* **2010**, *32*, 1062–1067. [CrossRef]

33. Bax, C.; Sironi, S.; Capelli, L. How can odors be measured? An overview of methods and their applications. *Atmosphere* **2020**, *11*, 92. [CrossRef]

34. Vera, P.; Uliaque, B.; Canellas, E.; Escudero, A.; Nerin, C. Identification and quantification of odorous compounds from adhesives used in food packaging materials by headspace solid phase extraction and

headspace solid phase microextraction coupled to gas chromatography-olfactometry-mass spectrometry. *Anal. Chim. Acta* **2012**, *745*, 53–63. [CrossRef] [PubMed]

35. Williams, J.; Ringsdorf, A. Human odour thresholds are tuned to atmospheric chemical lifetimes. *Philos. Trans. R. Soc. Lond B Biol. Sci.* **2020**, *375*, 20190274. [CrossRef] [PubMed]

36. Cometto-Muniz, J.E.; Cain, W.S. Nasal pungency, odor, and eye irritation thresholds for homologous acetates. *Pharmacol. Biochem. Behav.* **1991**, *39*, 983–989. [CrossRef]

37. Cain, W.S.; Lee, N.S.; Wise, P.M.; Schmidt, R.; Ahn, B.H.; Cometto-Muniz, J.E.; Abraham, M.H. Chemesthesis from volatile organic compounds: Psychophysical and neural responses. *Physiol. Behav.* **2006**, *88*, 317–324. [CrossRef]

38. van Thriel, C.; Schäper, M.; Kiesswetter, E.; Kleinbeck, S.; Juran, S.; Blaszkewicz, M.; Fricke, H.H.; Altmann, L.; Berresheim, H.; Brüning, T. From chemosensory thresholds to whole body exposures-experimental approaches evaluating chemosensory effects of chemicals. *Int. Arch. Occup. Environ. Health* **2006**, *79*, 308–321. [CrossRef] [PubMed]

39. Nagata, Y. Measurement of odor threshold by triangle odor bag method. *Odor Meas. Rev.* **2003**, *118*, 118–127.

40. Khan, R.M.; Luk, C.-H.; Flinker, A.; Aggarwal, A.; Lapid, H.; Haddad, R.; Sobel, N. Predicting odor pleasantness from odorant structure: Pleasantness as a reflection of the physical world. *J. Neurosci.* **2007**, *27*, 10015–10023. [CrossRef]

41. DIN EN 13725. *Stationary Source Emissions—Determination of Odour Concentration by Dynamic Olfactometry and Odour Emission Rate from Stationary Sources,* German and English version prEN 13725:2019; Beuth: Berlin, Germany, 2019. [CrossRef]

42. van Thriel, C.; Kiesswetter, E.; Schäper, M.; Blaszkewicz, M.; Golka, K.; Juran, S.; Kleinbeck, S.; Seeber, A. From neurotoxic to chemosensory effects: New insights on acute solvent neurotoxicity exemplified by acute effects of 2-ethylhexanol. *Neurotoxicology* **2007**, *28*, 347–355. [CrossRef] [PubMed]

3

Real-Time Monitoring of Indoor Air Quality with Internet of Things-Based E-Nose

Mehmet Taştan * and Hayrettin Gökozan

Department of Electronic, Turgutlu Vocational School, Manisa Celal Bayar University, 45400 Manisa, Turkey
* Correspondence: mehmet.tastan@cbu.edu.tr.

Abstract: Today, air pollution is the biggest environmental health problem in the world. Air pollution leads to adverse effects on human health, climate and ecosystems. Air is contaminated by toxic gases released by industry, vehicle emissions and the increased concentration of harmful gases and particulate matter in the atmosphere. Air pollution can cause many serious health problems such as respiratory, cardiovascular and skin diseases in humans. Nowadays, where air pollution has become the largest environmental health risk, the interest in monitoring air quality is increasing. Recently, mobile technologies, especially the Internet of Things, data and machine learning technologies have a positive impact on the way we manage our health. With the production of IoT-based portable air quality measuring devices and their widespread use, people can monitor the air quality in their living areas instantly. In this study, e-nose, a real-time mobile air quality monitoring system with various air parameters such as CO_2, CO, PM_{10}, NO_2 temperature and humidity, is proposed. The proposed e-nose is produced with an open source, low cost, easy installation and do-it-yourself approach. The air quality data measured by the GP2Y1010AU, MH-Z14, MICS-4514 and DHT22 sensor array can be monitored via the 32-bit ESP32 Wi-Fi controller and the mobile interface developed by the Blynk IoT platform, and the received data are recorded in a cloud server. Following evaluation of results obtained from the indoor measurements, it was shown that a decrease of indoor air quality was influenced by the number of people in the house and natural emissions due to activities such as sleeping, cleaning and cooking. However, it is observed that even daily manual natural ventilation has a significant improving effect on air quality.

Keywords: internet of things; e-nose; indoor air quality; smart home; ESP32

---

1. Introduction

In recent years, air pollution has been a major environmental problem and a global concern that has exceeded recommended national limits. Air pollution has negative effects on human health and ecosystems, as well as affecting the world's climate [1]. Air pollution can be classified as internal or external air pollution, depending on where the activities take place [2]. Outdoor air pollution occurs in an open environment, covering the entire atmosphere. Fossil fuels used to meet the energy needs of factories, industries and vehicles are the main activities contributing to agricultural and mining outdoor air pollution. The external air pollutants are mainly composed of nitrogen oxides (NOx), nitrogen dioxide (NO_2), sulfur dioxide (SO_2), ozone (O_3), carbon monoxide (CO), hydrocarbons and particulate matter (PM) of different particle sizes. Indoor air pollution, found in offices, hospitals, schools, libraries, entertainment areas, gymnasiums, public transport vehicles, etc., is classified as the pollution of the air of indoor areas [3]. Major indoor air pollutants include NOx, SO_2, O_3, CO, carbon dioxide (CO_2), volatile and semi-volatile organic compounds VOCs, PM, radon and microorganisms. The air quality index includes an internationally adopted parameter for assessing air quality.

Ground-level ozone reflects five pollution standards including PM, carbonic oxide, SO_2 and NO_2 [4]. Indoor air quality (IAQ) is very important for many people who spend most of their lives in closed spaces, such as the elderly, disabled, infants and patients [5]. Indoor air pollution occurs due to household activities and products used in these activities. Home cleaning products and paint materials emit toxic chemicals into the air and cause air pollution. Exposure to air pollution, especially indoor air pollution, is one of the largest environmental health risk factors and is directly related to millions of premature deaths worldwide each year [6]. Air pollution comes second on the list of deaths caused by non-contagious health reasons [7].

In the United States, indoor and outdoor air quality is regulated by the Environmental Protection Agency (EPA). The EPA has shown that indoor pollutant levels may be up to 100 times higher than the level of external pollutants and poor air quality is one of the five most important environmental risks threatening public health [8]. PM, a complex mixture of solid and liquid particles of organic and inorganic substances suspended in the air, is considered the most common pollutant. Air pollution from PM ($PM_{2.5}$ and PM_{10}) poses a significant health threat to people living in cities. The health-damaging PM, which can penetrate deep into the lungs, contributes to the risk of developing cardiovascular diseases as well as lung cancer [9]. In addition to health problems affecting the low IAQ nervous system, its effects are associated with other long-term diseases [10]. Simple interventions, such as natural ventilation provided by homeowners, have significant positive effects on IAQ. However, consideration of the household's thermal comfort and climate conditions in natural ventilation is crucial [11]. Furthermore, power losses due to manually regulated natural ventilation should be considered [12]. Therefore, indoor real-time monitoring of IAQ is very important to detect unhealthy conditions [13].

Today, cities face interesting challenges and problems to meet socio-economic development and quality of life goals, and the concept of "smart cities" responds to these challenges. The smart city is directly related to a strategy to reduce the problems caused by rapid urbanization [14]. The IoT is a network where physical objects are linked to each other or to larger systems. This network collects large amounts of data from different devices we use in our daily life and converts it into utilizable information [15]. The rapidly emerging IoT concept supports many different areas and applications including health, education, agriculture, industry and environmental monitoring. Real-time environmental monitoring and analysis is an important area of research for IoT [16]. IoT provides remarkable features to smart cities which improve environmental quality control, innovative real-life solutions and services. The smart home is an indispensable element of smart cities. In the future, smart homes will be fully managed through smartphone applications and will include IoT wearable devices [17] that are supported by microsensors. To date, several have been conducted in the literature with the aim of establishing real-time monitoring solutions for air quality analysis.

Numerous IoT-based air quality monitoring systems using micro-sensors for data collection have been proposed, including open source technologies for data processing and transmission [18]. According to an IAQ analysis [19], it has been reported that indoor $PM_{2.5}$ pollution mainly comes from outdoor sources. In another study, three typical activities such as a having fireplace in the house, cooking with kitchen appliances and toasting were investigated, and indoor pollutant levels such as $PM_{2.5}$, PM_{10} and VOC were found to be above the limit values [20]. Monitoring of the PM concentration for 63 participants between the ages of 18–65 years in the Perth metropolitan area indicates that the PM concentration increases the number of heartbeats by 4–6 beats per minute (bpm) and has a significant effect on systolic blood pressure (SBP) [21]. An IoT-based IAQ measurement system was created using the WEMOS D1 mini microcontroller and the PMS5003 PM sensor. This system allows the household individual to intervene for ambient assisted living (AAL) [22].

In this study, a low-cost, portable, IoT-based and real-time monitoring system that can measure ambient air quality with a range of sensors is proposed. For the proposed system, an Android interface has been designed primarily by the Blynk platform. Then, with the embedded architecture ESP32 module, the controller unit of the system is created. Using this IoT controller with an internal Wi-Fi module, all measured air quality data are displayed in the mobile app and these data are stored on a cloud server. The mobile interface provides users with numerical and graphical data on contaminating gas concentrations, temperature and humidity. The proposed e-nose measurement system sends a notification to the users via the mobile application if any gas concentration levels reach health-threatening values. In this way, households can take measures to reduce gas concentrations when necessary. The data can be viewed graphically through the mobile interface which allows users to observe the effects of activities such as sleep, cleaning and cooking on the gas concentration. In addition, the IoT-based e-nose air quality measurement system has a low-cost (about \$100), easy-to-install and open-source feature produced by a DIY approach.

2. Related Work

Due to the fast advancement in IoT and sensor techniques, interest in air quality measurements continues to grow. In [23] an environmental monitoring system was propose that included the sensors Raspberry Pi and MICS-4514, which enabled the assessment both indoors and outdoors at a university campus of multiple environmental parameters such as temperature, humidity, light, noise level, CO, NO_2. A proposed system for detecting environmental contamination [24] used various environmental detectors to identify temperature, humidity, ambient light, gas sensors and PM. All data measured by the system using the sensors STM32f4xx and Sharp GP2Y1010 for PM detection, and TGS5342 for CO, are stored in the internal storage and on an Internet server over the Wi-Fi network. Using the ATmega328AVR controller, DHT22 temperature sensor, Sharp GP2Y1010AU0F dust sensor and UVM-30A UV sensors, an Integrated Environmental Monitoring System (IEMS) [25] was proposed to detect the microenvironment. In another study [26], the Nano Environmental Monitoring System (nEMoS) was proposed using an IoT-based indoor environmental quality (IEQ) assessment system created using an Arduino Uno module and low-cost sensors such as DHT22. To determine the quality of packaged products, an IoT-based measuring system was intended, including low-cost pressure, temperature, humidity, gas sensors (BME680, DHT22 and MQ5), Arduino and XBee wireless module [27]. In the study, in which iAQ, an air quality surveillance system based on IoT architecture, was proposed for AAL [28], the wireless sensor nodes (WSN), Gateway and Android user interface provided environmental data such as temperature, humidity and CO_2 to users through the mobile application. In the IoT-based system [29] developed for real-time IAQ surveillance, multiple gases such as CO_2, NO_2, ethanol, methane and propane were detected using ESP8266 as the controller and sensors MICS-6814 and MICS-6814 as the detection unit. This system, based on open source technology, has a mobile phone application that transmits real-time notifications to users. The solution is based on the IoT concept and is entirely wireless in a study that presented an IAQ surveillance solution that can measure temperature, humidity, PM_{10}, CO_2 and light intensity in real-time [30]. The Arduino UNO emitter utilizes the ESP8266 controller and the "ThingSpeak" open-source IoT platform to record wireless data to enable wireless Internet access. In this study, a real-time surveillance system based on IoT architecture for PM surveillance was presented [31]. The iDust system was created using open source technologies and low-cost sensors. System measurement data, consisting of a WEMOS D1 Wi-Fi controller and PMS 5003 dust sensor, are transferred to users via an IoT-based implementation.

3. Materials and Methods

Low air quality poses a significant health threat for individuals who spend most of their time indoors. Some pollutants such as tobacco smoke, CO, NO_2, formaldehyde, asbestos fibres, microorganisms and allergens are known to be closely related to health problems. Temperature and humidity monitoring are part of everyday life, but in the vast majority of buildings, real-time air quality monitoring is not performed. In this study, air quality measurement system with an IoT-based e-nose has been proposed for real-time, low-cost and easy-to-install air quality monitoring. With the proposed e-nose system the ambient temperature and humidity values are measured in real-time in addition to the polluting gases such as CO_2, CO, PM_{10} and NO_2. Information on the presence of these monitored gases in excessive quantities is transmitted to users via the mobile application as a notification. This is a completely wireless solution developed using the ESP32 module, which integrates the IEEE 802.11 b/g/n network protocol into the IoT architecture.

The architecture of the proposed e-nose system is given in Figure 1. A real-time air quality monitoring system provides information about the concentration of pollutants in the environment. It provides precise and detailed information about the air quality of the living environment and helps to plan interventions that lead to improved air quality. The e-nose air quality monitoring system in Figure 1 consists of two parts: The first part is the detection and communication unit consisting of an ESP32 microcontroller-based sensor array with built-in Wi-Fi; and the second is the Android/iOS-based mobile user interface.

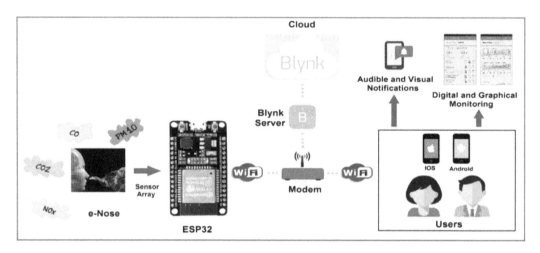

Figure 1. The proposed e-nose system architecture.

The ESP32 module with a built-in Wi-Fi module is used in the e-nose system created for monitoring air quality. The low-cost and high-performance 32-bit controller is frequently preferred in IoT applications. An ESP32 has a dual-core structure and has many internal modules such as Wi-Fi, Bluetooth, RF, IR, CAN, Ethernet module, temperature sensor, hall effect sensor and touch sensor needed for smart home applications. In the ESP32 module structure, the Harvard Tensilica Xtensa LX6 32-bit Dual Core features a processor capable of operating at up to 240 MHz. The detection unit includes the sensors GP2Y1010AU, MH-Z14, MICS-4514 and DHT22, which measure air quality parameters such as CO_2, CO, PM_{10}, NO_2, temperature and humidity.

The GP2Y1010AU is a dust sensor with an analogue output system. An infrared light-emitting diode (IRED) and a phototransistor are arranged across from each other. The IR beams reflected from dust entering the air chamber of the sensor is detected by the phototransistor and generates a corresponding voltage [32].

The MH-Z14A CO_2 sensor module uses the non-dispersive infrared (NDIR) principle. It measures between 0–5000 ppm, 5 ppm resolution with an accuracy of ±50 ppm. The sensor module sends the CO_2 concentration in three different output modes: Serial output (RS-232), analogue output and pulse width modulation (PWM) [33].

The MICS-4514 is mainly used for measuring emissions from automobile exhausts but is also used for measuring concentrations of gases such as NO_2, CO and hydrocarbons. The sensor has a built-in heating element and a micro-sensing diaphragm on the upper side. The MICS-4514 includes two sensor chips with independent heaters and delicate layers. One sensor chip detects oxidizing gases (OX) and the other sensor detects reducing gases (RED) [34].

The DHT22 consists of two parts: A thermistor temperature sensor and a capacitive humidity sensor. The DHT22 is an advanced sensor unit that provides a calibrated digital signal output. It is equipped with an 8-bit microcontroller and has a short response time. It has a relative error of ±0.5 °C in temperature measurement and ±2% rH in humidity measurement [35]. The electronic features of the sensors used in the IoT-based e-nose system are given in Table 1.

Table 1. Electronic features of the sensors in the IoT-based e-nose measuring system.

ID	Equipment Name	Types	Electronic Features
1	CO_2 gas sensor	MH–Z14 [36,37]	Detection range 0–10000 ppm; operating voltage: 4–6 V; accuracy: ± 50 ppm ±5%; resolution: 5 ppm; output Voltage: 0.4–2 V; operating temperature: 0–50 °C
2	NO_2, CO gas sensor	MICS-4514 [38–40]	Detection range 1–1000 ppm (CO); 0.05–5 ppm (NO_2); operating voltage: 4.9–5.1 V; operating temperature: −35–85 °C; heating current: 58 mA
3	Dust sensor	GP2Y1010AU [41–43]	Operating voltage: 5 V; output voltage: 0.9 (no dust)–3.4 V; operating current: max 20 mA; operating temperature: −10 to 65 °C, accuracy ±15%
4	Temperature and Humidity sensor	DHT22 [44–46]	Temperature range: −40 °C to 80 °C; humidity range: 0% to 100%; operating voltage: 3.5–5.5 V; operating current: 60 uA; output: serial; resolution: 0.1 °C and ±1 rH%; accuracy: ±0.5 °C and ±1 rH%; resolution: Temperature and humidity are 16-bit.

The images of the IoT-based e-nose system used in air quality measurements and formed from different sensors are displayed in Figure 2. The sensors used in the system are mounted in a 17 × 12 × 8 cm size sealed box per the measurement specifications. In addition to the sensors used for measurement, the box includes a printed circuit board, a fan for airflow to the dust sensor, and a power supply for the energy of the system.

Figure 2. Images of the IoT-based e-nose system.

The measurement system can work with a 12 V DC adapter, and additionally has a mobile use feature when paired with a power bank that can be connected to the 5 V DC USB port. This feature provides a great advantage for short-term measurements. When supplied from a 5 V power supply, the measuring device draws a current of 160 mA. With a 5000 mAh power bank, it has a measuring time of approximately 30 h. Using a wireless internet connection, it provides the opportunity to measure air quality from many common living areas such as parks, gardens, highways, industrial areas, sports fields, public transportation, cafes, restaurants, schools, hospitals.

The climate parameters and gas concentrations are a median of 12 measurements taken at 5 s intervals. This minimizes the effect of erroneous measurements brought about by faulty sensors. Users are notified by exceeding the designated gas concentration and climate parameter thresholds. Five-minute averages are calculated and delivered to users as notifications, preventing false warnings that may mislead users. In this e-nose system which measures air quality, no memory element is used for data recording. The received data are recorded directly to the Blynk cloud server via the mobile interface. The data are sent to the registered e-mail address of the user when requested. The status of the internet connection is checked before each data transmission. When not connected, data packets that cannot be temporarily transmitted are stored to ensure data integrity, and then, when the internet connection is re-established, these packets are sent to the cloud server using past time tags. In this way, data loss can be prevented by providing continuity in the data flow.

The front panel images of the developed Android-based mobile user interface are given in Figure 3. Control over mobile devices in IoT applications has become very common. There are many free options available for Android and iOS devices. Blynk is one of these applications and it is an IoT platform developed for iOS and Android applications that enables management of different controllers such as Raspberry Pi, ESP8266, ESP32, chipKIT, Intel, LeMarker, Onion Omega, SparkFun and STM32. Using the Blynk cloud server service, digital data such as temperature, humidity, current, voltage measurements and control systems are stored and can be easily accessed at any time. The Blynk graphical components (widgets) allow real-time clock and calendar (RTCC) features to be used.

Figure 3. Mobile user interface developed for the e-nose system.

Table 2 shows the cost table of the IoT-based e-nose measurement system. The total cost of the software and hardware components required to install the system is approximately $100. Compared to most commercially accessible non-IoT-based home air quality meters, this cost is very economical and convenient.

Table 2. Cost table for IoT-based e-nose measurement system.

Component	Cost ($)
ESP-32 Controller	5
GP2Y1010AU Sensor	5
MH-Z14 Sensor	25
MICS-4514 Sensor	20
DHT22 Sensor	3
5V Fan	3
5V Power Sup.	2
PCB	3
Plastic Box	5
Cable, Socket	5
Power Supply/Bank	15
Arduino IDE	free
Blynk IoT Platform	5
Total Cost	**100 $**

4. Results and Discussion

For air quality analysis with IoT-based e-nose, data were collected in a residence for 4 days at one-minute intervals. The housing area where air quality measurements were taken is 150 m^2.

The room in which the measurements were made is 25 m² and the measurements were taken from a height of 1.5 m. The five-person house is heated by a central heating system. Although there were changes in the number of inhabitants during the day, there were always at least two people in the household. There was no air cleaning system in the house and the ventilation of the environment was done by manually opening the windows.

The doors in the household were not closed during the measurement and thus the air was homogeneous throughout the house. Ventilation was carried out once a day for a duration of one hour. Ventilation started at 11:30 on the first day and at 07:30 on the other days.

Figure 4 displays 4-day measurement graphs showing the relationships between humidity–CO_2, CO–CO_2, CO_2–NO_2, CO–NO_2, PM_{10}–CO and PM_{10}–NO_2.

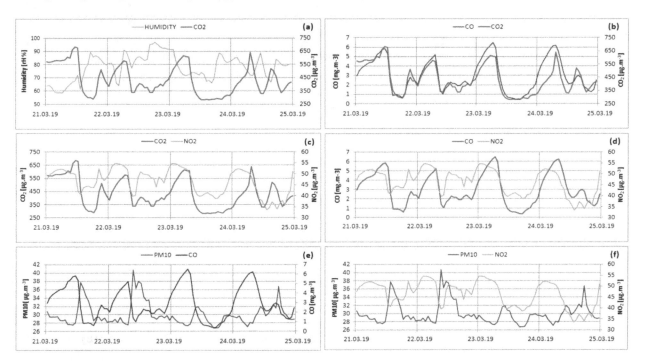

Figure 4. Relationship between gases during four-day measurement, (**a**) Humidity- CO_2, (**b**) CO-CO_2, (**c**) CO_2- NO_2, (**d**) CO-NO_2, (**e**) PM_{10}-CO, (**f**) CO-NO_2.

Figure 5 displays PM_{10}, CO_2, CO and NO_2 measurements over 4 days. When the graph is examined, it shows that the indoor PM_{10}, CO, NO_2 and CO_2 PM and gas concentrations vary depending on the number of household inhabitants and individual activities. It is seen that natural ventilation in the morning hours caused a decrease in CO, NO_2 and CO_2 gas concentrations. In addition, an increase in PM_{10} is observed in the morning as a result of increased activity in the household. These increased rates at the PM_{10} level are 27.1%, 32.2%, 14.4% and 12.5%, respectively.

Figure 6 displays the daily changes of temperature, humidity, PM_{10}, CO, CO_2 and NO_2 values. One-day time, period 1 (00:00–07:30), period 2 (07:30–16:30) and period 3 (16:30–00:00) are divided into three parts; period 1 sleeping of household inhabitants, period 2 during which the daily activities (cooking, cleaning, etc.) of the individuals in the household are performed, and the period 3 of eating and resting of the household inhabitants. The PM_{10} value seen in Figure 6a is in the order of 28.18 µg m^{-3}, 29.93 µg m^{-3} and 28.40 µg m^{-3} for periods 1-3, respectively. It can be seen that PM_{10} has the lowest value in period 1 when the household is asleep. The CO_2 values in the periods are as follows: period 1, 565 µg m^{-3}; period 2, 317 µg m^{-3}; and period 3 has a mean value of 297 µg m^{-3}. The value of CO_2 reaches the highest value in period 1 when the household inhabitants are asleep.

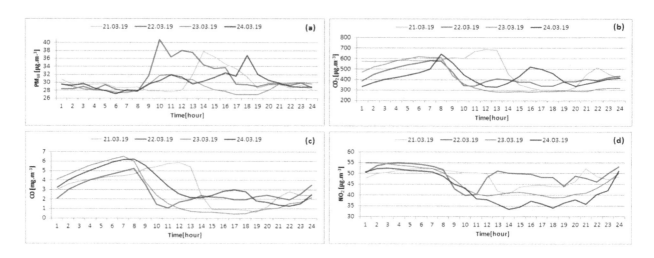

Figure 5. Daily changes of PM and gases, (**a**) PM_{10}, (**b**) CO_2, (**c**) CO, (**d**) NO_2.

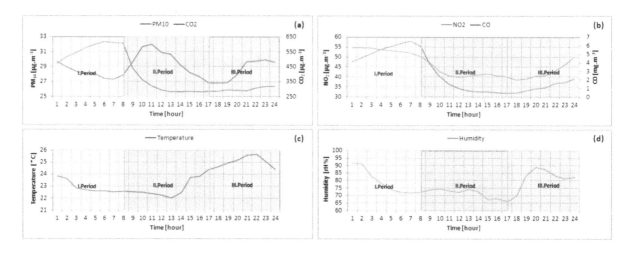

Figure 6. Hourly changes of climate parameters and gases values of day 3, (**a**) PM_{10} - CO_2, (**b**) NO_2-CO, (**c**) temperature, (**d**) humidity.

When Figure 6b is examined, it is seen that NO_2 concentration has average values of 53.15 µg m^{-3} for period 1, 41.59 µg m^{-3} for period 2 and 42.22 µg m^{-3} for period 3. It is seen that the NO_2 concentration value reached the highest value in the period 1, similar to CO_2. The CO concentration has an average of 5.48 mg m^{-3} in the period 1, 1.38 mg m^{-3} in the period 2 and 1.09 mg m^{-3} in the period 3. The CO value, and also CO_2 and NO_2 values, are observed to be highest in the period 1 while the household is asleep. Also, natural ventilation by opening the windows in the morning results in a significant reduction of CO, CO_2 and NO_2 concentrations.

As a result of natural ventilation, the indoor concentration of CO is seen to fall from 6.51 mg m^{-3} to 0.4 mg m^{-3}, NO_2 concentration from 54.7 µg m^{-3} to 38.6 µg m^{-3} and CO_2 concentration from 608 µg m^{-3} to 282 µg m^{-3}.

As a result of one hour of natural ventilation, the increased gas concentrations in the closed and unventilated environment resulted in a significant decrease of 93.8% in CO, 29.4% in NO_2 and 53.6% in CO_2.

Figure 7 shows the minimum, maximum and average values of daily gas concentrations. When the values given in Figure 7 are examined, it is seen that the largest variance is observed in the CO concentration. The CO values increased by nine times on day 1 and 16 times on day 3. Similarly, CO_2 concentration increased by 1.36 times on day 2. NO_2, with a lower change, increased by half on day 4 and PM_{10} concentration increased by half on day 2.

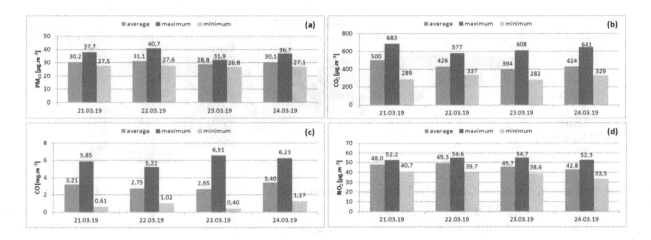

Figure 7. Minimum, maximum and average values of daily gas concentrations and PM, (**a**) PM_{10}, (**b**) CO_2, (**c**) CO, (**d**) NO_2

The daily average concentration values of indoor CO gas are as follows: 3.21, 2.75, 2.65 and 3.40 mg m^{-3}, daily concentrations of NO_2, 47.9, 49.2, 45.6 and 42.8 µg m^{-3}, CO_2, with mean values of 500, 426, 394 and 424 µg m^{-3} and PM_{10} 30.2, 31.1, 28.8 and 30.1 µg m^{-3}.

The correlation relationships between air quality dates have been estimated by the Konstanz Information Miner (KNIME) "Linear Correlation" algorithms. Figure 8 shows the linear correlation workflow of KNIME.

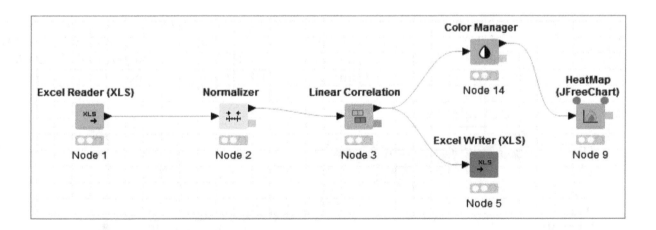

Figure 8. Workflow of Linear Correlation in KNIME application.

Tables 3–5 demonstrate the period 3 linear correlation matrices on 22.03.2019. Table 3 demonstrates the period 1 linear correlation matrix. There was a positive correlation between CO–CO_2, NO_2–humidity and NO_2–temperature at values of 0.943, 0.871, 0.755, respectively, during the period when the household was asleep. In addition, there were 0.979, 0.911 positive and 0.941 negative correlations between time–CO, time–CO_2 and time–NO_2, respectively. There was a strong negative correlation between climate parameters and values of CO and CO_2 and a strong positive correlation between climate parameters and values of NO_2. On the other hand, the correlation between PM10 and gas concentrations and climate parameters were minuscule. The findings indicate that in period 1 there was an elevated positive and negative correlation between gas concentrations and climatic parameters, which represents sleep hours.

Table 3. Linear correlation matrix for period 1.

	Time	CO	NO$_2$	CO$_2$	Temp	Humidity	PM$_{10}$		
Time	1	0.979	−0.941	0.911	−0.866	−0.944	−0.251		1
CO		1	−0.919	0.943	−0.907	−0.969	−0.254		
NO$_2$			1	−0.841	0.755	0.871	0.245		0
CO$_2$				1	−0.886	−0.918	−0.213		
Temp.					1	0.941	0.248		
Humidity						1	0.258		
PM$_{10}$							1		−1

Table 4 demonstrates the period 2 linear correlation matrix. There was a positive correlation between CO–CO$_2$. CO–NO$_2$ and NO$_2$–CO$_2$ at values of 0.966, 0.881, 0.861, respectively, and a negative correlation of 0.915 for temperature–humidity during this period, in which households are undertaking domestic activities (cooking, cleaning, washing, etc.). There were strong negative correlations between time, CO, CO$_2$, humidity and NO$_2$ at −0.844, −0.752, −0.735 and −0.669 during this period. In addition, this period's correlations between PM$_{10}$ value and gas levels and climate parameters were very low.

Table 4. Linear correlation matrix for period 2.

	Time	CO	NO$_2$	CO$_2$	Temp	Humidity	PM$_{10}$		
Time	1	−0.844	−0.669	−0.752	0.640	−0.735	−0.242		1
CO		1	0.881	0.966	−0.266	0.415	0.014		
NO$_2$			1	0.861	−0.098	0.254	−0.092		0
CO$_2$				1	−0.186	0.314	−0.020		
Temp.					1	−0.915	−0.327		
Humidity						1	0.317		
PM$_{10}$							1		−1

The linear correlation matrix provided in Table 5 belongs to period 3, in which families carry out activities such as eating and resting. There was a positive correlation between CO–NO$_2$ and CO–CO$_2$ at 0.833, 0.671, respectively, during this period. Furthermore, there was also a positive correlation between time–CO, time–NO$_2$, time–CO$_2$ and time–PM$_{10}$ at 0.968, 0.846, 0.622 and 0.407, respectively.

Table 5. Linear correlation matrix for period 3.

	Time	CO	NO$_2$	CO$_2$	Temp	Humidity	PM$_{10}$		
Time	1	0.968	0.846	0.622	0.234	0.304	0.407		1
CO		1	0.833	0.671	0.213	0.296	0.379		
NO$_2$			1	0.474	0.213	0.328	0.347		0
CO$_2$				1	0.051	0.001	0.192		
Temp.					1	0.629	0.230		
Humidity						1	0.169		
PM$_{10}$							1		−1

We can conclude that the cost of the proposed e-nose measurement system is very low compared to commercial products sold on the market. Moreover, the majority of domestic air quality meters available for commercial use measure only a restricted amount of air quality and climate parameters. In this system, CO, CO$_2$, NO$_2$, PM$_{10}$, and six parameters of temperature and humidity can be evaluated. Another significant advantage is that our measuring system is IoT-based and provides users with real-time data transfer through its mobile interface. The major drawback of the proposed e-nose system is that no comparison and calibration is conducted with any conventional measuring system. The measuring precision of the system is equivalent to the measuring precision indicated by manufacturers. To rectify this, in our future studies the measuring system will be calibrated first, then the sensor node created and longer-term readings made. Thus, the change in IAQ through individual activities will be disclosed more obviously.

5. Conclusions

Although many people spend most of their lives indoors, they have very limited information about the air quality in their environment. IAQ, which plays an especially important role in the health of children and the elderly, should be measured and necessary ventilation measures should be taken. In recent years, individual air quality measuring devices have been produced owing to the developing information communication technologies, Wi-Fi based microcontrollers and low-cost sensors.

In this study, an IoT-based personalized air quality measurement and monitoring system was proposed by using air quality sensors and DIY approach. The developed e-nose measurement system made data measurements at one-minute intervals and recorded these data to the cloud server. The measurement data can be monitored instantaneously via the Blynk mobile interface, and if the limit values are exceeded, the application sends a notification to the user to take the necessary measures.

According to the four-day measurement results, the following inferences have been obtained:

- IAQ is directly related to the number of people in the household and the activities carried out in the household.

- Activities such as cooking, sleeping, cleaning have a significant effect on CO, NO_2 and CO_2 gas concentrations.

- CO, NO_2, CO_2 gas concentration values of the house reach the highest level during the sleep period (period 1). During this period, PM_{10} concentration has the lowest value.

- PM_{10} concentration reaches the highest level in period 2, when daily routine tasks such as cleaning and house arrangement are carried out.

- When the daily maximum and minimum values were taken into consideration, there was a 16-fold daily maximum change on day 3 in CO concentration, with minimum values of 0.4 mg m^{-3} and maximum 6.51 mg m^{-3}.

- The largest change in PM_{10} concentration was achieved with a minimum of 27.6 μg m^{-3} and a maximum of 40.7 μg m m^{-3} on day 2, resulting in a 47% change.

- Although CO and CO had the lowest concentration in period 3, NO_2 had the lowest concentration in period 2.

- A rapid decrease in CO, NO_2, CO_2 gas concentrations was observed from the moment of natural ventilation, while a concurrent increase in PM_{10} was observed.

- In period 1, the highest positive correlations occurred between time–CO and CO–CO_2; the lowest negative correlations occurred between CO–humidity and time–humidity.

- In period 2, the largest positive correlations occurred between CO–CO_2 and CO–NO_2, whereas the smallest negative correlations occurred between temp–humidity and time–CO.

- The highest positive correlations between time–CO, time–NO_2 and CO–NO_2 occurred in period 3. There was no negative correlation between the weather parameters and gas concentrations during this period.

IAQ has been observed to change depending on the daily activities of the inhabitants. Low air quality is undoubtedly an important parameter that directly affects our health. Even simple measures such as opening only windows to reduce the concentration of harmful gases in the environment can significantly improve IAQ. The numerical data obtained show that the e-nose system is a feature that can contribute to a healthier living environment. In future studies, it is planned to model the effects of individual activities on IAQ by using the e-nose air quality measurement system used in this study.

Author Contributions: Conceptualization, M.T. and H.G.; methodology, H.G.; software, M.T.; validation, M.T.; formal analysis, M.T.; investigation, M.T. and H.G.; visualization, M.T. and H.G.; resources, H.G.; data curation, M.T.; writing—original draft preparation, M.T.; review and editing, H.G.

References

1. Sharma, A.; Mitra, A.; Sharma, S.; Roy, S. Estimation of Air Quality Index from Seasonal Trends Using Deep Neural Network. In Proceedings of the International Conference on Artificial Neural Networks, Island of Rhodes, Greece, 5–7 October 2018; pp. 511–521.

2. Leung, D.Y. Outdoor-indoor air pollution in urban environment: Challenges and opportunity. *Front. Environ. Sci.* **2015**, *2*, 69. [CrossRef]

3. De Gennaro, G.; Dambruoso, P.R.; Loiotile, A.D.; Di Gilio, A.; Giungato, P.; Tutino, M.; Porcelli, F. Indoor air quality in schools. *Environ. Chem. Lett.* **2014**, *12*, 467–482. [CrossRef]

4. Chen, M.; Yang, J.; Hu, L.; Hossain, M.S.; Muhammad, G. Urban healthcare big data system based on crowd sourced and cloud-based air quality indicators. *IEEE Commun. Mag.* **2018**, *56*, 14–20. [CrossRef]

5. Marques, G. Ambient Assisted Living and Internet of Things. In *Harnessing the Internet of Everything (IoE) for Accelerated Innovation Opportunities*; IGI Global: Hershey, PA, USA, 2019; pp. 100–115.

6. Wei, W.; Ramalho, O.; Mandin, C. Indoor air quality requirements in green building certifications. *Build. Environ.* **2015**, *92*, 10–19. [CrossRef]

7. Neira, M.; Prüss-Ustün, A.; Mudu, P. Reduce air pollution to beat NCDs: From recognition to action. *Lancet* **2018**, *392*, 1178–1179. [CrossRef]

8. Seguel, J.M.; Merrill, R.; Seguel, D.; Campagna, A.C. Indoor air quality. *Am. J. Lifestyle Med.* **2017**, *11*, 284–295. [CrossRef] [PubMed]

9. Kampa, M.; Castanas, E. Human health effects of air pollution. *Environ. Pollut.* **2008**, *151*, 362–367. [CrossRef]

10. Annesi-Maesano, I.; Baiz, N.; Banerjee, S.; Rudnai, P.; Rive, S.; Sinphonie Group. Indoor air quality and sources in schools and related health effects. *J. Toxicol. Environ. Health Part B* **2013**, *16*, 491–550. [CrossRef]

11. Tong, Z.; Chen, Y.; Malkawi, A.; Liu, Z.; Freeman, R.B. Energy saving potential of natural ventilation in China: The impact of ambient air pollution. *Appl. Energy* **2016**, *179*, 660–668. [CrossRef]

12. Chen, Y.; Tong, Z.; Wu, W.; Samuelson, H.; Malkawi, A.; Norford, L. Achieving natural ventilation potential in practice: Control schemes and levels of automation. *Appl. Energy* **2019**, *235*, 1141–1152. [CrossRef]

13. Jones, A.P. Indoor air quality and health. *Atmos. Environ.* **1999**, *33*, 4535–4564. [CrossRef]

14. Gökozan, H.; Taştan, M.; Sarı, A. *Smart Cities and Management Strategies*; 2017 Socio-Economic Strategies; Lambert Academic Publishing: Riga, Latvia, 2017; pp. 115–123, ISBN 978-3-330-06982-4.

15. Gubbi, J.; Buyya, R.; Marusic, S.; Palaniswami, M. Internet of Things (IoT): A vision, architectural elements, and future directions. *Future Gener. Comput. Syst.* **2013**, *29*, 1645–1660. [CrossRef]

16. Rathore, P.; Rao, A.S.; Rajasegarar, S.; Vanz, E.; Gubbi, J.; Palaniswami, M. Real-time urban microclimate analysis using internet of things. *IEEE Internet Things J.* **2018**, *5*, 500–511. [CrossRef]

17. Taştan, M. IoT Based Wearable Smart Health Monitoring System. *Celal Bayar Univ. J. Sci.* **2018**, *14*, 343–350. [CrossRef]

18. Pitarma, R.; Marques, G.; Caetano, F. Monitoring indoor air quality to improve occupational health. In *New Advances in Information Systems and Technologies*; Springer: Cham, Switzerland, 2016; pp. 13–21.

19. Yang, Y.; Liu, L.; Xu, C.; Li, N.; Liu, Z.; Wang, Q.; Xu, D. Source apportionment and influencing factor analysis of residential indoor PM2. 5 in Beijing. *Int. J. Environ. Res. Public Health* **2018**, *15*, 686. [CrossRef] [PubMed]

20. Canha, N.; Lage, J.; Galinha, C.; Coentro, S.; Alves, C.; Almeida, S. Impact of Biomass Home Heating, Cooking Styles, and Bread Toasting on the Indoor Air Quality at Portuguese Dwellings: A Case Study. *Atmosphere* **2018**, *9*, 214. [CrossRef]

21. Rumchev, K.; Soares, M.; Zhao, Y.; Reid, C.; Huxley, R. The association between indoor air quality and adult blood pressure levels in a high-income setting. *Int. J. Environ. Res. Public Health* **2018**, *15*, 2026. [CrossRef]

22. Wang, S.; Bolling, K.; Mao, W.; Reichstadt, J.; Jeste, D.; Kim, H.C.; Nebeker, C. Technology to Support Aging in Place: Older Adults' Perspectives. *Healthc. Multidiscip. Digit. Publ. Inst.* **2019**, *7*, 60. [CrossRef]

23. Alvarez-Campana, M.; López, G.; Vázquez, E.; Villagrá, V.; Berrocal, J. Smart CEI moncloa: An iot-based platform for people flow and environmental monitoring on a Smart University Campus. *Sensors* **2017**, *17*, 2856. [CrossRef]

24. Cho, H. An Air Quality and Event Detection System with Life Logging for Monitoring Household Environments. In *Smart Sensors at the IoT Frontier*; Springer: Cham, Switzerland, 2017; pp. 251–270.

25. Wong, M.; Yip, T.; Mok, E. Development of a personal integrated environmental monitoring system. *Sensors* **2014**, *14*, 22065–22081. [CrossRef]

26.   Salamone, F.; Danza, L.; Meroni, I.; Pollastro, M. A Low-Cost Environmental Monitoring System: How to Prevent Systematic Errors in the Design Phase through the Combined Use of Additive Manufacturing and Thermographic Techniques. *Sensors* **2017**, *17*, 828. [CrossRef] [PubMed]

27.   Popa, A.; Hnatiuc, M.; Paun, M.; Geman, O.; Hemanth, D.J.; Dorcea, D.; Ghita, S. An Intelligent IoT-Based Food Quality Monitoring Approach Using Low-Cost Sensors. *Symmetry* **2019**, *11*, 374. [CrossRef]

28.   Marques, G.; Pitarma, R. An indoor monitoring system for ambient assisted living based on internet of things architecture. *Int. J. Environ. Res. Public Health* **2016**, *13*, 1152. [CrossRef] [PubMed]

29.   Marques, G.; Pitarma, R. A cost-effective air quality supervision solution for enhanced living environments through the internet of things. *Electronics* **2019**, *8*, 170. [CrossRef]

30.   Marques, G.; Pitarma, R. Monitoring health factors in indoor living environments using internet of things. In Proceedings of the World Conference on Information Systems and Technologies, Madeira, Portugal, 11–13 April 2017; pp. 785–794.

31.   Marques, G.; Roque Ferreira, C.; Pitarma, R. A system based on the Internet of Things for real-time particle monitoring in buildings. *Int. J. Environ. Res. Public Health* **2018**, *15*, 821. [CrossRef] [PubMed]

32.   GP2Y1010AU0F: Compact Optical Dust Sensor. Available online: http://www.socle-tech.com/doc/IC%20Channel%20Product/sharp%20products.pdf (accessed on 11 June 2019).

33.   Zhengzhou Winsen Electronics Technology CO. Ltd. MH-Z14 CO2 Module Datasheet. 2013. Available online: http://www.futurlec.com/Datasheet/Sensor/MH-Z14.pdf (accessed on 17 June 2019).

34.   MICS-4514 Datasheet. Available online: http://files.manylabs.org/datasheets/MICS-4514.pdf (accessed on 11 June 2019).

35.   Aosong Electronics CO. Ltd. DHT22 Temperature and Humidity Sensor Datasheet. Available online: https://www.sparkfun.com/datasheets/Sensors/Temperature/DHT22.pdf (accessed on 13 June 2019).

36.   Hong, C.S.; Ghani, A.S.A.; Khairuddin, I.M. Development of an Electronic Kit for detecting asthma in Human Respiratory System. In Proceedings of the IOP Conference Series: Materials Science and Engineering, Yogyakarta, Indonesia, 7–8 December 2017; IOP Publishing: Bristol, UK, 2018; Volume 319, p. 012040.

37.   Bimaridi, A.; Putra, K.D.; Djunaedy, E.; Kirom, M.R. Assasment of Outside Air Supply for Split AC system–Part A: Affordable Instrument. *Procedia Eng.* **2017**, *170*, 248–254. [CrossRef]

38.   Suárez, J.I.; Arroyo, P.; Lozano, J.; Herrero, J.L.; Padilla, M. Bluetooth gas sensing module combined with smartphones for air quality monitoring. *Chemosphere* **2018**, *205*, 618–626. [CrossRef]

39.   McKercher, G.R.; Vanos, J.K. Low-cost mobile air pollution monitoring in urban environments: A pilot study in Lubbock, Texas. *Environ. Technol.* **2018**, *39*, 1505–1514. [CrossRef]

40.   Nguyen, T.N.T.; Ha, D.V.; Do, T.N.N.; Nguyen, V.H.; Ngo, X.T.; Phan, V.H.; Bui, Q.H. Air pollution monitoring network using low-cost sensors, a case study in Hanoi, Vietnam. In Proceedings of the IOP Conference Series: Earth and Environmental Science, Pan Pacific Hanoi, Vietnam, 23–24 January 2019; IOP Publishing: Bristol, UK, 2019; Volume 266, p. 012017.

41.   Liu, H.Y.; Schneider, P.; Haugen, R.; Vogt, M. Performance assessment of a low-cost PM2.5 sensor for a near four-month period in Oslo, Norway. *Atmosphere* **2019**, *10*, 41. [CrossRef]

42.   Budde, M.; El Masri, R.; Riedel, T.; Beigl, M. Enabling low-cost particulate matter measurement for participatory sensing scenarios. In Proceedings of the 12th International Conference on Mobile and Ubiquitous Multimedia, Lulea, Sweden, 2–5 December 2013; p. 19.

43.   Carminati, M.; Ferrari, G.; Sampietro, M. Emerging miniaturized technologies for airborne particulate matter pervasive monitoring. *Measurement* **2017**, *101*, 250–256. [CrossRef]

44.   Muangprathub, J.; Boonnam, N.; Kajornkasirat, S.; Lekbangpong, N.; Wanichsombat, A.; Nillaor, P. IoT and agriculture data analysis for smart farm. *Comput. Electron. Agric.* **2019**, *156*, 467–474. [CrossRef]

45.   Xie, J.; Gao, P.; Wang, W.; Xu, X.; Hu, G. Design of Wireless Sensor Network Bidirectional Nodes for Intelligent Monitoring System of Micro-irrigation in Litchi Orchards. *IFAC-PapersOnLine* **2018**, *51*, 449–454.

46.   Wijaya, D.R.; Sarno, R.; Zulaika, E.; Sabila, S.I. Development of mobile electronic nose for beef quality monitoring. *Procedia Comput. Sci.* **2017**, *124*, 728–735. [CrossRef]

4

Chemical Characterization of Electronic Cigarette (e-cigs) Refill Liquids Prior to EU Tobacco Product Directive Adoption: Evaluation of BTEX Contamination by HS-SPME-GC-MS and Identification of Flavoring Additives by GC-MS-O

Jolanda Palmisani [1,*], Carmelo Abenavoli [2], Marco Famele [2], Alessia Di Gilio [1,*], Laura Palmieri [1], Gianluigi de Gennaro [1] and Rosa Draisci [2]

[1] Department of Biology, University of Bari Aldo Moro, via Orabona 4, 70125 Bari, Italy; lapalmieri@libero.it (L.P.); gianluigi.degennaro@uniba.it (G.d.G.)

[2] National Institute of Health, National Centre for Chemicals, Cosmetic Products and Consumer Health Protection, Viale Regina Elena 299, 00161 Rome, Italy; carmelo.abenavoli@iss.it (C.A.); marco.famele@iss.it (M.F.); rosa.draisci@iss.it (R.D.)

* Correspondence: jolanda.palmisani@uniba.it (J.P.); alessia.digilio@uniba.it (A.D.G.);

Abstract: The present study focused on the determination of benzene, toluene, ethylbenzene and xylenes (BTEX) concentration levels in 97 refill liquids for e-cigs selected by the Italian National Institute of Health as representative of the EU market between 2013 and 2015 prior to the implementation of the European Union (EU) Tobacco Product Directive (TPD). Most of the e-liquids investigated (85/97) were affected by BTEX contamination, with few exceptions observed (levels below the limit of quantification (LOQ) of headspace-solid phase micro extraction-gas chromatography-mass spectrometry (HS-SPME-GC-MS) methodology). Across brands, concentration levels ranged from 2.7 to 30,200.0 µg/L for benzene, from 1.9 to 447.8 µg/L for ethylbenzene, from 1.9 to 1,648.4 µg/L for toluene and from 1.7 to 574.2 µg/L for m,p,o-xylenes. The variability observed in BTEX levels is likely to be related to the variability in contamination level of both propylene glycol and glycerol and flavoring additives included. No correlation was found with nicotine content. Moreover, on a limited number of e-liquids, gas chromatography-mass spectrometry-olfactometry (GC-MS-O) analysis was performed, allowing the identification of key flavoring additives responsible of specific flavor notes. Among them, diacetyl is a flavoring additive of concern for potential toxicity when directly inhaled into human airways. The data reported are eligible to be included in the pre-TPD database and may represent a reference for the ongoing evaluation on e-liquids safety and quality under the current EU Legislation.

Keywords: electronic cigarettes; flavoring additives; BTEX; contamination; headspace solid micro phase extraction; gas chromatography-olfactometry; human health; EU regulation

1. Introduction

Electronic cigarette (e-cig) use has increased extremely quickly worldwide over the last decade due to an intense marketing campaign aiming to advertise them as an aid to reducing and/or eliminating addiction to tobacco cigarette smoke [1]. Emerging in 2006 in China, e-cigs became widely available on the market throughout the world in 2008–2009. EU Commission public opinion surveys focused on the smoking attitudes of European citizens across 27 European Union (EU) member states highlighted that e-cig consumption increased from 7.2% to 11.6% between 2011 and 2014 and is expected to increase further [2]. Despite the claims of manufacturers and retailers advertising e-cigs as a healthier way to smoke nicotine and other chemicals in public places, to date reliable sociological data confirming the effectiveness of e-cigs use in changing smokers' behavior (e.g., smoking cessation and/or reduction) are not exhaustive enough to draw certain conclusions [3–5]. On one hand, public opinion surveys have provided data suggesting a relationship between e-cig consumption and quitting and significant reduction of traditional tobacco smoking [6]. On the other hand, however, scientific research still raises doubts regarding the role of e-cigs in smoking cessation and highlights the interchangeable and simultaneous use of e-cigs with tobacco cigarettes [7,8]. Moreover, a controversial debate is still ongoing within the scientific community on potential adverse effects on the health of both users and bystanders. Concerns about e-cig consumption, specifically related to e-liquids composition, are: (a) the potential inhalation exposure to chemicals of concern present in e-liquid formulations as contaminants of the main ingredients (i.e., aromatic hydrocarbons, aldehydes, PAHs, heavy metals); (b) the potential exposure to harmful by-products formed during the vaporization process; and (c) the unknown and unpredictable long-term health effects due to flavoring additive and main ingredient (i.e., glycerol and propylene glycol) inhalation exposure [9–12]. In view of the health-related concerns raised by the international scientific community and EU member states' competent authorities, specific provisions concerning e-cigs manufacture, labelling, and advertising were included in the EU Tobacco Products Directive 2014/40/EU (TPD), entered into force on May 2014 and fully implemented in EU countries between 2016 and 2018 [13]. E-liquids, available on the market in bottles or in replaceable cartridges, are basically a mixture of propylene glycol, glycerol, and water (the latter generally in smaller quantities). The inclusion of propylene glycol and glycerol in e-liquid formulations is common due to humectant and solvent properties, although the use of other chemicals, such as ethanol (EtOH), has been recently reported in literature [14]. This basic formulation may be enriched with nicotine (in variable and allowed quantities) and a wide selection of flavoring additives, in order to provide users a satisfying and enhanced sensory perception while vaping.

1.1. Flavoring Additives

It is estimated that several hundred flavoring chemicals are currently used for e-liquid formulations, allowing consumers to choose on the market among several flavors belonging to menthol, tobacco, fruit (i.e., cherry, blueberry, strawberry, apple), sweets (i.e., caramel, vanilla, liquorice, chocolate) categories, to mention the most popular ones [15,16]. Scientific reports on addictive behaviors highlighted the key role of flavors in vaping initiation, especially among young adults, and the resulting addiction along to nicotine [17]. The inclusion of flavoring additives in e-liquids is one of the most debated issues. They are approved in foods, beverages, and cosmetics and included in the Generally Recognized As Safe (GRAS) list of the Flavors and Extracts Manufacturers' Association (FEMA); therefore their use is intended through ingestion and dermal contact routes, not for direct inhalation. As a result, both short- and long-term effects due to inhalation exposure cannot be predicted. Due to the lack of epidemiological data able to elucidate the issue and to be reliable foundations for human risk assessment, precautionary measures have to be taken. Moreover, besides this general precautionary principle, specific flavorings are worthy of further attention for their potential toxicity. For instance, 2,3-butanedione (usually named diacetyl) has been widely used in the past in microwave popcorn in the USA with the purpose to generate, depending on the concentration, buttery and caramel tastes. It is a chemical mentioned in the GRAS list and approved in certain limits for ingestion, therefore it is used as additive in foods [18]. Due to its flavoring properties it is also used in the manufacturing process of e-liquid formulations and its presence has been documented in previous investigations carried out in EU Member States, raising concerns in the scientific community regarding potential health implications [19,20]. In this regard, recently published scientific papers based on epidemiological data collected over recent decades have revealed that inhalation exposure to diacetyl is likely related to increased risk of a specific lung disease called bronchiolitis obliterans [21,22]. The use of flavoring chemicals for e-liquid manufacture stimulated scientists to focus on safety and quality aspects of the formulations. As a result, the number of scientific publications on the chemical characterization of e-liquids in terms of flavoring additives has recently increasing. To cite the most recent studies, in 2017 Aszyk et al. carried out a comprehensive determination of flavoring additives on 25 e-liquid samples highlighting that limonene and benzyl acetate were the two most frequently detected [23]. In 2018 Girvalaki et al. reported findings from qualitative and quantitative analysis performed on 122 of the most commonly sold e-liquids in 9 EU member states. Among the 293 flavoring chemicals identified, menthol was the most frequently detected compound, regardless the overall e-liquid flavor [24]. Specific flavoring chemicals with known respiratory irritant properties or identified as inhalation toxicants were detected in other studies in relevant amounts, i.e., benzaldehyde by Kosmider et al., methyl cyclopentenolone and menthol by Vardavas et al., diacetyl and acetylpropionyl by Barhdadi et al. [19,25,26].

1.2. E-Liquids Contamination

The attention of scientists to the chemical composition of e-liquids has not only been aimed at the identification of flavoring chemicals, but also to address the issue concerning the potential presence of compounds of toxicological concern, such as volatile organic compounds (VOCs), due to main component contamination and low purity level of nicotine and flavors [11,27–30]. Among VOCs, aromatic hydrocarbons have attracted remarkable attention in view of a toxicity assessment of refill liquids due to the recognized carcinogenic properties of benzene, classified as carc. 1A according to EU CLP regulation [31]. Specific investigations were carried out to perform both qualitative and quantitative characterization in terms of VOCs of e-liquids commercially available on the EU market prior to the EU TPD implementation and after 2016, in order to verify the compliance of e-liquids distributed over EU countries with the TPD in force, in terms of both chemical composition and classification/labelling [24,32]. With specific regard to aromatic hydrocarbons, BTEX contamination has been detected in e-liquids available on extra-EU markets. Lim et al. highlighted the potential health hazards for e-cig users reporting the results of investigations made on 283 flavored liquids, 21 nicotine-content liquids, and 12 disposable cartridges [33]. BTEX coexisted in most of the investigated samples at relevant concentrations (e.g., benzene concentration ranging from 0.008 to 2.28 mg/L) and the contamination was hypothetically related to the use of petrogenic hydrocarbons in the extraction process of nicotine and flavors from natural plants. BTEX contamination of liquid formulations was also previously observed by Han et al. in a study aiming to assess VOCs levels in 55 refill liquids of 17 different brands available on the Chinese market [34]. Benzene and m,p-xylenes were found in all of the samples investigated, whilst ethylbenzene and toluene were detected with different frequencies. They all were present at comparable levels in the concentration range 1.10–17.31 µg/g. In view of the findings obtained to date on e-liquids composition in terms of a broad range of chemicals, reported above, it appears clear that the attention on the issue has to remain high to ensure that consumers' health is safeguarded and that compliance to safety and quality standards is guaranteed. On one hand there is the need for a comprehensive database referred to e-liquids both manufactured and imported in EU member states before the implementation of the TPD allowing us to define a pre-TPD baseline reference useful for comparison. On the other hand, ongoing investigations into e-liquids currently on the market are necessary to evaluate the effectiveness of TPD provisions in EU member states with regard to the manufacture and labelling of e-liquids, and to formulate further recommendations to policymakers.

1.3. Aim of the Present Study

The aim of the present study was to evaluate BTEX contamination across a representative group of refill liquids for e-cigs (n = 97) and to identify, in a selected sub-group (n = 5), the main flavoring additives responsible for the flavor/taste perceived. BTEX quantification was carried out applying headspace-solid phase micro extraction-gas chromatography-mass spectrometry (HS-SPME-GC-MS) methodology.

The identification of flavoring additives was performed applying a hydrid analytical-sensory technique, the gas chromatography-mass spectrometry-olfactometry (GC-MS-O). This research activity has been carried out in the context of a more comprehensive national project supported by the Italian Ministry of Health and coordinated by the National Institute of Health aimed to evaluate in a comprehensive manner potential risks related to e-cig consumption. The refill liquids investigated were selected through a preliminary survey and were considered representative of the EU market between 2013 and 2015, prior to the implementation of TPD in most of EU member states. Therefore, the data here reported are eligible to be included in the pre-TPD database on e-liquids manufactured and/or imported in EU and may represent a useful reference for the ongoing evaluation on e-liquid safety and quality under the current EU Legislation.

2. Materials and Methods

2.1. E-Liquids Selection

In the framework of the national research project, the Italian National Institute of Health carried out a preliminary survey allowing to identify the most popular brands of e-liquids manufactured and imported in EU and representative of the EU market between 2013 and 2015. Ninety-seven e-liquids of 12 different brands, with and without Nicotine and characterized by different flavors, were purchased online from EU manufacturers and importers in 10–30 ml plastic bottles, as sold commercially. More specifically, the selected e-liquids were manufactured in Italy (n = 45), China (n = 28), France (n = 8), UK (n = 8), Germany (n = 4), and the USA (n = 4). E-liquid composition in terms of propylene glycol, glycerol, water content (expressed in %), nicotine content (expressed as mg/ml or mg/g) as well as characteristic flavor is reported in Table 1, as declared on product label. E-liquids belonging to different brands and within the same brand were classified with progressive letters and number, respectively (sample ID in Table 1). Moreover, three identical e-liquids in terms of brand, basic composition, flavor and nicotine content (e.g., samples 10, 11 and 12 C) belong to different production batches. Nicotine-containing e-liquids were 59 with variable content (11,14,16 and 18 mg/ml and 11, 18 mg/g), as reported on the product label. The remaining 38 e-liquids were declared nicotine-free. Most of the investigated e-liquids were flavored and may be included in the following typical flavor categories: tobacco (48), mint (17), sweets/candy (11), spicy (7), fruits (3), coffee (3), and alcohol (3). Before analysis, all e-liquids were properly stored at room temperature and kept away from direct sunlight, as recommended on the product label.

2.2. BTEX Determination by HS-SPME-GC-MS Analysis

2.2.1. Standards and Reagents

The reference standard benzene (99.96%), toluene (99.93%), ethylbenzene (≥ 99.90%), p-xylene (99.90%), and benzene-d_6 (99.99%), the latter used as internal standard (IS), were purchased from Sigma Aldrich. The reagents methanol and propylene glycol used for the preparation of standard/calibration solutions, as well as blank and samples solutions of a purity grade of more than 99%, were purchased from Sigma Aldrich.

2.2.2. Standards and Calibration Solutions

For each compound, internal standard included, two standard stock solutions were preliminarily prepared. The first set of standard solutions (S1) was prepared diluting reference standards in methanol at a concentration of about 9×10^7 µg/L. The second set of standard solutions (S2) was prepared diluting S1 solutions in methanol (1:100 dilution) to obtain a concentration of about 9×10^5 µg/L.

Table 1. E-liquids composition and information: manufacturing country, % of the main components, characteristic flavor and nicotine content (expressed as mg/mL or mg/g).

Sample ID	Manufacturing Country	Propylen Glycol (%)	Glycerol (%)	Water (%)	Other (%) *	Flavor	Nicotine (mg/mL, ** mg/g)
1-A	China	–	–	–	–	coca cola	18
2-A	China	–	–	–	–	kiwi	18
3-A	China	–	–	–	–	Davidoff-tobacco	11
4-A	China	–	–	–	–	Green USA mix-tobacco	11
5-A	China	–	–	–	–	cigar	11
1-B	Italy	–	–	–	–	cuban cigar	18
2-B	Italy	–	–	–	–	natural	18
3-B	Italy	–	–	–	–	mint	0
4-B	Italy	–	–	–	–	tobacco USA	18
5-B	Italy	–	–	–	–	Virginia blend tobacco	0
6-B	Italy	–	–	–	–	natural	0
7-B	Italy	–	–	–	–	coffee	0
8-B	Italy	–	–	–	–	anise	0
9-B	Italy	–	–	–	–	cuban cigar	0
10-B	Italy	–	–	–	–	rhum	0
11-B	Italy	–	–	–	–	biscuit	0
12-B	Italy	–	–	–	–	anise	18
13-B	Italy	–	–	–	–	liquirice	18
14-B	Italy	–	–	–	–	biscuit	18
15-B	Italy	–	–	–	–	tobacco USA	0
16-B	Italy	–	–	–	–	mint	18
17-B	Italy	–	–	–	–	Virginia blend tobacco	18
1-C	Italy	50	40	5–10	–	Virginia blend tobacco	0
2-C	Italy	50	40	5–10	–	Virginia blend tobacco	18
3-C	Italy	50	40	5–10	–	Virginia blend tobacco	0
4-C	Italy	50	40	5–10	–	basic flavor	18
5-C	Italy	50	40	5–10	–	basic flavor	0
6-C	Italy	50	40	5–10	–	anise	18
7-C	Italy	50	40	5–10	–	mint	0
8-C	Italy	50	40	5–10	–	mint	18
9-C	Italy	50	40	5–10	–	biscuit	0
10-C	Italy	50	40	5–10	–	biscuit	18
11-C	Italy	50	40	5–10	–	biscuit	18
12-C	Italy	50	40	5–10	–	biscuit	18
13-C	Italy	50	40	5–10	–	cuban cigar	0
14-C	Italy	50	40	5–10	–	cuban cigar	18
15-C	Italy	50	40	5–10	–	tobacco USA	0
16-C	Italy	50	40	5–10	–	tobacco USA	18
17-C	Italy	50	40	5–10	–	rum	0
18-C	Italy	50	40	5–10	–	cognac	0
19-C	Italy	50	40	5–10	–	coffee	0
20-C	Italy	50	40	5–10	–	liquirice	18
1-D	Italy	–	–	–	–	mint	0
2-D	Italy	–	–	–	–	mint	14
3-D	Italy	–	–	–	–	black tobacco	0
4-D	Italy	–	–	–	–	black tobacco	14
5-D	Italy	–	–	–	–	Virginia blend tobacco	0

Table 1. *Cont.*

Sample ID	Manufacturing Country	Propylen Glycol (%)	Glycerol (%)	Water (%)	Other (%) *	Flavor	Nicotine (mg/mL, ** mg/g)
6-D	Italy	-	-	-	-	Virginia blend tobacco	14
7-D	Italy	-	-	-	-	tobacco	0
8-D	Italy	-	-	-	-	tobacco	14
1-E	China	80	20	-	-	cuban cigar	0
2-E	China	80	20	-	-	cuban cigar	16
3-E	China	80	20	-	-	Davidoff-tobacco	0
4-E	China	80	20	-	-	Davidoff-tobacco	16
5-E	China	80	20	-	-	Virginia blend tobacco	0
6-E	China	80	20	-	-	Virginia blend tobacco	16
1-F	China	>45	<12	<1	<42	almond	11**
2-F	China	>45	<12	<1	<42	bubble gum	11**
3-F	China	>45	<12	<1	<42	cigar	11**
4-F	China	>45	<12	<1	<42	cigar	18**
5-F	China	>45	<12	<1	<42	cinnamon	11**
6-F	China	>45	<12	<1	<42	coffee	11**
7-F	China	>45	<12	<1	<42	Davidoff-tobacco	11**
8-F	China	>45	<12	<1	<42	Davidoff-tobacco	18**
9-F	China	>45	<12	<1	<42	lemon	11**
10-F	China	>45	<12	<1	<42	Marlboro cigarettes	11**
11-F	China	>45	<12	<1	<42	Marlboro cigarettes	18**
12-F	China	>45	<12	<1	<42	mint	11**
13-F	China	>45	<12	<1	<42	tobacco	11**
14-F	China	>45	<12	<1	<42	tobacco	18**
15-F	China	>45	<12	<1	<42	fruits	11**
16-F	China	>45	<12	<1	<42	Virginia blend tobacco	11**
17-F	China	>45	<12	<1	<42	Virginia blend tobacco	18**
1-G	France	>80	<20	-	-	American blend-tobacco	0
2-G	France	>80	<20	-	-	American blend-tobacco	18
3-G	France	>80	<20	-	-	Virginia blend tobacco	0
4-G	France	>80	<20	-	-	Virginia blend tobacco	18
5-G	France	>80	<20	-	-	Habanos cigar-tobacco	0
6-G	France	>80	<20	-	-	Habanos cigar-tobacco	18
7-G	France	>80	<20	-	-	mint	0
8-G	France	>80	<20	-	-	mint	18
1-H	United Kingdom	50	50	-	-	tobacco	0
2-H	United Kingdom	50	50	-	-	tobacco	18
1-I	United Kingdom	>80	<20	-	-	Virginin Leaf - tobacco	0
2-I	United Kingdom	>80	<20	-	-	Virginin Leaf - tobacco	18
3-I	United Kingdom	>80	<20	-	-	mint	0
4-I	United Kingdom	>80	<20	-	-	mint	18
5-I	United Kingdom	>80	<20	-	-	cuban cigar	0
6-I	United Kingdom	>80	<20	-	-	cuban cigar	18
1-L	Germany	50	50	-	-	chocolate/vanille	0
2-L	Germany	50	50	-	-	chocolate/vanille	18
3-L	Germany	50	50	-	-	mint/herbs	0
4-L	Germany	50	50	-	-	mint/herbs	18
1-M	USA (California)	50	50	-	-	mint/vanilla/chocolate	0
2-M	USA (California)	50	50	-	-	mint/vanilla/chocolate	18
1-N	USA	20	80	-	-	thin mint	0
2-N	USA	20	80	-	-	thin mint	18

* Other components declared on the product label: tobacco essential and leaf oil, nicotine from tobacco leaf, plant extracts, trace level compounds; ** nicotine concentration expressed as mg/g; (-) means that information was not provided on product label.

Starting from S2 and with subsequent dilution with methanol, five solutions for each compound were prepared in the concentration range, approximately 20.0–450.0 µg/L (S3–S7). In order to simulate e-liquid basic composition, five matrix-matched calibration solutions for each compound were prepared by adding 100 µl of the corresponding S3–S7 solutions and 100 µl of benzene-d_6 solution (S2 set) in a headspace (HS)-vial containing 1 ml of laboratory-made liquid (90% propylene glycol, 10% water). Similarly, a blank solution was also prepared by adding 100 µl of benzene-d_6 solution (S2 set) and 100 µl of methanol in a HS-vial containing 1 ml of laboratory-made liquid (90% propylene glycol, 10% water). Both blank and matrix-matched calibration solutions were used for calibration, resulting in five concentration levels for each compound in the dynamic range between limit of quantification (LOQ) value and 45.0 µg/L.

2.2.3. Sample Preparation

Sample preparation prior analysis required the dilution of an aliquot of refill liquid (1 ml) with 100 µl of methanol and 100 µl of IS solution. The dilution with a proper solvent is fundamental to avoid inhomogeneous samples due to the difficulty in sampling exact volumes of high viscosity fluids [35].

2.2.4. HS-SPME-GC-MS Method Conditions and Performance Characteristics

The collection of BTEX in the volatile fraction of both calibration and sample solutions was carried out in 20-ml HS vials with magnetic screw caps provided with polytetrafluoroethylene (PTFE)/silicone septa (Agilent Technologies). BTEX were collected through adsorption onto the polydimethylsiloxane (PDMS) stationary phase-coated fused silica fiber (thickness 100 µm, length 1 cm) introduced into the sample vial. The PDMS fiber was left in the vial for 30 s at 50 °C. Mechanical stirring was performed for 5 s with a stirring speed of 500 rpm. Analyses were performed using a gas chromatograph (7890B Agilent Tecnologies, Santa Clara CA, USA) equipped with an automated sampler (Pal System, CTC Analytics AG, Zwingen, Switzerland), a split/splitless injector and a single-quad mass spectrometer (5977A Agilent Technologies, Santa Clara CA, USA). Once incubation was completed, the heated gas-tight syringe containing the fiber was automatically transferred into the GC injector via the automated sampler and BTEX were thermally desorbed at 250 °C for 300 s and injected into the GC column in split injection mode (split ratio 1:10). Separation was performed on capillary column semivolatiles, 30 m × 0.25 mm, i.d. 0.25 µm film thickness (Phenomenex). Helium (purity ≥ 99.999%) was applied as carrier gas at a constant flow rate of 1 ml/min. The GC oven temperature program used for optimal separation was: 40 °C for 2 min, ramped 8 °C/min up to 80 °C, then ramped 60 °C/min up to 250 °C. Transfer line and ion source temperatures were kept at 260 °C and 270 °C, respectively. The mass spectrometer was operated in electron impact (EI) ionization mode (70 eV). Identification of BTEX was based on comparison of the obtained mass spectra with those included in the National Institute of Standards and Technology (NIST) library (MassHunter software) and considered positive by library search match >800 for both forward and reverse matching. Further criteria for compounds identification were: (a) the matching of relative retention times (t_R) with those of the authentic standards within the allowed deviation of ± 0.05 min; and (b) the matching of ion ratios collected with those of the authentic standards within a tolerance of ± 20%. Quantification was performed in a selected ion monitoring mode (SIM). One quantifier ion and two qualifier ions were selected for each compound on the basis of their selectivity and abundance: 79 m/z as quantifier ion and 51 and 39 m/z as qualifier ions for benzene; 91 m/z as quantifier ion and 65 and 39 m/z as qualifier ions for toluene; and 91 m/z as quantifier ion and 106 and 51 m/z as qualifier ions for ethylbenzene and xylenes. Five point matrix-matched calibration curves were constructed for quantification (r^2 > 0.995) reporting compound/benzene-d_6 quantifier ion peak areas ratio vs amount ratio. Calibration curves were in the range 2.6–41.6 µg/L for benzene, 2.7–43.2 µg/L for toluene and xylenes isomers and 2.8–44.8 µg/L for ethylbenzene. The xylenes isomers were quantified on the basis of p-xylene response factor (e.g., p-xylene calibration curve) and reported as sum in Table 2. Chromatograms of a blank sample and a sample spiked with the BTEX standard solution (calibration level 3) were compared in Figure S1 (Supplementary Material, Figure S1). The main performance characteristics of the HS-SPME-GC-MS method were also evaluated. Linearity was calculated on the basis of three sets of replicates for each calibration level on three different days. As for the results, all matrix-matched calibration curves were linear over the set concentration ranges: relative accuracy (%) for each point was within the $\pm$ 5% of the expected concentrations, and all coefficients of determination (r^2) were >0.995. Selectivity/specificity was assessed directly onto the chromatograms obtained from the blank and from spiked matrices. The occurrence of possible extra peaks was tested by monitoring in SIM mode qualifier and quantifier ions characteristic for each investigated compound onto the blank matrix chromatograms, within the retention time window expected for the analyte elution. Limit of detection (LOD) and LOQ values were assessed in the spiked matrix by determining the lowest concentration of the analytes that resulted in a signal-to-noise (S/N) ratio of $\geq$ 3 and $\geq$ 10, respectively. LOD values were 1.4 µg/L for benzene and toluene, 1.5 µg/L for xylenes, and 1.6 µg/L for ethylbenzene. LOQ values were 2.6 µg/L for benzene, 2.7 µg/L for toluene and xylenes and 2.8 µg/L for ethylbenzene. Repeatability expressed as intra-day coefficients of variation (CV%) was evaluated on a set of results (n = 6 replicates) obtained for each analyte at three validation levels (i.e., LOQ values; 10.4 µg/L for benzene, 10.8 µg/L for toluene and xylenes, 11.2 µg/L for ethylbenzene; 41.6 µg/L for benzene, 43.2 µg/L for toluene and xylenes and 44.8 µg/L for ethylbenzene). Intra-day CV% values were 1.2–4.5% for benzene, 1.2–9.9% for toluene, 3.2–10.9% for ethylbenzene and 2.8–11.4% for xylenes. Intermediate precision (expressed as inter-day CV%) and recovery were calculated by analyzing the series within the three different days (n = 18 replicates). Inter-day CV% values were 5.1–15.3% for benzene, 6.6–10.0% for toluene, 8.8–14.6% for ethylbenzene and 9.4–15.4% for xylenes. Finally, recoveries were in the range of 96.6–113.0%.

2.3. Identification of Flavoring Additives by GC-MS-O Analysis

GC-MS-O methodology was revealed to be a powerful approach for accurate identification of volatile odor-active compounds in high-level complexity matrices through coupling traditional chromatographic analysis with human sensory perception [36–38]. For this reason, GC-MS-O methodology was applied in the present study, allowing us to accurately identify, on a limited number of e-liquids, the odor-active compounds responsible for the overall flavor perceived or of specific flavor notes.

2.3.1. Sample Selection and Preparation

The e-liquids subjected to the in-depth investigation were e-liquids with ID A 1-5 manufactured in China, with medium-high nicotine content and characterized by flavors covering different categories, from tobacco to fruits (Table 1). The aforementioned e-liquids were chosen for further study on the basis of collected data from BTEX investigation that highlighted high level of contamination. Moreover, during the preliminary survey and e-liquid selection made by the National Institute of Health, the brand A was already considered worthy of particular attention due to previous precautionary seizing actions made by Italian authorities and financial police. The preparation of the gaseous sample for GC-MS-O analysis starting from e-liquid formulation involved the use of the Adsorbent Tube Injector System device (ATIS™, Supelco). Before gaseous sample preparation, 250 µl of each e-liquid was preliminarily diluted, adding 250 µl of methanol, resulting in a solution with final volume of 500 µl. An aliquot (100 µL) of the obtained solution was injected by a syringe through the septum of the ATIS injection glassware and the volatile fraction was conveyed by ultrapure air flow (50 mL/min) into a collecting bag (Nalophan®), connected at the outlet of the injection glassware, resulting in a gaseous sample with a final volume of 2 L. The temperature, controlled by a thermometer inserted into the heating block, was set at 120°C. As a result, only the volatile fraction was collected into the bag, avoiding the vaporization of the high-boiling point fraction composed by propylene glycol and glycerol that would have resulted in two broad chromatographic peaks in the GC chromatogram.

2.3.2. GC-MS-O Analysis Conditions

The VOCs collected were analyzed using an air sampler-thermal desorber integrated system (UNITY 2™Markes International Ltd, Llantrisant, UK) connected to a gas chromatograph (7890 Agilent Technologies, Santa Clara CA, USA) equipped with an Olfactory Detection Port (ODP 3 Gerstel GmbH&Co, Mülheim an der Ruhr, Germany) and a single-quad mass spectrometer (5975 Agilent Technologies, Santa Clara CA, USA). The collection of VOCs onto the sorbent-pack focusing trap at −10°C of the desorption system UNITY2™ was performed by connecting the Nalophan bag to the inlet port of the automated air sampling device. The cold trap was flash heated to 300 °C and the compounds were transferred via the heated transfer line (200 °C) to the GC column and to the ODP port. The chromatographic separation was performed on a HP5-MS capillary column (30m × 250μm × 0.25μm). Carrier gas (Helium) flow was controlled by constant pressure and equal to 1.7 ml/min. The GC oven temperature program was set as follows: from 37 °C up to 100 °C at 3.5 °C/min (ramp 1); and from 100 °C up to 250 °C at 15 °C/min (ramp 2). After the GC separation, the column flow was split into two parts (ratio 1:1), one part was connected to the MS detector and the other one to ODP. The transfer line connecting the GC column and MS detector was kept at 250 °C. The mass spectrometer was operated in electron impact (EI) ionization mode (70eV) in the mass range 20–250 m/z. The effluent from the capillary column was connected to the ODP port through an uncoated transfer line (deactivated silica capillaries), constantly heated to prevent compounds condensation. Two trained panelists, one male and one female (24 years old), were asked to sniff in the conical ODP simultaneously with the GC run, indicating exactly when they start and stop perceiving the odor and providing a qualitative description of the odor (using suitable descriptors) [36] and odor intensity based on an intensity scale from 0 (no odor perceived) to 4 (strong odor). Auxiliary air (make-up gas) was added to the GC effluent to prevent the assessors' nose mucous membranes drying, which may potentially cause discomfort, especially in extended analysis sessions. The panelists involved in the present study had previously been selected according to a standardized procedure used for the panel selection in Dynamic Olfactometry, the official methodology for odor emissions assessment standardized by a European technical law (EN 13725/2003) [39]. The standardized procedure provides for individuals with average olfactory perception sensitivity that constitute a representative sample of the human population. The screening was performed evaluating the response to the most used reference gas, 1-butanol. Only assessors who fulfilled predetermined repeatability and accuracy criteria were selected as panelists. The identification of flavoring additives and other VOCs in e-liquid formulation was performed by comparing the mass spectra obtained with those listed in the NIST library (Agilent Technologies). It was considered valid when the confidence rating of mass spectra comparison was superior or equal to 95%. The attribution was further confirmed using the retention times of authentic compounds. Before GC-MS-O sessions, panelists were asked to carry out preliminary sensory tests by sniffing and vaping the liquid formulations. This preliminary approach revealed to be useful in appreciating discrepancies between the flavors reported on e-liquid labels and the overall flavor perceived by panelists' noses and mouths (see Section 3.2 in results section).

3. Results

3.1. Quantitative Analysis: BTEX Contamination of the Investigated E-Liquids

Single and total BTEX concentrations, expressed in μg/L, are reported in Table 2. As shown, most of the e-liquids investigated in the present study (85/97) were revealed to be affected, to a lesser or greater extent, by BTEX contamination. Only a few exceptions were observed with BTEX levels below the LOQ of the analytical methodology applied. Across all of the brands investigated (ID A-N, Table 1), concentration levels ranged from 2.7 μg/L to 30,200.0 μg/L for benzene, from 1.9 μg/L to 447.8 μg/L for ethylbenzene, from 1.9 μg/L to 1,648.4 μg/L for toluene and, finally, from 1.7 μg/L to 574.2 μg/L for m,p,o-xylenes. HS-SPME-GC-MS analysis of e-liquids with ID A (1-5), manufactured in China, highlighted a relevant contamination by BTEX with concentration levels up to four order of magnitude higher than those determined in all the other investigated e-liquids, regardless of the manufacturing country and the chemical composition. More specifically, within brand A, benzene concentration levels ranged from 7,200.0 µg/L (sample 4-A) to 30,200.0 µg/L (sample 3-A), toluene concentration levels ranged from 764.4 µg/L (sample 1-A) to 1,648.4 µg/L (sample 4-A), ethylbenzene concentration levels ranged from 187.9 µg/L (sample 1-A) to 447.8 µg/L (sample 4-A) and, finally, m,p,o-xylenes concentration levels ranged from 201.8 µg/L (1-A) to 574.2 µg/L (sample 5-A). Moreover, making a comparison among samples ID A in terms of BTEX total concentration, it is possible to observe that 3-A shows the highest BTEX total concentration, equal to 32,151.1 µg/L. The comparison between samples ID A with all the other samples under investigation (ID B-N) revealed that benzene concentrations in 1-5 A samples were between one and four orders of magnitude higher than those determined in all the other e-liquids. Moreover, toluene concentrations in 1-5 A samples were up to three order of magnitude higher than those determined in all the other e-liquids, whilst ethylbenzene and m,p,o-xylenes were up to two order of magnitude higher. Benzene concentrations in 1–5 A samples were higher than toluene concentrations (from 4 to 22 times higher), a finding that was not observed for all the other samples characterized by toluene concentrations higher than benzene concentrations, with very few exceptions. To mention some examples, e-liquids with ID E and F manufactured in China showed toluene concentrations ranging from 20.7 µg/L to 96.2 µg/L and from 6.8 µg/L to 385.9 µg/L, respectively, in both cases one up to two order of magnitude higher than benzene concentrations. As already mentioned, some of the samples investigated were not affected by BTEX contamination. It is possible to observe that in most of the samples C (i.e., 1,2,3,8,10,12,14,15 and 17) and in samples 5D, G5, G6 the presence of BTEX was not detected at all with all concentration levels below the LOQ of the analytical methodology applied. Therefore the samples with ID C manufactured in Italy were revealed to be the highest quality e-liquids among all the tested samples. On the contrary, across samples with ID B-N, the highest BTEX total concentrations were associated with samples belonging to the batch with ID F (manufacture country China) with samples 12-F and 17-F showing the highest values, equal to 739.2 µg/L and 743.8 µg/L respectively. Therefore, it is possible to state that the highest BTEX contamination was observed in e-liquids belonging to two different brands (A and F), both of Chinese origin. Another important observation is that the highest BTEX total concentrations observed for most of the brands were associated with e-liquids characterized by mint flavor (brands B, F and L) and tobacco flavor (brands D, E, F and I).

Table 2. Benzene, toluene, ethylbenzene and xylenes (BTEX) concentration (expressed in µg/L) in the investigated e-liquids.

E-Liquid ID	Flavor	Benzene (µg/L)	Ethylbenzene (µg/L)	Toluene (µg/L)	m,p,o-Xylenes (µg/L)	BTEX Total (µg/L)
1-A	coca cola	11,000.0	187.9	764.4	201.8	12,154.1
2-A	kiwi	16,700.0	305.1	902.5	388.6	18,296.2
3-A	Davidoff-tobacco	30,200.0	295.8	1,331.7	323.6	32,151.1
4-A	Green USA mix-tobacco	7200.0	447.8	1,648.4	559.1	9,855.3
5-A	cigar	12,900.0	442.0	1,566.0	574.2	15,482.2
1-B	cuban cigar	<LOQ	<LOQ	5.9	<LOQ	5.9
2-B	natural	<LOQ	2.8	4.4	3.6	10.8
3-B	mint	2.7	39.0	42.8	77.3	161.8
4-B	tobacco USA	<LOQ	<LOQ	3.3	<LOQ	3.3
5-B	Virginia blend tobacco	<LOQ	<LOQ	3.6	<LOQ	3.6
6-B	natural	<LOQ	<LOQ	4.4	<LOQ	4.4
7-B	coffee	<LOQ	<LOQ	3.2	<LOQ	3.2
8-B	anise	<LOQ	<LOQ	3.7	<LOQ	3.7
9-B	cuban cigar	<LOQ	<LOQ	4.6	<LOQ	4.6
10-B	rhum	<LOQ	<LOQ	6.3	<LOQ	6.3
11-B	biscuit	<LOQ	<LOQ	4.0	<LOQ	4.0
12-B	anise	<LOQ	3.1	4.9	<LOQ	8.0
13-B	liquirice	<LOQ	<LOQ	4.3	4.6	8.9
14-B	biscuit	<LOQ	<LOQ	4.1	<LOQ	4.1
15-B	tobacco USA	<LOQ	<LOQ	3.3	<LOQ	3.3
16-B	mint	3.4	37.3	38.8	80.7	160.2
17-B	Virginia blend tobacco	<LOQ	2.8	3.3	<LOQ	6.1
1-C	Virginia blend tobacco	<LOQ	<LOQ	<LOQ	<LOQ	/
2-C	Virginia blend tobacco	<LOQ	<LOQ	<LOQ	<LOQ	/
3-C	basic flavor	<LOQ	<LOQ	<LOQ	<LOQ	/

Table 2. *Cont.*

E-Liquid ID	Flavor	Benzene (µg/L)	Ethylbenzene (µg/L)	Toluene (µg/L)	m,p,o-Xylenes (µg/L)	BTEX Total (µg/L)
4-C	basic flavor	<LOQ	4.0	7.3	18.8	30.1
5-C	anise	5.2	<LOQ	2.7	4.0	11.9
6-C	anise	<LOQ	<LOQ	8.3	<LOQ	8.3
7-C	mint	<LOQ	<LOQ	3.0	<LOQ	3.0
8-C	mint	<LOQ	<LOQ	<LOQ	<LOQ	/
9-C	biscuit	4.5	<LOQ	<LOQ	<LOQ	4.5
10-C	biscuit	<LOQ	<LOQ	<LOQ	<LOQ	/
11-C	biscuit	<LOQ	<LOQ	3.1	4.3	7.4
12-C	biscuit	<LOQ	<LOQ	<LOQ	<LOQ	/
13-C	cuban cigar	<LOQ	<LOQ	10.0	5.4	15.4
14-C	cuban cigar	<LOQ	<LOQ	<LOQ	<LOQ	/
15-C	tobacco USA	<LOQ	<LOQ	<LOQ	<LOQ	/
16-C	tobacco USA	<LOQ	<LOQ	7.9	<LOQ	7.9
17-C	rum	<LOQ	<LOQ	<LOQ	<LOQ	/
18-C	cognac	4.2	<LOQ	<LOQ	<LOQ	4.2
19-C	coffee	<LOQ	<LOQ	29.4	<LOQ	29.4
20-C	liquirice	<LOQ	<LOQ	5.0	<LOQ	5.0
1-D	mint	<LOQ	<LOQ	22.5	<LOQ	22.5
2-D	mint	<LOQ	<LOQ	29.1	<LOQ	29.1
3-D	black tobacco	<LOQ	<LOQ	11.4	<LOQ	11.4
4-D	black tobacco	<LOQ	<LOQ	10.9	<LOQ	10.9
5-D	Virginia blend tobacco	<LOQ	<LOQ	<LOQ	<LOQ	/
6-D	Virginia blend tobacco	2.7	<LOQ	5.2	26.4	34.3
7-D	tobacco	<LOQ	8.0	4.1	22.1	34.2
8-D	tobacco	<LOQ	<LOQ	7.9	19.9	27.8
1-E	cuban cigar	<LOQ	<LOQ	75.2	4.9	80.1
2-E	cuban cigar	<LOQ	7.3	96.2	9.8	113.3
3-E	Davidoff-tobacco	<LOQ	<LOQ	36.5	6.2	42.7
4-E	Davidoff-tobacco	<LOQ	4.9	73.3	9.2	87.4
5-E	Virginia blend tobacco	<LOQ	<LOQ	20.7	8.6	29.3
6-E	Virginia blend tobacco	6.7	6.9	25.9	15.1	54.6
1-F	almond	260.6	3.0	154.0	13.9	431.5
2-F	bubble gum	12.0	37.4	121.3	148.2	318.9
3-F	cigar	17.3	64.0	81.5	50.8	213.6
4-F	cigar	120.6	80.8	334.9	110.1	646.4
5-F	cinnamon	23.2	6.4	102.0	4.4	136.0
6-F	coffee	<LOQ	<LOQ	6.8	<LOQ	6.8
7-F	Davidoff-tobacco	113.6	19.5	212.0	85.4	430.5
8-F	Davidoff-tobacco	11.6	3.0	38.6	17.6	70.8
9-F	lemon	<LOQ	<LOQ	20.5	5.5	26.0
10-F	Marlboro cigarettes	13.8	<LOQ	35.4	11.4	60.6
11-F	Marlboro cigarettes	18.9	5.1	44.0	15.9	83.9
12-F	mint	7.8	111.9	326.5	293.0	739.2
13-F	tobacco	67.2	12.5	151.4	46.0	277.1
14-F	tobacco	9.1	2.8	31.8	12.6	56.3
15-F	fruits	<LOQ	<LOQ	14.6	5.9	20.5
16-F	Virginia blend tobacco	12.2	7.1	31.1	14.2	64.6
17-F	Virginia blend tobacco	176.6	40.8	385.9	140.5	743.8
1-G	American blend-tobacco	<LOQ	<LOQ	5.1	<LOQ	5.1
2-G	American blend-tobacco	<LOQ	<LOQ	4.3	<LOQ	4.3
3-G	Virginia blend tobacco	<LOQ	<LOQ	4.5	17.5	22.0
4-G	Virginia blend tobacco	<LOQ	<LOQ	11.9	3.9	15.8
5-G	Habanos cigar-tobacco	<LOQ	<LOQ	<LOQ	<LOQ	/
6-G	Habanos cigar-tobacco	<LOQ	<LOQ	<LOQ	<LOQ	/
7-G	mint	<LOQ	4.0	8.5	24.5	37.0
8-G	mint	<LOQ	3.0	9.4	18.4	30.8
1-H	tobacco	<LOQ	6.4	<LOQ	7.8	14.2
2-H	tobacco	<LOQ	<LOQ	<LOQ	2.8	2.8
1-I	Virginin Leaf - tobacco	<LOQ	<LOQ	3.5	4.3	7.8
2-I	Virginin Leaf - tobacco	<LOQ	<LOQ	4.6	<LOQ	4.6
3-I	mint	<LOQ	<LOQ	5.0	<LOQ	5.0
4-I	mint	<LOQ	<LOQ	3.2	<LOQ	3.2
5-I	cuban cigar	<LOQ	<LOQ	4.0	<LOQ	4.0
6-I	cuban cigar	<LOQ	<LOQ	2.7	<LOQ	2.7
1-L	chocolate/vanille	<LOQ	<LOQ	4.0	<LOQ	4.0
2-L	chocolate/vanille	<LOQ	<LOQ	4.0	2.8	6.8
3-L	mint/herbs	<LOQ	<LOQ	2.8	7.8	10.6
4-L	mint/herbs	<LOQ	<LOQ	4.8	4.5	9.3
1-M	mint/vanilla/chocolate	<LOQ	<LOQ	13.4	5.3	18.7
2-M	mint/vanilla/chocolate	<LOQ	4.1	27.3	15.4	46.8
1-N	thin mint	<LOQ	<LOQ	29.0	10.1	39.1
2-N	thin mint	<LOQ	<LOQ	10.0	<LOQ	10.0

3.2. GC-MS-O Qualitative Analysis: Identification of Flavoring Additives

The sensory evaluation report by GC-MS-O analysis of e-liquids ID A 1–5 is shown in Table 3. Molecular formula, CAS number and retention time (TR), expressed in minutes, of identified odor-active compounds, as well as the intensity of the odor perceived and the associated qualitative description provided by both trained panelists, are reported. GC/MS-O analysis of the sample 1-A with labelled flavor Coca cola allowed to distinctly identify 4 odor-active compounds: ethoxyethane, 2-ethoxybutane, camphene, and γ-terpinene. In more detail, the integration of chromatographic data with sensory perception revealed that the first odorous stimulus perceived by both assessors with intensity 3 (clear odor) and qualitatively described with the descriptor 'sweet' was associated with ethoxyethane eluted at 2.8 min. The odor-active compounds 2-ethoxybutane and γ-terpinene, eluted at 4.8 and 18.1 min respectively, were associated with the characteristic flavor of coca cola beverage and related to the overall flavor perceived during the preliminary odor test with the refill liquid. More specifically, 2-ethoxybutane was perceived by both panelists with intensity 3 and described as coca cola-like flavor while γ-terpinene was perceived by both the panelists with intensity 2, described with the descriptor 'bitter' and referred to the bitter aftertaste of coca cola. Another odor-active compound detected at the olfactory port and chromatographically identified was camphene, perceived by both assessors with intensity 2 and associated with citrus and fresh notes. The odor-active compounds 2-ethoxybutane, camphene and γ-terpinene are all classified by FEMA as flavoring agents with a specific flavor profile. 2-ethoxybutane is associated with the flavor profile 'floral' while camphene and γ-terpinene to the flavor profile 'camphor/oil' and 'bitter/citrus' respectively. Other sources e.g., The Good Scents Company (TGSC) Information System reports a more detailed flavor profile of camphene including minty, fresh, woody and citrus notes depending on the concentration confirming, in part, the assessors' olfactory perception. Finally, as shown in Table 3, two odorous stimuli although distinctly perceived at the olfactory port approximately at 8.9 and 21.3 min were not identified due to chromatographic peaks not sufficiently intense to allow accurate identification. The lack of clear correspondence between sensory perception and chromatographic data highlights that, despite the potentialities of GC-MS-O technique, in certain cases the sensory perception of human nose is more sensitive than the analytical detection as reported by Plutowska et al., 2008 [40]. The GC-MS-O analysis of e-liquid 2-A with the characteristic kiwi flavor resulted in the identification of seven odor-active compounds. Most of the odorous stimuli were qualitatively described by assessors with the odor descriptors 'sweet' and 'fruity'. The odor-active compounds identified, in order of chromatographic elution, were: ethoxy ethane (sweet, 2.8 min), ethyl acetate (aromatic/alcoholic, 3.6 min), 2-ethoxybutane (sweet/fruity, 4.8 min), methyl butanoate (fruity, 5.3 min), ethyl butanoate (fruity, 7.6 min), ethyl 2-methyl butanoate (fruity, 9.3 min), and methyl hexanoate (fruity, 12.3 min). Two were in common with e-liquid 1-A, i.e., ethoxy ethane and 2-ethoxy butane perceived by both assessors with intensity 3 and 2, respectively. The esters methyl butanoate, ethyl butanoate, ethyl 2-methylbutanoate, and methyl hexanoate are odor-active compounds with fruity attributes and represent a characteristic portion of the volatile aroma profile of fruits. They are also classified by FEMA as flavoring agents and are primarily used to impart fruity flavor in foods and beverages. Ethyl acetate is also included in the FEMA list of flavoring agents (with specification as food additive, carrier solvent) but its flavor profile is based on aromatic, brandy, and grape odor notes. Among the 'sweet' and 'fruity' odorous stimuli, both the assessors clearly indicated the one associated with the characteristic kiwi flavor, with odor intensity equal to 3. Comparing GC-MS results with the sensory response provided by both the panelists, ethyl 2-methyl butanoate was identified as the odor-active compound responsible of the kiwi flavor of the refill. This specific ester has been already identified in previous investigations by GC-MS and GC-MS-O as the key contributor of the aroma profile of several fruits such as pineapples [41], strawberries [42], cranberries [43] and melons [44]. The preliminary sensory tests (e.g., sniffing and vaping) on e-liquids

3-A, 4-A and 5-A, performed by both the panelists before GC-MS-O analytical sessions, allowed to appreciate a significant discrepancy between the flavor reported on the label and the overall flavor perceived. E-liquids 3-A and 4-A labels 'Davidoff' and 'Green USA mix' referred to tobacco brands whilst e-liquid 5-A label reported 'cigar' flavor. In all three cases, the overall flavor coming from e-liquids vaporization should have simulated the characteristic notes of the tobacco leaves aroma (i.e., woody, leather). Instead, the qualitative description provided by both assessors highlighted that the overall e-liquids flavors were dominated by sweet and caramel-like notes with the only exception of e-liquid 5-A that in addition was characterized by distinct woody notes. GC/MS-O analysis of sample 3-A ('Davidoff' flavor) allowed to confirm the role of ethoxyethane in giving the formulation a characteristic sweet and pleasant flavor. Moreover, the odor-active compound found to be the key contributor to the caramel notes of the overall flavor was 2,3-butanedione (or diacetyl), whose relevance as a flavoring additive will be deeply discussed in Section 4. Similarly to samples 1-A and 2-A, other odorous stimuli perceived approximately at 17.6 and 20.4 min and resembling tobacco flavor were associated with low intensity chromatographic peaks and, as a result, the tentative attribution was not allowed. At this regard, it has been already highlighted in Tierney et al., 2015 that the majority of tobacco flavored liquids were found to contain confectionary flavor chemicals instead of tobacco extracts therefore it is likely that the flavor chemicals pattern (i.e., benzyl alcohol, vanillin, ethylacetate, maltol) included in the formulations for resembling tobacco flavor is not necessarily what is expected to be found in a tobacco extract [45]. Considerations made for sample 3-A are relevant also for sample 4-A ('Green USA mix' flavor). Ethoxyethane and diacetyl were also detected in sample 4-A and associated, similarly with sample 3-A, to sweet and caramel-like flavor notes respectively. In addition, ethoxybutane was identified and associated with sweet flavor notes. The attribution for other odorous stimuli perceived during the GC/MS run, approximately at 8.7, 11.6 and 17.6 min (the latter similarly with sample 3-A), was not successful due to low intensity chromatographic peaks. More specifically, in addition to tobacco-like flavor, hearbaceous and grass/mint notes were perceived by assessors and this perception was considered reliable taking into account that, at least in principle, the formulation 'Green USA mix' should have simulated menthol-tobacco cigarettes and its characteristic menthol and herbaceous flavor notes. A comprehensive list of flavoring additives was obtained for sample 5-A ('cigar' flavor). Ethoxyethane and 2-ethoxybutane (both perceived with intensity 2) were confirmed as key contributors for sweet flavor notes while diacetyl (perceived with intensity 3) responsible for the caramel-like flavor.

An interesting GC-MS-O outcome, allowing us to characterize the odor profile of the sample 5-A in a more distinctive way, was the identification of three odor-active compounds, perceived with odor intensity ranging from 1 to 2: α-terpinene (woody, 16.3 min), α-phellandrene (woody, 18.3 min), and α-terpinolene (woody/pine, 19.4 min). They all are classified as flavoring agents by FEMA: the associated flavor profile varies from woody, fresh, citrus, and spice notes in the case of α-phellandrene to pine flavor notes in the case of α-terpinolene. Their inclusion in the liquid formulation is therefore related to the intention of enriching the overall flavor profile of the product with woody and pine flavor notes with the purpose to simulate as closely as possible the cigar flavor. Finally, the integration of sensory perception and GC-MS chromatographic data failed in the identification of the odor-active compound perceived by both evaluators as responsible for the tobacco-like and burnt flavor, similar to what was previously observed for the sample 3-A.

Table 3. Gas chromatography-mass spectrometry-olfactometry (GC-MS-O) report: identified odor-active compounds with specification of molecular formula, CAS number, retention time (TR, min), odor description and intensity.

Sample ID	Compound Identified	Molecular Formula	CAS Number	Retention Time (TR, min)	Odor Description Panelist 1/Panelist 2	Odor Intensity Panelist 1/Panelist 2
1-A (Coca cola)	ethoxyethane	$(C_2H_5)_2O$	60-29-7	2.8	sweet/sweet	3/3
	2-ethoxybutane	$C_6H_{14}O$	2679-87-0	4.8	Coca cola-like/sweet	3/3
	?			8.9	alcohol/sweet	2/1
	camphene	$C_{10}H_{16}$	79-92-5	13.3	citrus/citrus,fresh	2/2
	γ-terpinene	$C_{10}H_{16}$	99-85-4	18.1	bitter,citrus/coca cola-like,bitter	2/2
	?			21.3	pungent/no response	2/0
2-A (Kiwi)	ethoxyethane	$(C_2H_5)_2O$	60-29-7	2.8	sweet/sweet	3/3
	ethyl acetate	$C_4H_8O_2$	141-78-6	3.6	aromatic/alcoholic	1/1
	2-ethoxybutane	$C_6H_{14}O$	2679-87-0	4.8	sweet and fruity/sweet and fruity	2/2
	methylbutanoate	$C_5H_{10}O_2$	623-42-7	5.3	fruity/fruity	2/2
	ethylbutanoate	$C_6H_{12}O_2$	105-54-4	7.6	fruity/fruity	3/3
	ethyl 2-methylbutanoate	$C_7H_{14}O_2$	7452-79-1	9.3	kiwi-like/kiwi-like	3/3
	methyl hexanoate	$C_7H_{14}O_2$	106-70-7	12.3	fruity/fruity	2/2
	?			20.9	sweet/sweet	2/2
3-A (Davidoff)	ethoxyethane	$(C_2H_5)_2O$	60-29-7	2.8	sweet/sweet	2/2
	2,3-butanedione	$C_4H_6O_2$	431-03-8	3.3	caramel/caramel,sweet	3/3
	?			11.7	sweet/uncertain response	2/?
	?			17.6	tobacco-like/tobacco-like	1/1
	?			20.4	tobacco,burnt/tobacco, burnt	2/2
4-A (Green USA Mix)	ethoxyethane	$(C_2H_5)_2O$	60-29-7	2.8	sweet/sweet	3/2
	2,3-butanedione	$C_4H_6O_2$	431-03-8	3.3	caramel/caramel,sweet	2/2
	2-ethoxybutane	$C_6H_{14}O$	2679-87-0	4.8	sweet/sweet	2/2
	?			8.7	grass/mint	2/2
	?			11.6	herbaceous/herbaceous	2/2
	?			17.6	tobacco-like/tobacco-like	2/2
5-A (Cigar)	ethoxyethane	$(C_2H_5)_2O$	60-29-7	2.8	sweet/sweet	2/2
	2,3-butanedione	$C_4H_6O_2$	431-03-8	3.3	caramel/caramel,sweet	3/3
	2-ethoxybutane	$C_6H_{14}O$	2679-87-0	4.8	sweet/sweet	2/2
	α-terpinene	$C_{10}H_{16}$	99-86-5	16.3	woody/woody	2/2
	α-phellandrene	$C_{10}H_{16}$	99-83-2	18.2	woody/woody, spice	1/1
	terpinolene	$C_{10}H_{16}$	586-62-9	19.4	woody/woody, pine	2/2
	?			20.4	tobacco-like,burnt/tobacco,burnt	2/2

4. Discussion

4.1. Discussion on BTEX Results

HS-SPME-GC-MS analysis of 97 e-liquids highlighted BTEX contamination. Experimental data obtained suggest that, during the period 2013–2015, contaminated e-liquids were commercially available on the EU market, particularly e-liquids imported into EU member states and manufactured in China. Taking into account all of the data obtained, no correlation was found between BTEX contamination levels and nicotine content, nor nicotine presence. The variability observed in BTEX contamination levels from one brand to another one is therefore likely to be related to the variability in contamination level of the basic components (i.e., propylene glycol and glycerol) and/or the flavoring additives included. In addition, the variability in BTEX contamination levels observed within the same brand is likely to be related to the flavoring additives used, and in the specific case of samples 10, 11 and 12 C, given the same flavor and nicotine content, to the contamination of basic components used in the production process of different batches. According to Regulation (EC) No 1272/2008 on Classification, Labelling and Packaging of substances and mixtures (CLP), benzene, toluene, ethylbenzene and m,o,p-xylenes are included in Annex VI, Table 3. Benzene is classified as carcinogenic for humans (Carc. 1A, H350: May cause cancer by inhalation), mutagenic (Muta. 1B, H340: May cause genetic defects), and represents a hazard when inhaled (Asp. Tox 1, H304: May be fatal if swallowed and enters airways; STOT RE 1, H372: causes damage to organs through prolonged and repeated exposure) [31]. Toluene is classified as reprotoxic (Repr. 2, H361d: Suspected of damaging the unborn child) and represents a hazard when inhaled (Asp.Tox 1, H304: May be fatal if swallowed and enters airways). Ethylbenzene and xylenes are both classified as follows: Acute tox. 4, H332: harmful if inhaled. Given all the information on toxicity classification reported above, more attention has necessarily to be paid to benzene, a human mutagenic and genotoxic carcinogen, detected in some e-liquids at high concentration levels. Therefore, an in-depth analysis of potential health effects due to inhalation exposure to benzene is due. Epidemiological studies over the years have provided evidence of a causal relationship between chronic inhalation exposure to benzene and serious adverse health effects and diseases, from non-cancer health effects (i.e., hematologic diseases and/or functional aberrations of immune, nervous, endocrine systems) to cancer (i.e., myeloid leukemia, non-Hodgkins lymphoma) [46]. Numerous studies have demonstrated that benzene metabolites, especially p-benzoquinone, are involved in the progression from cytotoxicity to carcinogenicity, as they activate oxygenated radical species able to cause DNA damage [47]. It has been estimated that approximately 50% of the quantity of inhaled benzene is adsorbed into the human body. Once introduced into the human body through the respiratory apparatus, benzene is preferentially adsorbed in fat-rich tissues (i.e., fat and bone marrow), owing to its lipophilic nature. Great concern about potential health hazards has been historically linked to occupational exposure (where higher benzene concentrations than in general environments are likely to be encountered) but knowledge on the issue, acquired over the years, has led the scientists and epidemiologists to be more and more focused on health effects induced by long term exposure of the general population to low concentrations of benzene. Although benzene is recognized as a 'non-threshold carcinogen' on the basis of the assumption that any exposure may result in some increase of risk, in the present study the carcinogenic risk related to the inhalation exposure to benzene resulting from the consumption of e-liquids affected by the highest contamination (brand A) has been estimated.

As reported in the results section, across all 97 e-liquids tested, benzene concentration levels ranged from 2.7 µg/L (in samples 3-B and 6-D, both produced in Italy) to 30,200.0 µg/L (sample 3-A produced in China). This means that, if we consider the daily average consumption of e-liquids by a regular vaper approximately equal to 3 ml per day [48], the total amount of benzene potentially inhaled by the vaper within one day would have ranged from 0.0081 µg to 90.6 µg. For the most contaminated Chinese brand (brand A) the total amount of daily inhaled benzene with 3 ml e-liquid consumption would have varied in the range 21.6–90.6 µg. Taking into account a regular vaper represented by an adult person with an average body weight of 60 kg, the daily consumption of brand A e-liquids would result in benzene exposure of 0.00036–0.00151 mg/kg/day. A carcinogenic risk assessment for benzene may be performed comparing the estimated exposure with derived minimal effect level (DMEL) value, representing the level of exposure expressed as mg/kg/day below which the risk level of cancer is considered tolerable/acceptable (indicative tolerable risk level is 10^{-5} associated with a life-time risk for cancer of 1 per 100000 exposed individuals). The DMEL value for benzene, derived from reference values reported on Integrated Risk Information System (IRIS) website of United States Environmental Protection Agency (USEPA), is 0.0000182 mg/kg/day [49]. The comparison exposure-DMEL allows to point out that the daily consumption of Chinese e-liquids belonging to brand A would have resulted in a serious inhalation exposure scenario for active users with a risk level of cancer that is not acceptable. These results are of particular concern, also in light of the World Health Organization (WHO) guidelines for indoor air quality, published in 2010, where it is clearly stated that 'no safe level of exposure to benzene can be recommended' and that 'from a practical standpoint, it is expedient to reduce exposure levels to as low as possible' reducing or eliminating activities and materials that may release it [50].

4.2. Discussion of Flavoring Additives Results

Among the flavoring additives identified, diacetyl is certainly worthy of an in-depth analysis. Diacetyl is a volatile α-diketone and is a natural constituent of many regularly consumed foods (i.e., dairy products, fruits, coffee). Due to its flavor characteristics, it is widely used in the food manufacturing industry as a flavoring additive. It is added to a wide selection of foods and beverages to mainly impart butter and caramel taste and smell, depending on the concentration used. Its use in the food manufacturing industry is approved by competent governmental bodies such as U.S. Food and Drug Administration (U.S. FDA) and the National Institute for Occupational Safety and Health (NIOSH) and is currently authorized in EU member states according to EU Regulation No 872/2012. The potential risks for consumers health associated with the dietary exposure have been deeply evaluated over the years. As a result of safety evaluations, diacetyl has been determined to be 'generally recognized as safe' (e.g., GRAS) by the FEMA Expert Panel, and has been included in the FEMA GRAS list of authorized flavoring substances [51]. The European Food Safety Authority was also asked to take a position on the issue and the final opinion was that, on the basis of the safety evaluations carried out so far, the use of diacetyl in food is of no safety concern for humans. In this regard, however, it is important to point out that toxicological evaluations used to approve and support diacetyl as a flavoring additive in foods are related to ingestion, and therefore do not provide assurance of safety when other routes of exposure are involved, such as inhalation. In the early 2000s, concerns were raised with respect to potential toxicity for humans associated with inhalation exposure to diacetyl following the reported cases of a severe obstructive lung disease in diacetyl-exposed workers at microwave popcorn manufacturing plants in USA [52]. Preliminary evidence of an association between the occupational exposure to diacetyl and adverse effects on human respiratory apparatus has been reported by Kreiss et al., from a decline in respiratory function to development of a rare irreversible lung disease characterized by fixed airflow obstruction, called bronchiolitis obliterans [52]. Extensive scientific research on diacetyl has been carried out from then both confirming preliminary hypothesis on exposure-occurrence of lung disease association and adding new relevant scientific data [53]. Recently published papers have highlighted both neurotoxicity and impairment of cilia function in human airway epithelium [54,55]. Therefore, in light of the knowledge progressively acquired, the inclusion of diacetyl as flavoring additive in the manufacturing process of liquid formulations for e-cigs has rapidly become a much-debated issue in the scientific community due to foreseeable toxicological implications from direct inhalation exposure. In reaction to this, a prompt response came from e-liquids manufacturers with the replacement of diacetyl with 2,3-pentanedione (acetylpropionyl), an α-diketone showing similar flavor properties, but this option was soon revealed to be unsuccessful when scientific data on acetylpropionyl toxicity started to be published [56]. Our findings, although related to a limited number of samples, are in line with the results obtained in previous investigations highlighting the presence of diacetyl in e-liquids commercially available in EU member states in the pre-TPD implementation period and with characteristic flavors appealing to teenagers and young adults [19,20,30]. Farsalinos et al., 2015 analyzed both liquid and aerosol matrices of a total number of 159 samples purchased from 36 manufacturers and retailers in 7 different countries. Diacetyl was found in 74% of the samples investigated and in a large proportion of sweet-flavored e-liquids, with similar concentrations in both liquid and aerosol. The simultaneous presence of acetylpropionyl also suggested that, instead of being used as a replacement, acetylpropionyl is often used in conjunction with diacetyl. Further, the authors highlighted that, for 47% of diacetyl-containing e-liquids, the daily exposure level (μg/day) for vapers could be higher than NIOSH-defined safety limits for occupational exposure. Barhdadi et al. investigated 12 flavored e-liquids by applying the HS/GC-MS method, properly developed for the screening and quantification of diacetyl and acetylpropionyl in e-liquids. The samples were provided by the Belgium Federal Agency for Medicinal and Health Products and collected either upon inspections in vaping shops or through seizure activity by Belgian authorities in the period 2013–2015, similar to the present study. The authors reported that only two sweet-flavored e-liquids contained measurable amounts of diacetyl and the determined concentrations were 6.04 μg/g and 98.84 μg/g. Finally, 42 e-liquids selected from among the 14 most popular brands dominating both the USA and EU markets in 2013 were investigated by Varlet et al. in terms of chemical and biological constituents. Diacetyl was detected in three e-liquids, two of them characterized by tobacco flavors and one by candy flavor. Similarly to Farsalinos et al., comparison with the NIOSH safety limit was made, revealing that one tobacco flavored e-liquid that resulted diacetyl-positive could lead to exposure higher the recommended limit. Although approximate for estimating risk for e-cig users, the use of occupational exposure limits is affected by several limitations [19,57]. This approach has raised some resistance, mainly because occupational safety limits for toxicants, for instance for diacetyl, have been set for workers not for the general population and are related to inhalation exposure scenarios not applicable to e-cigs users. According to the authors' knowledge, other two studies carried out by Allen et al. in 2017 and Omayie et al. in 2019 have raised concerns about diacetyl, confirming its inclusion as flavoring additive in refill liquids for e-cigs (diacetyl detected in 39 of 51 tested refills and in 150 of 277 samples, respectively), but in both cases the investigated samples were considered dominating the current extra-EU market and therefore are not representative of the EU market before the implementation of TPD. To summarize, our findings on diacetyl, although related to a limited number of e-liquids manufactured in China and commercially available in the EU during the period 2013-2015, are in line with the results obtained in other investigations made on larger sets of samples representative of the EU market at that time. The only discrepancy on diacetyl presence detectable among the studies performed before the TPD implementation was reported by Girvalaki et al. in 2018. The authors evaluated the chemical composition of 122 e-liquids selected among the most commonly sold brands in 9 EU member states in mid-2016 before the TPD implementation. The result of this comprehensive investigation was a list of 177 compounds detected (e.g., flavoring additives and other VOCs), the majority with associated Globally Harmonized System of Classification and labeling of Chemicals (GHS) health hazard statements. Diacetyl, however, was not detected in the samples tested, and therefore not included in the list. This discrepancy between Girvalaki et al. and the other abovementioned studies may be related or to the different period of e-liquids selection (2013–2015 versus 2016), although both periods were before TP -implementation, when the first actions aimed to the progressive replacement/elimination of diacetyl started to be made on a voluntary basis by some EU manufacturers and importers, or it simply reflects the potential heterogeneity due to the multitude of samples commercially available on the EU market in the period of reference. To date, following the implementation of TPD in most EU member states in 2016, both manufacturers and importers are obliged to submit a notification to competent authorities reporting detailed information on refill liquids (Article 20) [13]. The notification must report the list of all the ingredients (including flavoring additives) contained in e-liquid formulations for e-cigs available on the market and indication of related quantities as well. It must be noted, however, that according to TPD, the use of diacetyl is neither explicitly prohibited nor subjected to restriction. In addition, due to difficulty in defining a typical inhalation exposure scenario fitting all vapers habits (high variability in daily e-liquid amount consumed), there is no scientific consensus on the maximum allowed level of diacetyl in e-liquids. Therefore, to date, diacetyl use as a flavoring additive in e-liquids remains an open issue, suggesting not only that quality controls remain necessary, even in e-liquids labelled as diacetyl-free, but also that the potential solution at the EU level to ensure that e-liquids supplied to consumers are safe is to follow the direction of some EU member states that proposed the ban of diacetyl and other flavoring additives of concern [58].

5. Conclusions

In the present paper, results from a study on the chemical characterization of levels of BTEX in 97 e-liquids, representative of the EU market between 2013 and 2015 prior the implementation of TPD in most EU member states, are reported. To our knowledge, there have been very few studies focused on BTEX analysis in refill fluids and cartridges for e-cigs commercially available on the EU market in the pre-TPD implementation period. Therefore, although the e-liquids investigated may not be representative of the current EU market, our findings may represent a useful reference for the ongoing evaluation on the effectiveness of e-liquid safety and quality requirements under the current legislative framework. Most of the e-liquids investigated were revealed to be affected, to a lesser or greater extent, by BTEX contamination. Few exceptions were observed (12 of 97 samples). High variability in BTEX total concentration level was observed from one brand to another, ranging from 2.7 µg/L to 32,151.1 µg/L. The contamination is likely to be related to the contamination of propylene glycol and glycerol, and/or the flavoring additives used. No correlation was found between BTEX concentration levels and nicotine content/presence. Moreover, it was estimated that an inhalation exposure of very high concern would have occurred for active users vaping the most contaminated e-liquids (brand A), characterized by high concentration levels (7,200–30,200 µg/L) of benzene, a known human carcinogen. Our findings, therefore, point out that higher quality ingredients should have been used and that quality control on the formulations should have been applied prior their introduction on the EU market in 2013–2015 period. Further investigations carried out on a limited number of e-liquids aimed at the identification of flavoring additives through GC-MS-O application confirmed, in the reference period of the present study, the use of diacetyl, a flavoring additive approved for foods but associated with the onset of a severe lung disease when inhaled. This finding is in line with results obtained by other investigations made in the same period on a larger number of e-liquids sold in EU, highlighting the use of diacetyl in the e-liquid manufacturing industry due to poor awareness of the potential harm to humans. There are now sufficient toxicological data on the potential adverse effects of diacetyl and other flavoring chemicals when directly inhaled into the human airways, and therefore harmonized regulation at EU level on flavoring additives use in e-liquids, resulting in ban or restriction, should be fully addressed, in order to ensure health protection.

Author Contributions: Conceptualization: J.P., R.D. and G.d.G.; methodology: J.P., C.A., M.F. and L.P.; investigation: J.P., C.A., M.F. and L.P.; data curation: J.P., C.A., M.F., A.D.G. and L.P.; writing—original draft preparation: J.P., A.D.G. and L.P.; supervision: R.D. and G.d.G. project administration: C.A., M.F., G.d.G. and R.D.; funding acquisition: G.d.G. and R.D. All authors have read and agreed to the published version of the manuscript.

References

1.   Dockrell, M.; Morrison, R.; Bauld, L.; McNeill, A. E-cigarettes: prevalence and attitudes in Great Britain. *Nicotine Tob. Res.* **2013**, *15*, 1737–1744. [CrossRef] [PubMed]

2.   European Commission. Directorate-General for Health and Food Safety and coordinated by the Directorate-General for Communication. Special Eurobarometer 429. Attitudes of Europeans towards Tobacco and Electronic Cigarettes. 2015. Available online: https://ec.europa.eu/public_opinion/index_en.htm. (accessed on 20 September 2019).

3.   Kitzen, J.M.; McConaha, J.L.; Bookser, M.L.; Pergolizzi, J.V.; Taylor, R.; Raffa, R.B. e-Cigarettes for smoking cessation: Do they deliver? *J. Clin. Pharm. Ther.* **2019**, *44*, 650–655. [CrossRef] [PubMed]

4.   Manzoli, L.; Flacco, M.E.; Ferrante, M.; La Vecchia, C.; Siliquini, R.; Ricciardi, W.; Marzuillo, C.; Villari, P.; Fiore, M.; Gualano, M.R.; et al. Cohort study of electronic cigarette use: effectiveness and safety at 24 months. *Tob. Control.* **2016**, *26*, 284–292. [CrossRef] [PubMed]

5.   Dai, H.; Leventhalab, A.M. Association of electronic cigarette vaping and subsequent smoking relapse among former smokers. *Drug Alcohol Depend.* **2019**, *199*, 10–17. [CrossRef]

6.   Farsalinos, K.; Romagna, G.; Tsiapras, D.; Kyrzopoulos, S.; Voudris, V. Characteristics, Perceived Side Effects and Benefits of Electronic Cigarette Use: A Worldwide Survey of More than 19,000 Consumers. *Int. J. Environ. Res. Public Health* **2014**, *11*, 4356–4373. [CrossRef]

7.   Goniewicz, M.L.; Leigh, N.J.; Gawron, M.; Nadolska, J.; Balwicki, Ł.; McGuire, C.; Sobczak, A. Dual use of electronic and tobacco cigarettes among adolescents: a cross-sectional study in Poland. *Int. J. Public Health* **2015**, *61*, 189–197. [CrossRef]

8.   Soneji, S.; Barrington-Trimis, J.L.; Wills, T.A.; Leventhal, A.M.; Unger, J.B.; Gibson, L.; Yang, J.; Primack, B.A.; Andrews, J.A.; Miech, R.A.; et al. Association Between Initial Use of e-Cigarettes and Subsequent Cigarette Smoking Among Adolescents and Young Adults. *JAMA Pediatr.* **2017**, *171*, 788–797. [CrossRef]

9.   Famele, M.; Palmisani, J.; Ferranti, C.; Abenavoli, C.; Palleschi, L.; Mancinelli, R.; Fidente, R.M.; De Gennaro, G.; Draisci, R. Liquid chromatography with tandem mass spectrometry method for the determination of nicotine and minor tobacco alkaloids in electronic cigarette refill liquids and second-hand generated aerosol. *J. Sep. Sci.* **2017**, *40*, 1049–1056. [CrossRef]

10. Geiss, O.; Bianchi, I.; Barahona, F.; Barrero-Moreno, J. Characterisation of mainstream and passive vapours emitted by selected electronic cigarettes. *Int. J. Hyg. Environ. Health* **2015**, *218*, 169–180. [CrossRef]

11. Flora, J.W.; Meruva, N.; Huang, C.B.; Wilkinson, C.T.; Ballentine, R.; Smith, D.C.; Werley, M.S.; McKinney, W.J. Characterization of potential impurities and degradation products in electronic cigarette formulations and aerosols. *Regul. Toxicol. Pharmacol.* **2016**, *74*, 1–11. [CrossRef] [PubMed]

12. Omaiye, E.E.; McWhirter, K.J.; Luo, W.; Tierney, P.A.; Pankow, J.F.; Talbot, P. High concentrations of flavor chemicals are present in electronic cigarette refill fluids. *Sci. Rep.* **2019**, *9*, 2468. [CrossRef] [PubMed]

13. European Parliament. *Directive 2014/40/EU of the European Parliament and of the Council of 3 April 2014 on the Approximation of the Laws, Regulations and Administrative Provisions of the Member States Concerning the Manufacture, Presentation and Sale of Tobacco and Related Products and Repealing Directive 2001/37/EC*; European Parliament: Brussels, Belgium, 2014.

14. Poklis, J.L.; Wolf, C.E.; Peace, M.R. Ethanol concentration in 56 refillable electronic cigarettes liquid formulations determined by headspace gas chromatography with flame ionization detector (HS-GC-FID). *Drug Test. Anal.* **2017**, *9*, 1637–1640. [CrossRef] [PubMed]

15. Aszyk, J.; Kubica, P.; Woźniak, M.; Namiesnik, J.; Wasik, A.; Kot-Wasik, A. Evaluation of flavour profiles in e-cigarette refill solutions using gas chromatography-tandem mass spectrometry. *J. Chromatogr. A* **2018**, *1547*, 86–98. [CrossRef] [PubMed]

16. Nguyen, N.; McKelvey, K.; Halpern-Felsher, B. Popular Flavors Used in Alternative Tobacco Products Among Young Adults. *J. Adolesc. Health* **2019**, *65*, 306–308. [CrossRef] [PubMed]

17. Landry, R.L.; Groom, A.; Vu, T.-H.T.; Stokes, A.C.; Berry, K.M.; Kesh, A.; Hart, J.L.; Walker, K.L.; Giachello, A.L.; Sears, C.G.; et al. The role of flavors in vaping initiation and satisfaction among U.S. adults. *Addict. Behav.* **2019**, *99*, 106077. [CrossRef] [PubMed]

18. EFSA Panel on Food Additives and Nutrient Sources added to Food (ANS); Younes, M.; Aggett, P.; Aguilar, F.; Crebelli, R.; Dusemund, B.; Filipič, M.; Frutos, M.J.; Galtier, P.; Gott, D.; et al. Re-evaluation of propane-1,2-diol (E 1520) as a food additive. *EFSA J.* **2018**, *16*, e05235. [CrossRef]

19. Barhdadi, S.; Canfyn, M.; Courselle, P.; Rogiers, V.; Vanhaecke, T.; Deconinck, E. Development and validation of a HS/GC–MS method for the simultaneous analysis of diacetyl and acetylpropionyl in electronic cigarette refills. *J. Pharm. Biomed. Anal.* **2017**, *142*, 218–224. [CrossRef]

20. Farsalinos, K.; Kistler, K.A.; Gillman, I.G.; Voudris, V. Evaluation of Electronic Cigarette Liquids and Aerosol for the Presence of Selected Inhalation Toxins. *Nicotine Tob. Res.* **2014**, *17*, 168–174. [CrossRef]

21. Shibamoto, T. Diacetyl: Occurrence, Analysis, and Toxicity. *J. Agric. Food Chem.* **2014**, *62*, 4048–4053. [CrossRef]

22. Egilman, D.; Schilling, J.H. Bronchiolitis obliterans and consumer exposure to butter-flavored microwave popcorn: a case series. *Int. J. Occup. Environ. Health* **2012**, *18*, 29–42. [CrossRef]

23. Aszyk, J.; Woźniak, M.; Kubica, P.; Kot-Wasik, A.; Namiesnik, J.; Wasik, A. Comprehensive determination of flavouring additives and nicotine in e-cigarette refill solutions. Part II: Gas-chromatography–mass spectrometry analysis. *J. Chromatogr. A* **2017**, *1517*, 156–164. [CrossRef]

24. Girvalaki, C.; Tzatzarakis, M.; Kyriakos, C.N.; Vardavas, A.I.; Stivaktakis, P.D.; Kavvalakis, M.; Tsatsakis, A.; Vardavas, C.I. Composition and chemical health hazards of the most common electronic cigarette liquids in nine European countries. *Inhal. Toxicol.* **2018**, *30*, 361–369. [CrossRef] [PubMed]

25. Kosmider, L.; Sobczak, A.; Prokopowicz, A.; Kurek, J.; Zaciera, M.; Knysak, J.; Smith, D.; Goniewicz, M.L. Cherry-flavoured electronic cigarettes expose users to the inhalation irritant, benzaldehyde. *Thorax* **2016**, *71*, 376–377. [CrossRef] [PubMed]

26. Vardavas, C.I.; Girvalaki, C.; Vardavas, A.; Papadakis, S.; Tzatzarakis, M.; Behrakis, P.; Tsatsakis, A. Respiratory irritants in e-cigarette refill liquids across nine European countries: A threat to respiratory health? *Eur. Respir. J.* **2017**, *50*, 1701698. [CrossRef] [PubMed]

27. Cheng, T. Chemical evaluation of electronic cigarettes. *Tob. Control.* **2014**, *23*, 11–17. [CrossRef]

28. Lee, M.-H.; Szulejko, J.; Kim, K.-H. Determination of carbonyl compounds in electronic cigarette refill solutions and aerosols through liquid-phase dinitrophenyl hydrazine derivatization. *Environ. Monit. Assess.* **2018**, *190*, 200. [CrossRef]

29. Hess, C.A.; Olmedo, P.; Navas-Acien, A.; Goessler, W.; Cohen, J.E.; Rule, A.M. E-cigarettes as a source of toxic and potentially carcinogenic metals. *Environ. Res.* **2016**, *152*, 221–225. [CrossRef]

30. Varlet, V.; Farsalinos, K.; Augsburger, M.; Thomas, A.; Etter, J.-F. Toxicity Assessment of Refill Liquids for Electronic Cigarettes. *Int. J. Environ. Res. Public Health* **2015**, *12*, 4796–4815. [CrossRef]

31. *Regulation (EC) no 1272/2008 of the European Parliament and of the Council of 16 December 2008 on Classification, Labelling and Packaging of Substances and Mixtures, Amending and Repealing Directives 67/548/EEC and 1999/45/EC, and Amending Regulation (EC) No 1907/2006*; European Parliament: Brussels, Belgium, 2008.

32. Girvalaki, C.; Vardavas, A.; Tzatzarakis, M.; Kyriakos, C.N.; Nikitara, K.; Tsatsakis, A.M.; I Vardavas, C. Compliance of e-cigarette refill liquids with regulations on labelling, packaging and technical design characteristics in nine European member states. *Tob. Control.* **2019**. [CrossRef]

33. Lim, H.-H.; Shin, H.-S. Determination of volatile organic compounds including alcohols in refill fluids and cartridges of electronic cigarettes by headspace solid-phase micro extraction and gas chromatography–mass spectrometry. *Anal. Bioanal. Chem.* **2016**, *409*, 1247–1256. [CrossRef]

34. Han, S.; Chen, H.; Zhang, X.; Liu, T.; Fu, Y. Levels of Selected Groups of Compounds in Refill Solutions for Electronic Cigarettes. *Nicotine Tob. Res.* **2015**, *18*, 708–714. [CrossRef]

35. Famele, M.; Ferranti, C.; Abenavoli, C.; Palleschi, L.; Mancinelli, R.; Draisci, R. The Chemical Components of Electronic Cigarette Cartridges and Refill Fluids: Review of Analytical Methods. *Nicotine Tob. Res.* **2014**, *17*, 271–279. [CrossRef]

36. Giungato, P.; Di Gilio, A.; Palmisani, J.; Marzocca, A.; Mazzone, A.; Brattoli, M.; Giua, R.; De Gennaro, G. Synergistic approaches for odor active compounds monitoring and identification: State of the art, integration, limits and potentialities of analytical and sensorial techniques. *TrAC Trends Anal. Chem.* **2018**, *107*, 116–129. [CrossRef]

37. De Gennaro, G.; Amenduni, A.; Brattoli, M.; Massari, F.; Palmisani, J.; Tutino, M. Chemical Characterization of Odor Active Volatile Organic Compounds Emitted from Perfumes by GC/MS-O. *Environ. Eng. Manag. J.* **2016**, *15*, 1963–1969. [CrossRef]

38. Brattoli, M.; Cisternino, E.; de Gennaro, G.; Giungato, P.; Mazzone, A.; Palmisani, J.; Tutino, M. Gas Chromatography Analysis with Olfactometric Detection (GC-O): An Innovative Approach for Chemical Characterization of Odor Active Volatile Organic Compounds (VOCs) emitted from a Consumer Product. *Chem. Eng. Trans.* **2014**, *40*, 121–126.

39. CEN (Committee for European Normalization) EN13725. *Air Quality Determination of Odour Concentration by Dynamic Olfactometry*; CEN (Committee for European Normalization): Brussels, Belgium, 2003.

40. Plutowska, B.; Wardencki, W. Application of gas chromatography–olfactometry (GC–O) in analysis and quality assessment of alcoholic beverages—A review. *Food Chem.* **2008**, *107*, 449–463. [CrossRef]

41. Zheng, L.-Y.; Sun, G.-M.; Liu, Y.-G.; Lv, L.-L.; Yang, W.-X.; Zhao, W.-F.; Wei, C.-B. Aroma Volatile Compounds from Two Fresh Pineapple Varieties in China. *Int. J. Mol. Sci.* **2012**, *13*, 7383–7392. [CrossRef] [PubMed]

42. Kim, Y.-H.; Kim, K.-H.; Szulejko, J.; Parker, D. Quantitative Analysis of Fragrance and Odorants Released from Fresh and Decaying Strawberries. *Sensors* **2013**, *13*, 7939–7978. [CrossRef]

43. Zhu, J.; Chen, F.; Wang, L.; Xiao, Z.; Chen, H.; Wang, H.; Zhu, J. Characterization of the Key Aroma Volatile Compounds in Cranberry (Vaccinium macrocarponAit.) Using Gas Chromatography–Olfactometry (GC-O) and Odor Activity Value (OAV). *J. Agric. Food Chem.* **2016**, *64*, 4990–4999. [CrossRef]

44. Aubert, C.; Bourger, N. Investigation of Volatiles in Charentais Cantaloupe Melons (Cucumis meloVar.cantalupensis). Characterization of Aroma Constituents in Some Cultivars. *J. Agric. Food Chem.* **2004**, *52*, 4522–4528. [CrossRef]

45. A Tierney, P.; Karpinski, C.D.; E Brown, J.; Luo, W.; Pankow, J.F. Flavour chemicals in electronic cigarette fluids. *Tob. Control.* **2015**, *25*, e10–e15. [CrossRef]

46. Bahadar, H.; Mostafalou, S.; Abdollahi, M. Current understandings and perspectives on non-cancer health effects of benzene: A global concern. *Toxicol. Appl. Pharmacol.* **2014**, *276*, 83–94. [CrossRef]

47. Atkinson, T.J. A review of the role of benzene metabolites and mechanisms in malignant transformation: Summative evidence for a lack of research in nonmyelogenous cancer types. *Int. J. Hyg. Environ. Health* **2009**, *212*, 1–10. [CrossRef] [PubMed]

48. Van Gucht, D.; Adriaens, K.; Baeyens, F. Online Vape Shop Customers Who Use E-Cigarettes Report Abstinence from Smoking and Improved Quality of Life, But a Substantial Minority Still Have Vaping-Related Health Concerns. *Int. J. Environ. Res. Public Health* **2017**, *14*, 798. [CrossRef] [PubMed]

49. United States Environmental Protection Agency (USEPA). Integrated Risk Information System (IRIS) Website. Available online: https://www.epa.gov/iris (accessed on 28 March 2020).

50. World Health Organization (WHO). *Indoor Air Quality Guidelines: Selected Pollutants*; WHO Regional Office for Europe: Geneva, Switzerland, 2010.

51. FEMA. Diacetyl FEMA 2370. 2018. Available online: https://www.femaflavor.org/flavor-library/diacetyl (accessed on 1 October 2019).

52. Kreiss, K.; Gomaa, A.; Kullman, G.; Fedan, K.; Simoes, E.; Enright, P.L. Clinical Bronchiolitis Obliterans in Workers at a Microwave-Popcorn Plant. *N. Engl. J. Med.* **2002**, *347*, 330–338. [CrossRef]

53. Kullman, G.; Boylstein, R.; Jones, W.; Piacitelli, C.; Pendergrass, S.; Kreiss, K. Characterization of Respiratory Exposures at a Microwave Popcorn Plant with Cases of Bronchiolitis Obliterans. *J. Occup. Environ. Hyg.* **2005**, *2*, 169–178. [CrossRef] [PubMed]

54. Park, H.-R.; O'Sullivan, M.; Vallarino, J.; Shumyatcher, M.; Himes, B.E.; Park, J.-A.; Christiani, D.C.; Allen, J.G.; Lu, Q. Transcriptomic response of primary human airway epithelial cells to flavoring chemicals in electronic cigarettes. *Sci. Rep.* **2019**, *9*, 1400. [CrossRef] [PubMed]

55. Das, S.; Smid, S. Small molecule diketone flavorants diacetyl and 2,3-pentanedione promote neurotoxicity but inhibit amyloid β aggregation. *Toxicol. Lett.* **2019**, *300*, 67–72. [CrossRef] [PubMed]

56. Holden, V.K.; Hines, S.E. Update on flavoring-induced lung disease. *Curr. Opin. Pulm. Med.* **2016**, *22*, 158–164. [CrossRef]

57. Allen, J.G.; Flanigan, S.S.; Leblanc, M.; Vallarino, J.; Macnaughton, P.; Stewart, J.H.; Christiani, D.C. Flavoring Chemicals in E-Cigarettes: Diacetyl, 2,3-Pentanedione, and Acetoin in a Sample of 51 Products, Including Fruit-, Candy-, and Cocktail-Flavored E-Cigarettes. *Environ. Health Perspect.* **2016**, *124*, 733–739. [CrossRef]

58. MHRA (Medicines and Healthcare products Regulatory Agency). *Discussion Paper on Submission of Notifications under Article 20 of Directive 2014/40/EU: Chapter 6- Advise on Ingredients in Nicotine-Containing Liquids in Electronic Cigarettes and Refill Containers*; MHRA (Medicines and Healthcare products Regulatory Agency): London, UK, 2016.

5

Study of Passive Adjustment Performance of Tubular Space in Subway Station Building Complexes

Junjie Li [1], Shuai Lu [2,*], Qingguo Wang [3], Shuo Tian [1] and Yichun Jin [1]

[1]   School of Architecture and Design, Beijing Jiaotong University, Beijing 100044, China;
      lijunjie@bjtu.edu.cn (J.L.); 17121721@bjtu.edu.cn (S.T.); 18121731@bjtu.edu.cn (Y.J.)

[2]   School of Architecture and Urban Planning, Shenzhen University, Shenzhen 518060, China

[3]   China Design and Research Group, Beijing 100042, China; wangqg@cadg.cn

*   Correspondence: Lyushuai@szu.edu.cn.

Abstract: The stereo integration of subway transportation with urban functions has promoted the transformation of urban space via extensive two-dimensional plans to intensive three-dimensional development. As sustainable development aspect, it has posed new challenges for the design of architectural space to be better environmental quality and low energy consumption. Therefore, subway station building complexes with high-performance designs should be a primary focus. Tubular space is a very common spatial form in subway station building complexes; it is an important space carrier for transmitting airflow and natural light. As such, it embodies the advantages of effectively utilizing natural resources, improving the indoor thermal and light environments, refining the air quality, and reducing energy consumption. This research took tubular space, which has a passive regulation function in subway station building complexes as its research object. It firstly established a scientific and logical method for verifying the value of tubular space by searching causal relationships among the parameterized building space information factors, occupancy satisfaction elements, physical environment comfort aspects, and climate conditions. Secondly, based on the actual field investigation, a database of physical environment performance data and users' subjective satisfaction information was collected. Through the fieldwork results and analysis, the research thirdly concluded that the potential passive utilization of tubular space in subway station building complexes can be divided into two aspects: improvement in comfort level itself and utilization of climate between natural or artificial. Finally, three typical integrated design method for tubular spaces exhibiting high levels of performance and low amounts of energy consumption in subway station building complexes was put forward. This interdisciplinary research provides a design basis for subway station building complexes seeking to achieve high levels of performance and low amounts of energy consumption.

Keywords: passive space design; tubular space; physical building environment; fieldwork test; subway station building complex

---

Highlights:

(1) Construction of a multi-criteria analysis framework to analyze the passive adjustment performance of tubular space in subway station building complexes;

(2) Establishment of a database of physical environment performance and occupants' subjective satisfaction, based on actual field investigations;

(3) Development of an integrated design idea for tubular spaces in subway station building complexes that displays a high level of performance and low amount of energy consumption as the target orientation; and

(4) Proposal of three typical design concepts for compound tubular space.

1. Introduction

1.1. Research Background

With the rapid evolution of urban construction, the Transit-Oriented Development (TOD) mode has gradually formed a new organizational model for urban public spaces [1–3]. With the expansion of city subway, subway station building complexes have also entered a period of reinvention [4,5]. The stereo integration of subway transportation with urban functions has promoted the transformation of urban spaces from extensive two-dimensional plans, to intensive three-dimensional development [6]. Due to China's rapidly-advancing urbanization, the demand for sustainable development is becoming more and more urgent [7,8], and the issues of improving occupant comfort and reducing environmental load must be optimized [9,10]. The significant flow rate of people mainly in pass-through mode has led to lower environmental quality in above- and underground spaces at the junctions in subway station, and this may directly affect occupant comfort [11] and health [12]. In addition, large and complex public buildings tend to occupy a significant proportion of a city's energy consumption, threatening the sustainable development of human living environments [13,14].

1.2. Passive Design of Tubular Space in Subway Station Building Complexes

Passive design, which affects the sustainability of architecture from the prototype stage onward, is an important aspect of green building design [15]. Passive building design does not rely on active system equipment, but it does depend on a strong capacity for climate adaptability and self-adjustment, which creates a harmonious indoor coexistence of people and the outdoor environment [16,17]. Passive architecture describes buildings that are designed to cope with climate factors by providing enduring and natural comfortable indoor conditions [18,19]. The term "passive" conveys the idea of self-defense or self-protection of users in architectural design, with respect to the local natural environment [20,21]. A quality passive design avoids the possibility of high levels of energy consumption, saving up to 50% over traditional methods [22]. Therefore, the architectural prototype generally determines the degree of sustainability of the building.

Tubular space includes horizontal and vertical corridors in buildings, usually in slender shape, such as ventilation shafts, patios and lighting tubes, and tunnel corridors for connection. Tubular space occupies an important proportion of buildings in subway station building complexes, and its passive regulation has not been deeply investigated [23]. Subway station building complexes are affected by the characteristics of the mode of space utilization, wherein it is very common to use tubular space in ground-level and underground spaces (as shown in Figure 1), including patios and lighting tubes to improve natural lighting efficiency. Other uses include ventilation shafts for improving the indoor thermal environment and air quality, a station's traffic tubes for connecting ground-level and underground stations, and tunnels for traffic transmission. Tubular space can be seen as a "communications device" that transmits people, mass, and energy to different spaces [24]. This space type is a passive adjustment strategy located between the external and interior environments of the building. It uses natural energy sources (such as wind, solar energy, and rainwater) and the natural environment to regulate microclimates and improve the indoor atmosphere. In subway station building complexes, tubular space has the potential to play an important role in passive adjustment performance, especially with regards to natural lighting and ventilation, passive cooling, etc., to optimize comfort and user satisfaction with the indoor space, and greatly reduce the energy consumption of the building's operating phase [25].

Figure 1. Examples of tubular spaces in subway station building complexes.

In their preliminary analyses of spatial design and climatic contradiction factors, some scholars have considered the particularity of using underground spaces over ground-level spaces, with respect to climate [26,27]. From the perspective of the physical environment of underground spaces, the degree of thermal and light comfort are of great importance. The comfort level of subway transit spaces and vehicle interiors has been verified by the coupling of actual tests with digital simulations [28]. After six years of actual tests of underground civil air defense space, Yong Li argued for a suitable acceptable thermal temperature range for underground areas [29]. In recent years, scholars have paid more attention to occupant health and placed a greater emphasis on ventilation and air quality, by conducting typological studies, and control and defense research on pollution and particulate matter (PM). Min Jeong Kim and others have proposed ventilation systems that can improve the platform PM10 levels and reduce ventilation energy, as compared to manual systems [30]. Practices based on this theory can be found as early as in ancient Rome, where ancient architects used tubular space to create more comfortable living environments. For instance, the underground corridor is a very good example of a kind of air cooling system in use at this time [31]. In terms of modern urban architecture, Hikarie Shibuya, a Japanese transportation complex designed as an integration concept that considers the passive utilization relationship between subway station buildings and above-ground structures, solved the problem of subway station lighting and ventilation [32]. In the retrofitting of the Les Halles area of Paris, the utilization of tubular space was adopted in underground spaces to provide natural lighting [33].

1.3. Objective of this Study

This study addressed a variety of forms of tubular space in city subway station building complexes, screening those spaces for passive adjustment potential in order to study the effect on the comfort level of the indoor environment and overall energy consumption. Through a fieldwork evaluation of building performance and occupant satisfaction in actual built projects, this research conducted a quantitative performance evaluation of passive architectural design strategies for tubular spaces. Through an analysis of the current situation and excavation of the spatial potential, this work determined the passive adjustment performance effects for subway station building complexes in terms of sustainable development, therefore providing a basis for improving architectural design methods to show higher levels of performance and lower amounts of energy consumption. This research pursued the following three objectives: (1) provide a basis for design by analyzing the types of passive function and specific variables for the technical strategies employed by subway station building complexes; (2) test the passive adjustment effects of subway station building complexes and improve the authenticity and objectivity of the designs via actual and effective environmental monitoring evaluation; and (3) explore typical further-optimized strategy models for the compound tubular space systems of subway station building complexes, and provide guidance for design optimization.

2. Methodology

This research was based on the dual perspectives of architecture and the built environment. With regards to architectural design, this work produced a space prototype and deconstructed the factors affecting the passive adjustment performance of architecture. According to a comparison of the factors that influence the quality of indoor buildings and actual built environments, a comprehensive evaluation was made of the passive adjustment performance of tubular spaces in subway station building complexes [34]. First, the factors that affect the passive adjustment performance were analyzed. Then, according to the analytic factors and taking the urban Beijing subway station building complexes as an example, a long-term physical environment test was carried out. The subject was a subway station building complex with a typical amount of tubular space. This research focused on the physical environment, as tube as the passive function of potential space and its influence on surrounding functional areas. It included actual measurement results such as the thermal conditions, air ventilation, lighting environments, indoor air quality, occupant satisfaction and comfort, and other subjective feedback. Through this objective investigation of the physical environment and subjective feedback of the occupants' degree of comfort, the problems with objective space were able to be studied and analyzed, and the potential for spatial optimization put forth from the perspective of passive adjustment performance. Finally, the database established through this research assisted in highlighting the design goals for tubular space in subway station building complexes. A model for three typical kinds of composite tubular spaces was constructed with the goal of achieving high levels of performance and low amounts of energy consumption. Therefore, this research method was divided into the following four steps. (as shown in Figure 2)

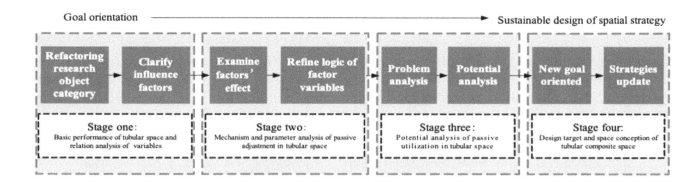

Figure 2. Research methodology route.

2.1. Stage One: Factor Analysis of the Effect of Passive Design on Tubular Space

Passive design belongs to the category of sustainable development in architecture, and is an essential part of addressing three factors: the environment, society, and the economy [35]. Tubular space is a typical passive spatial design strategy that coordinates contradictions among these three factors and architecture, leading building construction in a more positive direction. Architecture is a carrier of the climate and its human occupants; building space and outdoor climate conditions can be seen as the reasons for indoor physical environments and occupant satisfaction [36]. This research is based on an AHP (Analytic Hierarchy Process) methodology which decomposes complex issues into several group factors, and compares those factors to one another to determine their relative importance [37,38]. It adopted the method of factor quantification analysis to classify climate conditions, building spaces, physical environmental comfort, and occupancy satisfaction. This logical framework was established through measurements and simulations; the influences therein were determined by a correlation analysis, based on the acquired data. Therefore, the passive design of tubular space was divided into four factor groups: spatial parameters, climate parameters, the degree of physical environmental comfort, and occupancy satisfaction. (as shown in Figure 3)

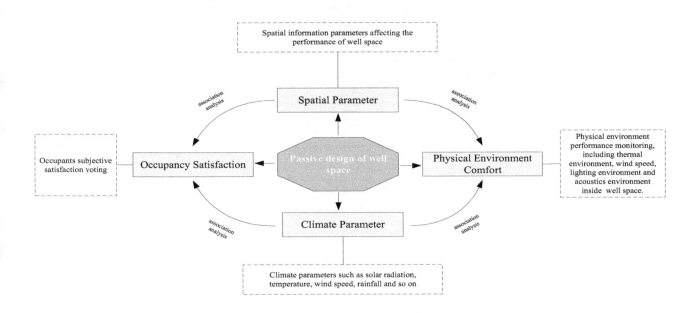

Figure 3. Analysis of the factors influential on tubular space in subway station building complexes.

According to authors pervious research in passive space design [36,39], the building space can be recognized as four categories including "shape", "mass", "quantity", and "connection". This research also adopts this research framework that the factor groups were further quantified and subdivided. The spatial parameters were again divided into four sub-factors: geometric dimensions, interface properties, and internal and external related categories (as shown in Table 1). Researchers have made useful contributions to indoor environment comfort research [40]. For example, Professor Fanger highlighted six elements that influence the comfort level of an indoor thermal environment (the mean radiation temperature, air temperature, relative humidity, air speed, clothing insulation, and metabolic rate) to form the Predicted Mean Vote (PMV) model [41]. LEED [42] in USGBC and China ESGB [43] lists lighting environment evaluation factors (including natural light, artificial light, view, and control) and offers clear standard indicators. As the tubular space can have a significant wind effect that affects the indoor air flow, while most tubular space exists underground, and the outdoor noise environment has little influence on the interior, therefore, outdoor climate parameters can be divided into four additional sub-factors: thermal environment, lighting environment, air quality, and wind speed. Corresponding to the outdoor climatic conditions, the indoor physical environment also includes four sub-factors: thermal environment, lighting environment, air quality, and air velocity. There are many factors involved in occupancy satisfaction. The method of semantic differential (SD) analysis [44] was used to divide the influencing factors into eight aspects: thermal conditions, humidity, light, air quality, air velocity, ease of use, cleanliness and maintenance, and overall environmental satisfaction (as shown in Table 1).

Table 1. Factor analysis of the passive adjustment effect of tubular space.

Factor Group	Factors	Parameter Acquisition Method	Parameter Unit
Spatial parameter	geometric dimensions (L:W:H)	distance measurement	m
	interface property (U-value)	material thermal performance calculation	$W/(m^2 \cdot K)$
	internal related categories (visitor flow rate)	statistics	N/h
	external related categories (outdoor, platform, commercial, none)	judgment	N/a
Climate parameter	thermal environment (temperature, relative humidity)	measurement	°C, %
	lighting (illuminance)		Lux
	air quality (PM2.5, PM10, CO_2)		$\mu g/m^3$, ppm
	wind speed		m/s
Degree of comfort with the physical environment	thermal environment (temperature, relative humidity)		°C, %
	lighting (illuminance)		Lux
	air quality (PM2.5, PM10, HCHO, CO_2)		$\mu g/m^3$, ppm
	air velocity, wind temperature		m/s, °C
Occupancy satisfaction	thermal comfort	occupant survey	Vote score [−3~3]
	humidity		
	air quality		
	lighting		
	ventilation		
	ease of use		
	cleanliness and maintenance		
	overall environmental quality satisfaction		

2.2. Stage Two: Field Survey

Corresponding with the public factors affecting the passive function of tubular space, the actual field investigation involved spatial drawings, monitoring the indoor and outdoor physical environments, and determining occupants' subjective levels of satisfaction. The relationship between buildings and people, especially in terms of the healthiness of the indoor environment, has a significant influence on human survival and sustainable development. Since the normal operating phase tends to be from 6:00 a.m. to 10:00 p.m., the opening time of building complexes (offices or businesses) is usually included within that time frame. Therefore, the survey had a clear research plan regarding a day cycle time, from 6:00 a.m. to 10:00 p.m., including two rush hours where there was peak human flow. The physical environment test called for the selection of a typical space, such as tube entrance, middle tubular space, connection point between a subway and complex building, station hall, or subway platform where the long-term physical environment could be monitored. The data were collected every five minutes. The physical quantities included nine parameters: temperature, humidity, illuminance, CO_2 concentration, PM2.5, PM10, HCHO, air velocity, and wind temperature. The test contents including long-term consecutive days of outdoor temperatures which measurement interval was 5 min, temperature measurements for each (selected) test point which measurement from 6:00 a.m. to 10:00 p.m. for each typical day; measurement interval was 5 min.

In addition, since piston wind can affect the interior tubular space and connection points of urban complexes during subway operation, and the aerodynamic forces of piston wind may be usable as a source of renewable energy [45], the fieldwork test also included instantaneous wind speed changes at the subway platform level. The observation frequency was 3 s, with a train cycle of arrival, stay, and departure (as shown in Table 2).

Table 2. Building the physical environment fieldwork test framework.

Measurement Items		Parameter Type	Test Content	Properties of the instruments
Thermal environment	outdoor temperature test	temperature	°C	Portable infrared temperature meter, Biaozhi GM700, Range: −50~700 °C, Resolution: 0.1 °C
	indoor temperature test for each (selected) test point			
	indoor humidity test for each (selected) test point	humidity	%	
Lighting	outdoor luminance test	luminance	lux/daylight factor %	Portable luminance meter, Reggiani DT-1301, Range: 0~50,000 Lux, Resolution: 1 Lux
	indoor luminance test for each (selected) test point			
IAQ	outdoor CO$_3$ concentration test	CO$_2$ concentration	ppm	Portable and self-record CO$_2$ meter, TJHY-EZY-1, Range: 0~5000 ppm, Resolution: 1 ppm
	indoor CO$_2$ concentration test for each (selected) test point			
	outdoor PM2.5/10 concentration test	PM2.5/10 concentration	μg/m^3	Portable air quality meter, temopt LKC-1000S+, Range: 0~999 mg/m^3, Resolution: 0.01 mg/m^3
	indoor PM2.5/10 concentration test for each (selected) test point			
	indoor HCHO concentration test for each (selected) test point	HCHO	g/cm^3	
Ventilation	indoor air velocity test for each (selected) test point	air velocity	m/s	Self-record instrument for wind velocity and wind temperature, TJHY-FB-1, Range: 0~10 M/S, Resolution: 0.01 M/S
	indoor air temperature of each (selected) test point	air temperature	°C	Self-record instrument for environment, TJHY-HCZY-1, Range: 0~5000 ppm, Resolution: 1 ppm

The object of this investigation was the tubular spaces in a subway station building complexes, so the function and shape of the spaces were simpler and more explicit than other common building spaces. The design of the subjective questionnaire focused on the degree of occupancy comfort and the space's influence on human health during short stays. In general, the questionnaire included three categories: satisfaction with the physical environment; space satisfaction votes, such as ease of use, cleanliness, and maintenance; and overall satisfaction with the environment's quality.

In order to improve the efficiency of the investigation, spatial drawings, subjective satisfaction research, and studies of comfort related to the objective physical environment were all conducted. The characteristics of the subway space in utilization mode were special in that mostly the area was designed for a rapidly-passing crowd who would be present only for a short stay. The connection spaces often fluctuated in terms of the physical environment, and users' moods and physical health conditions likely varied considerably. Therefore, the method of subjective investigation also needed to be more diverse than in other similar types of research. The questionnaire was collected mainly by three means: website-based and on-site surveys, and on-site interviews. The purpose of the website-based questionnaire was to avoid misunderstandings related to temporally-subjective factors, and dispel individual elements through the long-term accumulation of memory. For the satisfaction and self-reported productivity questions, the survey used a 7-point semantic differential scale with endpoints of "very dissatisfied" and "very satisfied." For the purposes of comparison, the scale was assumed to be roughly linear, with ordinal values for each of the points that ranged from −3 (very dissatisfied) to +3 (very satisfied) and 0 as the neutral midpoint [46]. The on-site interviews were with specifically-selected occupants who stayed in the space for a long period of time, such as retail vendors, subway station operators, commercial building security, cleaning and maintenance staff, etc. The research methods were not rigidly obeyed for prescribed problems and formats, and were altered via conversations to help the researchers understand the respondent's age, cultural and economic background, space satisfaction, environmental problems, etc (as shown in Table 3).

Table 3. Occupancy satisfaction voting framework.

Test Items		Parameter Type	Test Content	Test Content
Physical environment satisfaction	thermal comfort	vote	web-based survey/ fieldwork-based survey/ human perception test	7-point scale [−3,−2,−1,0,1,2,3] very dissatisfied to very satisfied
	humidity			
	air quality			
	lighting			
	ventilation			
Space satisfaction	ease of use	vote	web-based survey/ fieldwork-based survey/ human perception test	7-point scale [−3,−2,−1,0,1,2,3] SD of feelings about space's atmosphere
	cleanliness and maintenance			
Overall environmental quality satisfaction		vote	web-based survey/ fieldwork-based survey/ human perception test	7-point scale [−3,−2,−1,0,1,2,3] very dissatisfied to very satisfied

2.3. Stage Three: Problem and Analysis of the Spatial Potential

Stage two set up a framework for a comprehensive system that focused on three factors: architecture, humans, and the environment. Based on the conclusions made during that stage, the passive strategy factors affecting the research object were classified and recombined to analyze the functional characteristics of different positions of space that can be found throughout a subway station building complexes. According to the research by Margarita N. Assimakopoulos regarding the thermal environment in Greek subways [47], Teresa Moreno and colleagues' work on airborne particulate matter in the Barcelona subways, and John Burnett [48] and associates' investigation and analysis of the lighting environment in the Hongkong metro in China [49], obvious problems such as high humidity, low thermal comfort, and poor air quality and lighting environments in subway halls and platform spaces all emerged as worthy of further research. Many scholars also argued for energy-saving strategies in subway systems by means of passive ventilation designs for complete ventilation systems, for instance by developing ventilation systems in subway stations that could control indoor air pollutants [50,51]. The core of the third stage of this study includes two aspects. First, through an investigation and analysis of the status quo, the existing environmental problems were extracted and a design strategy put forward to resolve certain issues. Secondly, through statistical data, the researchers discovered the potential capacity of passive space, and identified design opportunities that could improve comfort, health, and energy efficiency.

2.4. Stage Four: Set up New Target Orientation and Space Update

The fourth step of this research put forward the spatial design goal of tubular space in subway station building complexes, from the perspective of sustainable development. Certain space assumptions and a particular design procedure for the compound tubular space were promoted to provide guidance for the optimized design.

Through data and problem analyses and potential excavation, this research searched typical models of complex tubular space complex systems that would be applicable to subway station building complexes. The key point was to determine an applicable and feasible space utilization model that could provide a design basis. The researchers put special emphasis on viable applications for potential natural resources in subway spaces, such as tunnel and piston wind, pull shafts, lighting, and landscape tubes, that could be further integrated into the design. Typical composite tubular spaces can be in the form of a tube tunnel (a solar chimney composite space system), combined wind tunnel (a displacement ventilation complex space system), combined active and passive ground source heat pump (wind tunnel composite tubular space system), hot air shaft ventilation, or lighting composite space system. Figure 3 offers an overall view of the study.

3. Results and Discussion

3.1. Building Space Information Factors

Based on the above-mentioned factors, a multi-criteria evaluation method for tubular space was proposed. The survey selected five typical subway station building complexes in Beijing, including Xizhimen (W1), Haidianhuangzhuang (W2), Guomao (W3), Dawanglu (W4), and Wangfujing Stations (W5) (as shown in Figure 4). The selected five stations involved six main subway lines: #1, #2, #4, #10, #13, and #14. All of the surveyed stations were urban subway transit hubs. Xizhimen Station (W1) is where the #2, #4, and #13 subway lines converge; it is also home to three high-rise commercial office buildings and the Beijing subway station. Haidianhuangzhuang (W2), Guomao (W3), and Dawanglu Stations (W4) all are points of convergence for more than two subway lines, and offer connections with urban complexes. Wangfujing Station (W5) is located in the middle of Beijing, and is connected to the largest integrated commercial building in Asia. As such, it features a substantial people flow rate; the location is also of geographical importance (as shown in Table 4).

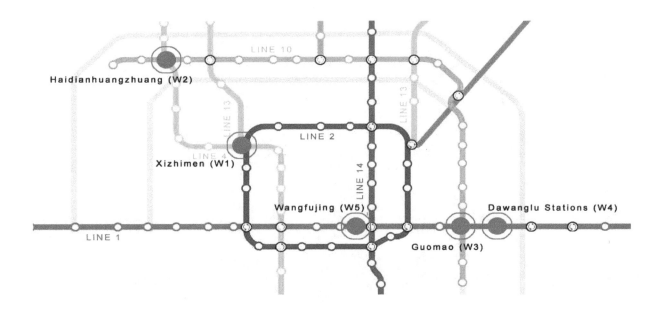

Figure 4. Five typical subway station building complexes in Beijing.

The test period was selected to be from 28th June 2017 to 20th July 2017, the highest-temperature time period in Beijing. It is a typical and continuous testing period of 3 weeks. The data excluded unstable factors which may conduct to instantaneous mutations data such as weather mutations, active equipment interference, people behavioral interference, misuse of testing instruments, etc, and used a mean value within the 3 weeks. The purpose of this research was to investigate the performance of the tubular space in the physical environment under the most unfavorable conditions in the summer climate.

Table 4. Survey object information.

No.	Station Name	City Complex	Building Function	Building Area (m²)	Subway Line	Number of Test Points	Type of Test Point
W1	Xizhimen	Cade mall	Commercial, office	89,000	#2, #4, #13	3	middle tubular space, tube entrance, platform layer
W2	Haidianhuangzhuang	Gate City mall	Commercial	47,000	#4, #10	3	middle tubular space, tube entrance
W3	Guomao	Yintai mall	Commercial, office	350,000	#1, #10	4	middle tubular space, tube entrance, platform layer, station hall
W4	Dawanglu	China Trade Center mall	Commercial, office	710,000	#1, #14	4	middle tubular space, tube entrance, platform layer, station hall
W5	Wangfujing	Oriental Plaza mall	Commercial	120,000	#1	4	middle tubular space, tube entrance, platform layer, station hall

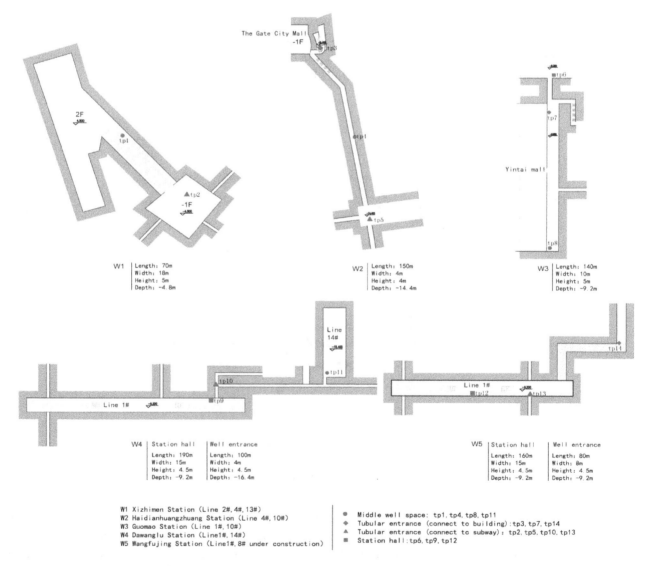

Figure 5. Test station space plan and test points location.

The first step in the fieldwork survey was to acquire the building information. This was performed in order to obtain the data supporting the tubular spaces. After the fieldwork test, the study first drew out five site plans and geometric scale parameters for the tubular space (as shown in Figure 5). The researchers then selected three or four test points for each site station; all tests contained 18 measuring points. Each test point had a certain representativeness. The W1 site contained three test points; tp1 was located in the middle of an above-ground glass corridor between the station and the complex, and was a middle tubular space.Tp4 and tp11 were located between the sites and underground associated tubes, which were all classified as middle tubular space. Tp8 was a link between a subway station and commercial building, which was also middle tubular space. Tp3, tp7, and tp14 belonged to the first kind of tube entrance space, and connected the commercial complex building at one end of the tube shaft. Tp2, tp5, tp10, and tp13 belonged to the second kind of tube entrance space, and connected to the station hall. Tp6, tp9, and tp12 were the test points at the subway station hall (as shown in Figure 5). Finally, tp15, tp16, tp17, and tp18 were the subway platform test points for W1, W3, W4, and W5, respectively.

Because the locations of the tubular spaces in each subway station building complexes were different, the environmental problems varied dramatically. Therefore, this research compared the physical environment parameters of the 18 measuring points, according to five types: middle tubular space (four test points), tube entrance (connection to building) (three test points), tube entrance (connection to subway) (four test points), station hall (four test points), and platform layer (four test points).

3.2. Field Survey Results and Analysis

3.2.1. Physical test results and analysis

Professor Fanger highlighted six elements that influence the comfort level of an indoor thermal environment (mean radiation temperature, air temperature, relative humidity, air speed, clothing insulation, and metabolic rate); all six are necessary to form the Predicted Mean Vote (PMV) model [41]. Tubular space in subway station building complexes is mostly underground, so the influence of radiation temperature can be ignored and people's metabolic rates can be set to the same level of 1.5 met [41]. During the test period, the hottest period in summer in Beijing was selected, so clothing level was chosen as 0.35 clo [41]. Data from this survey were collected according to eight physical parameters: environment temperature, humidity, illuminance, air velocity, PM2.5, PM10, and the HCHO and CO_2 concentrations at each point. The average result values are shown in Tables 5–9. According to the current national standards and norms, the typical range of thermal comfort is defined as between 16~28 °C [43], humidity comfort is between 30~60% [42], illumination should be no higher than 150 lux [52], and indoor air velocity in winter should be lower than 0.15 m/s and 0.25 m/s in summer. The concentrations of PM2.5 and PM10 should be lower than 75 $\mu g/m^3$ and 150 $\mu g/m^3$, respectively, according to the 24-hour average concentration limits of the two grades listed in the national standard [53]. The concentration of HCHO and CO_2 should be lower than 0.08 mg/m^3 [54] and 1000 ppm [55], respectively.

(1) Comfort analysis

Figures 6–9 are box diagrams of the physical test results for thermal conditions, lighting, IAQ, and ventilation in all five types of tubular space. The red zones show locations where the occupant comfort values were beyond the related comfort standard.

As regards the thermal environment, tp1 was a solar corridor space with a higher temperature than the other three points, which were at the boundary of the comfort zone (the point at which the human body would no longer enjoy thermal comfort). The tube lengths of tp4 and tp8 were 150 m and 140 m, respectively, and the humidity levels of the two test points significantly exceeded the standard. The highest reached 84.7%. The temperature in the space was lower than the human comfort level standard would deem acceptable, and the excessive humidity could easily cause mildew and affect users' health. As regards the lighting environment, the illuminance levels of tp4 and tp8 in the middle tubular space were not sufficient; the values did not reach the national standard requirement and thus could be hiding potential dangers. Due to the large amount of natural light at tp1, the average illuminance could reach 990 lux without artificial lighting. A better lighting environment would also improve the quality of the indoor environment. Past research results have shown that occupants become uncomfortable and can even experience headaches and chest tightness, causing their work ability to decline, when the CO_2 concentration is over 1000 ppm. In the test of CO_2 concentration for the middle tube, the two longer tube sat tp4 and tp8 had lower air quality; the highest CO_2 concentration was at tp4, which reached 1931 ppm. This is over two times the standard. The maximum PM2.5 concentration at tp8 reached 121 g/m^3; the outdoor concentration was 74.6 g/m^3. Thus, the disadvantages were greatly exacerbated. Although the air velocity values at tp4 and tp11 were slightly higher than the standard, moderate ventilation in a thermal environment can lower body surface temperature, taking away the sweat that collects on the human body's surface.

To sum up, the main problems as regards comfort in the middle tubular space were its high humidity, poor light environment, and low air quality.

Table 5. Average values for the physical test results in the middle tubular space.

Site No.	Test Point Number	Temperature (°C)	Humidity (%)	Illuminance (Lux)	Air Velocity (m/s)	PM2.5	PM10	HCHO	CO$_2$
W1	Tp1	32.1	47.7	990	0.196	47.9	67.2	0.022	681
W2	Tp4	28.1	70.2	72.0	0.265	45.4	64.3	0.030	1163
W3	Tp8	25.5	76.8	19.0	0.0	66.1	92.0	0.057	1000
W4	Tp11	28.3	57.4	153.6	0.307	52.0	73.1	0.044	892.1

Note: The data in the grey background indicate that it is beyond comfortable zone.

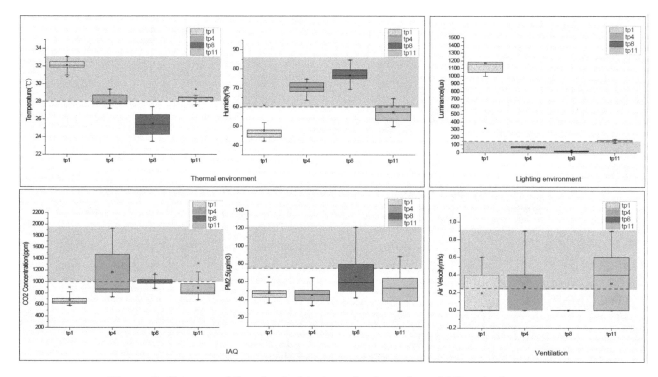

Figure 6. diagram of the physical test results from the middle tubular space.

Figure 7 shows the performance of the physical environment of the tube entrance space approach to one end of a shopping mall. All of the environment's temperatures and most of the humidity index values for the three test points exceeded the comfort zone. The maximum value of tp3 was 32.3 °C, while the mean outdoor temperature was 29 °C. The average illuminance levels of the three test points in the tube entrance (connection to building) did not reach the national standard; tp3 and tp7 were especially low. Both of the test points located at the indoor and outdoor junctions had light levels that would require human eyes a significant amount of time to adjust to the darkness. This can cause sensations of insecurity and discomfort when entering from a strong outdoor light environment. The concentration of CO_2 in these spaces was also high, with maximum values for tp3 and tp7 reaching 1761 ppm and 1690 ppm, respectively.

To sum up, because most of the tube entrance space was connected to a shopping mall and was close to the outdoors, the thermal environment and humidity levels were low, light problems were obvious, and concentrations of CO_2 were high.

Table 6. Average values for the physical test results from the tube entrance (connection to a building).

Site No.	Test Point Number	Temperature (°C)	Humidity (%)	Illuminance (Lux)	Air Velocity (m/s)	PM2.5	PM10	HCHO	CO_2
W2	Tp3	30.6	62.1	7.5	0.117	54.8	76.3	0.055	981.9
W3	Tp7	30.4	65.2	24.4	0.144	73.5	102.8	0.126	1308.2
W5	Tp14	28.9	52.9	147.9	0.604	32.2	45.0	0.047	928.9

Note: The data in the grey background indicate that it is beyond comfortable zone.

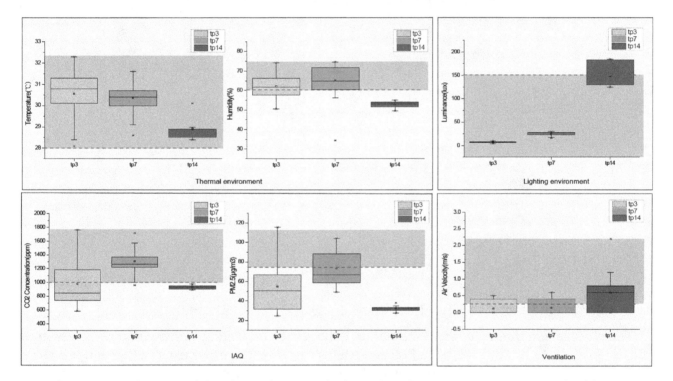

Figure 7. Box diagram of the physical test results from the tube entrance (connection to building).

According to Figure 8, the thermal and humidity environments at the tube entrance space near the subway were obvious. The temperatures at the four test points all exceeded the comfort level. The average temperature at tp5, with an elevation depth of 14.4 m, was 2.5 °C lower than that of tp10 and tp13, which were at an elevation depth of 9.2 m. The higher wind speed relieved the high humidity and CO_2 concentrations at the tube entrance space (connection to subway). Generally, the average humidity was lower than 70% and both the average and maximum values of CO_2 were greatly reduced. However, due to the influence of the subway piston wind, the wind speed presented a sinusoidal fluctuation and the direction of the wind speed changed periodically. Thus, the wind environment became the most unfavorable factor with regards to comfort.

To sum up, ventilation was one of the most detrimental comfort factors at the tube entrance space (connection to subway). It was affected by the piston wind so that cold air (from the air conditioning) was sometimes sucked out of the subway platform layer and hot air was pumped out of the tubular space. People who stayed at that location for long periods of time were frequently affected by two kinds of wind that had large temperature and direction differences. They expressed great discomfort and likely were experiencing threats to their health.

Table 7. Average values for the physical test results from the tube entrance (connection to subway).

Site No.	Test Point Number	Temperature (°C)	Humidity (%)	Illuminance (Lux)	Air Velocity (m/s)	PM2.5	PM10	HCHO	CO_2
W1	Tp2	30.8	50.0	154	0.365	53.5	75.9	0.019	688.4
W2	Tp5	28.6	66.0	32.5	0.859	52.7	73.6	0.019	840.5
W4	Tp10	31.3	68.3	107.3	1.73	107.6	151.1	0.021	920.8
W5	Tp13	31.1	70.0	91.2	0.683	149.2	201.7	0.026	802.4

Note: The data in the grey background indicate that it is beyond comfortable zone.

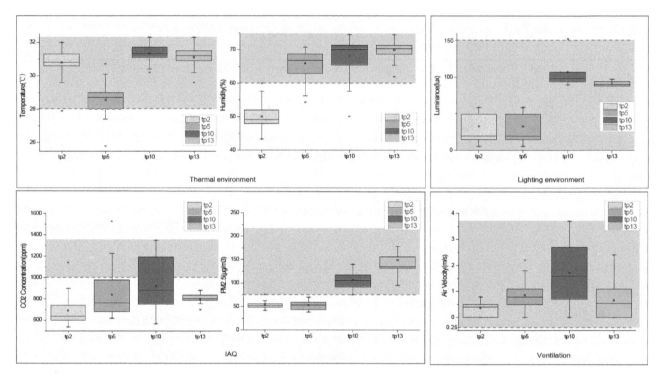

Figure 8. Box diagram of the physical test results at the tube entrance (connection to subway).

The physical environment test at the station hall focused on Beijing subway line #1, and included three test points. As can be seen from Figure 9, the environment temperature in the station hall was significantly higher than the comfort zone and 5 to 6 °C higher than the outdoor temperature at that time. The humidity in the station hall was also relatively high, but the light environment basically met the national standard. Due to a large number of people flowing through the middle of the station hall and the relatively longer tube lengths (130 m, 190 m, and 160 m), the CO_2 concentrations at the three points were correspondingly higher. The peak concentration of CO_2 at the tp6 measuring point reached 1833 ppm. Tp12 was located at a point with a smaller people flow rate, so its CO_2 concentration stayed within the standard range. However, due to the impact of the subway piston wind, there was a high level of PM pollution in the subway tunnel that was brought into the entrance hall, with a peak concentration of 160 g/m^3. Compared with the outdoor concentration 74.6 g/m^3, this was two times the outdoor concentration at that time. Thus, it can be seen that a large number of people were gathered at the station and hall levels, resulting in a higher concentration of CO_2. The station hall layer was directly connected to the train platform and had a higher PM concentration due to the influence of the piston wind from the subway.

In summary, influenced by the space size and flow rate, the physical environment of the hall at all stations was the worst, which is reflected in the high temperature and humidity, and poor air quality.

Table 8. Average values of the physical test results from the station hall.

Site No.	Test Point No.	Temperature (°C)	Humidity (%)	Illuminance (Lux)	Air Velocity (m/s)	PM 2.5	PM 10	HCHO	CO_2
W3	Tp6	33.5	62.2	144.1	0.724	68.6	96.0	0.038	1374.4
W4	Tp9	32.6	61.9	152.3	0.446	100.9	139.2	0.023	1153.6
W5	Tp12	31.2	71.4	92.0	0.172	119.6	167.2	0.021	922.5

Note: The data in the grey background indicate that it is beyond comfortable zone.

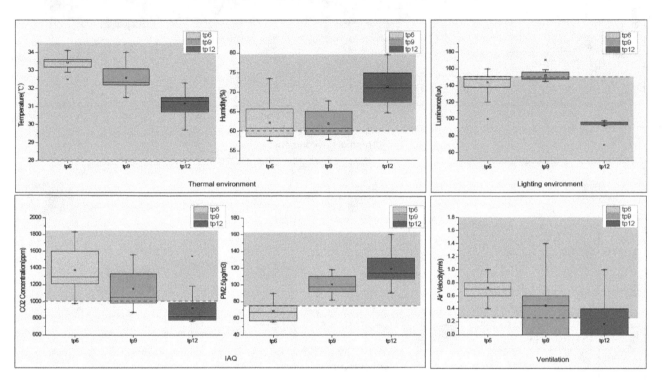

Figure 9. Box diagram of the physical test results from the station hall.

The physical environment of the platform layer is shown in Figure 10. Since the wind speed at the platform layer was significantly affected by the movement of the subway vehicles, it will be discussed in detail in the next chapter. The lighting environment met the national lighting standards. There were two key problems with comfort: the thermal environment and air quality. The temperature was generally too high; the highest value was from W4, which reached 34.5 °C. The values were 3.5 °C higher than the outdoor temperature at that time. For air quality, the most significant problem was the PM concentration. The PM2.5 data for almost all of the test sites were higher than the human body's comfort range; the PM10 concentration was too high at the W4 site as tube, reaching a maximum of 225.7 g/m^3, while the outdoor concentration was 83.2 g/m^3.

To sum up, the temperature and PM values were the key problems with comfort at the platform layer.

Table 9. Average values for the physical test results from the platform layer.

Site No.	Test Point No.	Temperature (°C)	Humidity (%)	Illuminance (Lux)	Air Velocity (m/s)	PM2.5	PM 10	HCHO	CO_2
W1	tp15	30.2	48	260.3	Instantaneous wind speed (as shown in Figure 12)	85.3	121.7	0.017	974.5
W3	tp16	31.6	64.3	263		98.5	138.3	0.019	867
W4	tp17	33.9	58.8	259.5		137.6	192.5	0.02	1005
W5	tp18	33.7	69.6	251.2		72.7	102.3	0.019	1283

Note: The data in the grey background indicate that it is beyond comfortable zone.

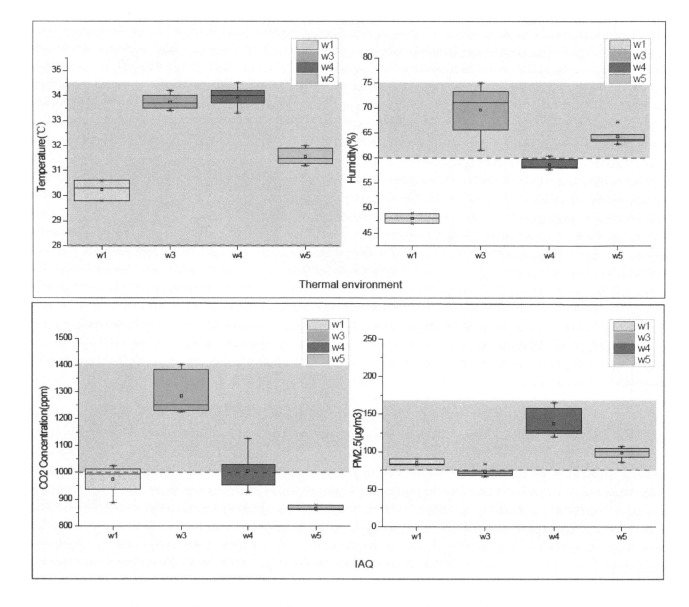

Figure 10. Box diagram of the physical test results from the platform layer.

(2) Changes in physical quantities with time and human flow
a: Temperature

The green curve in Figure 11 shows the fluctuation in outdoor temperature. Except for the aboveground space of tp1, the temperature at other points was not directly affected by solar radiation; therefore, the temperature curve hardly varied over time. The overall temperature environment at the middle of the tubular space was the best, followed by the tube entrance space near the shopping center. The third best was the tube entrance space near the subway measurement point. The thermal environment of the station hall was the worst, maintaining the highest temperature nearly the entire day. Changes in the flow rate had little influence on the thermal environment of the tubular space; the relative position was the decisive factor for the temperature there.

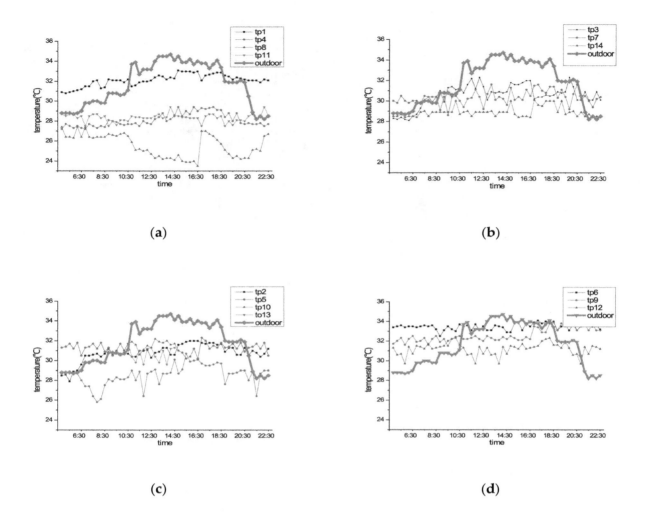

Figure 11. Comparison curves of hourly temperature data for each point during the monitoring day. (**a**) Middle tubular space; (**b**) Tube entrance (connection to building); (**c**) Tube entrance (connection to subway); (**d**) Station hall.

b: CO_2 concentration

In the diurnal variation curve for CO_2 concentration, there was almost no consistent fluctuation pattern for the same type of tubular space. From the overall curvilinear relation in Figure 12, it can be seen that there were two main trends. The first was the stable value for the whole day; the range of change was not large, as illustrated by test points tp8, tp13, and tp14. The second was the dramatic changes during two time periods, 8:00~9:00 a.m. and 4:30~7:30 p.m., which indicated that the concentration of CO_2 was significantly affected by indoor people flow, and the elevated concentration occurred during morning and evening peak hours. In addition, tp3, tp4, tp5, tp6, tp7, tp9, tp10, and tp12 exposed a lack of air adjustment capacity when a large number of people gathered together.

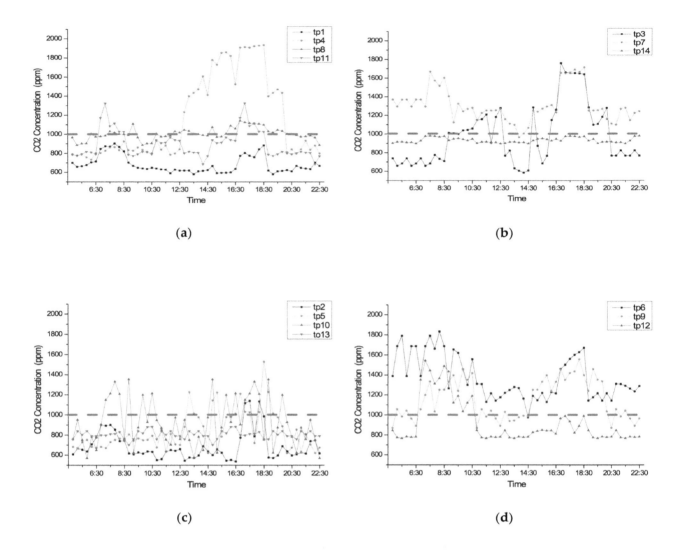

Figure 12. Comparison curves of hourly CO_2 concentration data for each point during the monitoring day. (**a**) Middle tubular space; (**b**) Tube entrance (connection to building); (**c**) Tube entrance (connection to subway); (**d**) Station hall.

c: PM2.5/PM10

As shown in Figure 13, the PM2.5 and PM10 data for each test point varied with time. This study took the middle of the tubular space to be the object of analysis. There were change rules over time for four test points: tp1, tp4, tp8, and tp11. The survey found that three points—tp1, tp4, and tp8—had a sudden increase in PM concentration at 11:00 a.m. and 6:00 p.m., almost two times that of other times. Tp11 was located in the tubular space near the #14 subway line, and was affected by the piston wind from the subway platform. The variations in PM concentration throughout the day were large, and the regularity presented was closely related to the subway's operation time.

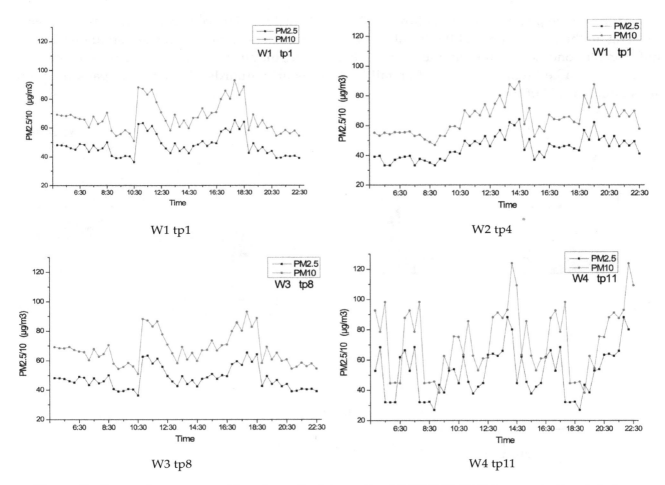

W1 tp1   W2 tp4

W3 tp8   W4 tp11

Figure 13. Comparison curves of hourly particulate matter 2.5 (PM2.5)/PM10 concentration data for each point during the monitoring day.

d: Wind speed

The test point height was 1.5 m (tp1, as shown in Figure 14a). One operating cycle for a train (arrival, stay, and departure) was around three minutes. Due to the piston effect caused by the train's operation, the tunnel wind reached remarkable levels. The wind speed when the train was arriving lasted for 15 s, with an average wind speed of 1.2 m/s. The stage during which the train stayed at the station lasted for 40~50 s (at the test point it was 48 s), with an average wind speed of 0.84 m/s. The train's departure stage lasted longer, about 120 s, with a maximum instantaneous wind speed of up to 3.6 m/s and an average wind speed of 1.79 m/s. During each three-minute cycle, the maximum wind speed was 3.6 m/s during the departure stage, while the minimum wind speed was 0 m/s during the period when the train stayed at the station (as shown Figure 14b). The wind direction was opposite during the arrival and departure stages. These two wind directions were off set when the train stayed at the station, presenting a brief state of calm. According to the coupling experiment in the Beijing subway conducted by Mingliang Ren and others, the maximum wind speed near the train could reach 7.5 m/s at the test point height of 2 m [56] (tp2, as shown in Figure 14a).

3.2.2. Occupancy satisfaction survey results and analysis

The subjective occupancy survey was comprised of two parts: on-site questionnaires and on-site interviews. The subjective questionnaire adopted a 7-point scale, where −3 corresponded with "Very Dissatisfied", 3 referred to the respondent being "Very Satisfied", and 0 was neutral (as shown in Table 10). The subjective questionnaire was issued at each test point (tp1 to tp14), 40 copies each, for a total of 560 copies. There was a total of 551 valid questionnaires completed, for a recovery rate of 98%. The test period was selected to be the same as the physical environment test, the highest temperature period for the Beijing area, from 28 June 2017, to 20 July 2017. The purpose of this research was to investigate the performance of the tubular space in terms of occupancy satisfaction under the most unfavorable conditions in the summer climate. In addition, the survey also included on-site interviews. Researchers talked with subway staff, retail traders, security guards, cleaners, and passers-by for a substantial period of time.

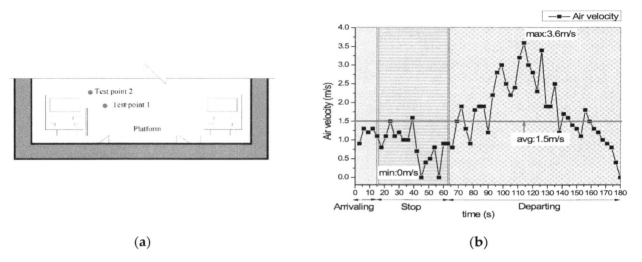

(a)   (b)

Figure 14. Wind speed curve for one operating cycle of a train. (**a**) Test point location; (**b**) Curve line.

Table 10. Questionnaire for the occupancy satisfaction survey.

Occupancy Satisfaction	Very Dissatisfied		Neutral		Very Satisfied		
Thermal comfort							
Humidity							
Air quality							
Lighting							
Ventilation	−3 ☐	−2 ☐	−1 ☐	0 ☐	1 ☐	2 ☐	3 ☐
Ease of use							
Cleanliness and maintenance							
Overall environmental quality satisfaction							

Table 11 shows the average data collected from the questionnaire. Corresponding with the analysis of the physical environment test results, this research also classified the 14 test points according to their location. There were a total of four types: middle tubular space, tube entrance (connection to building), tube entrance (connection to subway), and station hall.

Based on a histogram analysis of Figure 15, the researchers found that occupants had higher levels of satisfaction in the middle tubular space, and most of the collected data were positive. The problems with higher temperatures and poor lighting were clear in the physical environment test, but the subjective feelings of the users were not obvious. However, in terms of humidity, both the subjective questionnaire and the interviews reflected the occupants' discomfort.

The satisfaction results at the tube entrance (connection to building) fell into two categories. The first type was in relation to where the tube entrance connected the building to the outdoor space (tp3 and tp7). All indexes of satisfaction were low, especially in terms of thermal comfort, humidity, air quality, and convenience. The other type related to where the tube entrance connected to the underground building (tp14), which demonstrated positive advantages for all indicators.

Table 11. Results data from occupancy satisfaction survey.

Occupancy Satisfaction Vote	Tp1	Tp2	Tp3	Tp4	Tp5	Tp6	Tp7	Tp8	Tp9	Tp10	Tp11	Tp12	Tp13	Tp14
Thermal Comfort	−0.23	1.1	−0.9	1.3	−1.1	−1.8	−2.3	2.1	−2.1	−1.2	0.9	−2.2	−2.5	1.9
Humidity	−0.25	0.2	−1.5	−1.6	−2.5	−2.2	−2.2	−0.6	−2.4	−1.5	0.8	−2.3	−2.1	1.5
Air Quality	−0.15	−0.3	−1.2	0.2	−1.9	−2.6	−1.9	1.1	−1.5	−1.2	1.1	−2.3	−0.5	1.3
Lighting	2	−0.9	0.5	0.3	−0.8	0.2	−0.6	1.2	0.8	0.8	1.2	−0.5	0.2	1.5
Wind Comfort	−0.9	1.1	−0.6	0.5	−2.6	−0.5	−0.5	1.4	0.5	−1.6	1.2	−1.1	−2.4	−0.1
Ease of Use	−0.15	0.6	−2.1	1.1	0.2	0.9	1.1	0.9	0.6	−0.2	1.1	0.3	0.2	0.2
Cleaning and Maintenance	0.6	0.5	1.6	0.5	1.8	1.2	−1.6	1.3	0.7	0.2	1.3	0.2	0.3	0.3
Overall Environmental Quality	0.5	0.6	1.1	0.5	−0.9	0.4	−1.2	1.4	0.5	0.3	1.3	−1.2	−0.9	0.8
Overall	0.18	0.36	−0.39	0.35	−0.98	−0.55	−1.15	1.10	−0.36	−0.55	1.11	−1.14	−0.96	0.93

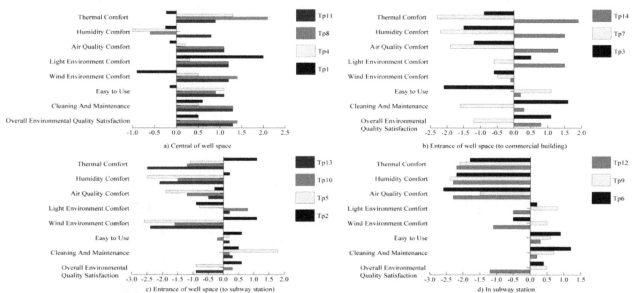

Figure 15. Results analysis of the occupancy satisfaction survey.

The most complaints from occupants were from the area near the subway. Both the space entrance near the subway and the subway platform level received lower occupancy satisfaction scores. The average thermal comfort of the tube entrance space near the subway was −1.85, the humidity score was −1.48, and air velocity comfort score was −1.38. Compared with the physical environment test results, the thermal environment of the tube entrance space near the subway had the greatest influence on the physiology and psychology of the human body. It was also the weakest point of the inner tubular space. In the physical environment test, the performance of the station hall was the worst, which is reflected in its high temperature, humid environment, and poor air quality. The averages of the three categories from the subjective survey data—thermal environment, humidity, and air quality—were as follows. Tp6 was −2.2, Tp9was −2, and tp12 was −2.3, which indicates extreme occupant dissatisfaction.

All of the questionnaire items can be sorted as follows: tp11 (middle tubular space) > tp8 (middle tubular space) > tp14 (tube entrance type1) > tp2 (tube entrance type2) > tp4 (middle tubular space) > tp1 (middle tubular space) > tp9 (station hall) > tp3 (tube entrance type1) > tp6 (station hall) = tp10 (tube entrance type2) > tp13 (tube entrance type2) > tp5 (tube entrance type2) > tp12 (station hall) > tp7 (tube entrance type 1).

3.2.3. Satisfaction–Comfort Matrix Results and Analysis

Figure 16 shows the results of the Satisfaction–Comfort Matrix in the overall 14 target test points. The horizontal axis of the matrix corresponds to the level of each comfort (thermal, lighting, ventilation, and air quality) with the physical environment inside of the tubular space. The data was calculated into a comfort percentage obtained from the test results, and divided into six classes from 0 to 100%. The vertical axis of the matrix corresponds to the subjective analysis of occupancy space satisfaction, following with the −3~3 score scale outlined in the above research method.

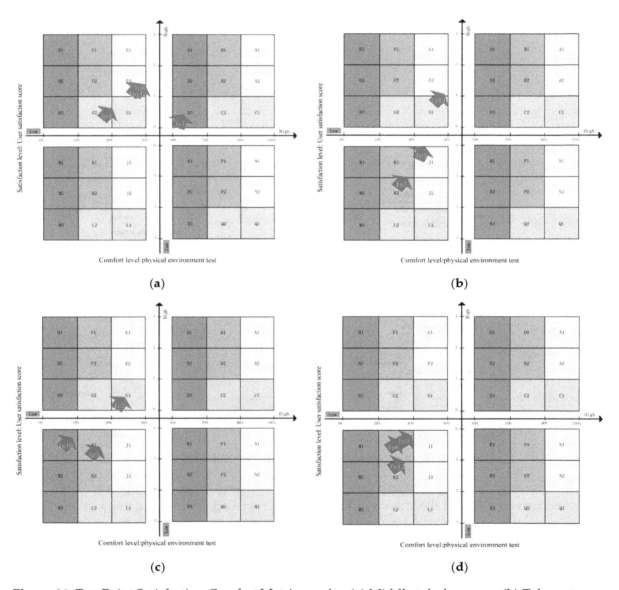

Figure 16. Test Point Satisfaction-Comfort Matrix results. (**a**) Middle tubular space; (**b**) Tube entrance (connection to building); (**c**) Tube entrance (connection to subway); (**d**) In subway station.

The Satisfaction–Comfort Matrix is divided into four quadrants. The results located in the first quadrant indicate that the animate space has a positive effect in terms of both satisfaction and comfort. The results located in the second quadrant mean it has a positive effect in terms of satisfaction but a negative effect in terms of comfort. The results located in the third quadrant imply neither satisfaction nor comfort. The results located in the fourth quadrant mean it has a negative effect in terms of satisfaction but a possible positive effect in terms of comfort [39]. Each quadrant's evaluation is divided into four grades (as shown in Figure 16).

3.3. Problem Analysis and Potential

According to the investigation of the objective physical environment and subjective comfort feelings, it can be concluded that the potential for passive utilization of tubular space includes two aspects: Improvement in comfort and utilization of climate.

3.3.1. Comfort Improvement from the Perspective of Passive Space Design

Tube entrance space plays a role in cohesion and intermediary conversion. However, it was revealed that in the Beijing subway station building complex in the underground tube entrance convergence space, the instantaneous wind speed was too high in winter and transition seasons; this seriously affected user comfort and the long-term health of security personnel. Combining a design for the complex building's tube entrance transitional buffer space with setting a reasonable wind barrier along the narrow tube entrance area, as tube as identifying transfer space in the underground connected tubular space, would help to avoid the local wind outlet issue (as shown in Figure 17).

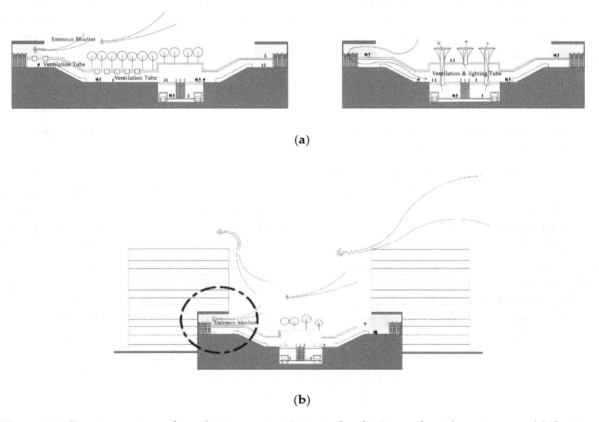

(a)

(b)

Figure 17. Passive strategy for relieving excessive wind velocity at the tube entrance. (**a**) Setting conversion space in the underground interconnected tubular space; (**b**) Transition buffer space at the tube entrance that combines with the complex building design.

The tubular space in the subway station area is always long and narrow. People in these locations experience high traffic flow, and the space lacks connection with the outdoor natural environment; this often results in poor air quality. Because it is a pass-through space, it is regularly neglected. Considering the complex building shape, setting aside areas such as air shafts, light tubes, atriums, and sunken courtyards would be an effective way of improving the air quality and indoor environment through space-based strategies.

3.3.2. Climate Utilization from the Perspective of Passive Space Design

However, tubular space as a means of connecting the external and internal environments has the advantage of transmitting natural resources. Effective natural resources can be transferred to the ground and underground spaces in subway transportation areas to solve comfort-based shortcomings in tubular spaces. These spaces can improve the quality of the area and reduce possible energy consumption during operation. In subway station building complexes, using aerodynamic, piston, and mixed ventilations, and exporting hot air to the outside through ventilation shafts or entrances can reduce energy consumption from air conditioning equipment, as tube as the area's levels of heat and humidity. Research has explored new modes for composite space systems, such as tubes with wind tunnels and solar chimneys [57], air supply tunnels that displace wind for better ventilation, active and passive combinations of wind tunnels and ground source heat pumps, and lighting from hot-press ventilation channels. These are all typical spatial patterns that can improve the utilization efficiency of passive tubular space. In addition, the shape of the tubular space has the advantage of introducing natural light. Natural light can improve the comfort level of the light environment and reduce lighting energy consumption. According to the actual test data, in a summer outdoor temperature of 28 °C, the air temperature in a space covered by four meters of soil can be maintained at 10 °C. When the outdoor temperature drops to −5 °C, the air temperature in that same underground area will remain stable at 10 °C [58]. In subway station building complexes, using geothermal energy with tubular spaces will not only improve the thermal environment quality and passive space design, but also transfer comfortable temperatures to other functional areas surrounding the tubular space, improving comfort and reducing the cost of operation.

3.4. Design Target and Space Conception of Tubular Composite Spaces from the Perspective of Sustainable Development

3.4.1. Passive Design for High Performance and Low Energy Consumption-Oriented Tubular Spaces in Subway Station Building Complexes

It was determined that there is often low levels of comfort and user satisfaction problems with tubular space. To design a new integrated subway station building complex both at home and abroad, one must continuously attempt to optimize the indoor environment and further coordinate the design strategy with the setting. Thus, this research studied tubular spaces in subway station building complex integration designs, forming a passive design method for tubular spaces that would offer high levels of performance and low amounts of energy consumption. This high level of performance is expressed in greater comfort and improved human health related to the indoor thermal environment, light and air quality, and user experience. Low energy consumption comes in the form of an energy-saving role in the building, such as by providing passive cooling, a fresh air supply, natural light, and wind energy utilization via the functional and shape advantages of tubular space (as shown in Figure 18).

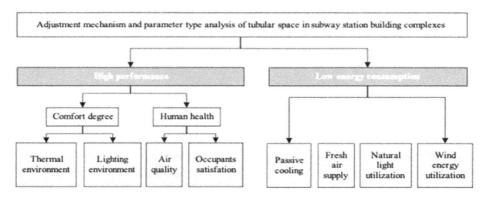

Figure 18. Type analysis of the passive parameters of high performance and low energy consumption-oriented tubular space in subway station building complexes.

3.4.2. Three Typical Design Concepts for Composite Tubular Space

An integrated design concept was presented in the study; it included two key aspects: the integration of an independent original design division that considered the organization of a single building, the underground station, and station space in the subway station building complex design process; and the integration of the space system with the goals of high performance and low energy consumption, to optimize the comprehensive performance of the composite space. Based on these issues, three design concepts for composite spaces are described below. The verification of their performance will be analyzed in detail in future research.

(a) Tubular complex space system for a wind tunnel and solar chimney (building atrium) for air filtration

The subway station building complex is a relatively complicated building space. Because of its large area, functional space coordination, and significant number of users, atriums are usually employed to organize the functional relationship. Vertically high atrium spaces can serve as a draft for solar chimneys. Combining them with a horizontal tunnel shaft design will help improve the quality of the indoor space's thermal environment and reduce the building's amount of energy consumption. In terms of the thermal environment, tunnel air temperature tends to be low, and combining the low temperature of tunnel air with air conditioning will reduce energy consumption in summer. Combining wind tunnels with the construction of solar chimneys and using the air dynamics principle of cold air sinking and hot air rising will benefit passive air circulation in buildings, exhausting high-temperature air and sucking low-temperature air from the tunnels. In addition, moderate thermal pressure ventilation in buildings is beneficial to the circulation of air and can reduce surface body temperature in summer, thus improving thermal comfort. However, due to the wind tunnels the air quality will remain low. Therefore, the composite tubular space system should be combined with an air filtration system to improve the air quality inside the tube and air circulation throughout the subway station building complex (as shown in Figure 19).

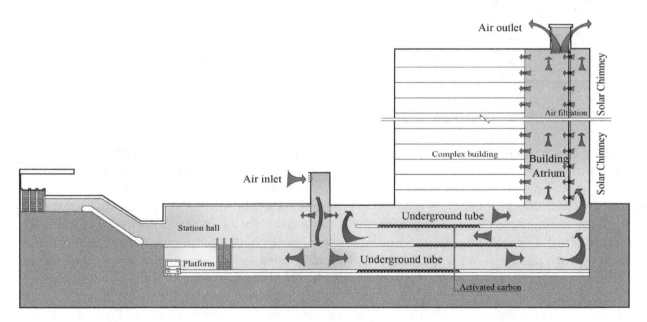

Figure 19. Sketch map of a complex tubular space system for wind tunnel and solar chimney air filtration.

(b) Complex tube path space system for lighting using hot air ventilation

The types of tubular space in a building can be divided into vertical, horizontal, and mixed. The advantage of a vertical tubular space is that it penetrates space in a vertical direction, which is beneficial for energy flowing from the top to the bottom, or vice versa. Natural lighting is a scarce resource in underground space, and thus underground areas consume much more energy for lighting than do general ground-level buildings. In most subway station building complex buildings, atrium spaces or pass heights are the most common organizational structures. The integrated design idea connects building complexes and subway station halls to provide natural lighting in tubular space, forming light tubes with the atrium space. In addition, vertical tubes in atriums can offer useful hot air ventilation. The low air temperature of tunnel wind can result in a substantial temperature difference from the higher air temperatures at the tops of atriums, which can lead low-temperature air into the building. This composite space system can also integrate the composite tubular space system of a wind tunnel with a solar chimney (atrium) for air filtration, and integrate natural light with the air filtration system, ventilation, and other functions resulting in a composite function space. This yields further improvement in the passive space adjustment effect (as shown in Figure 20).

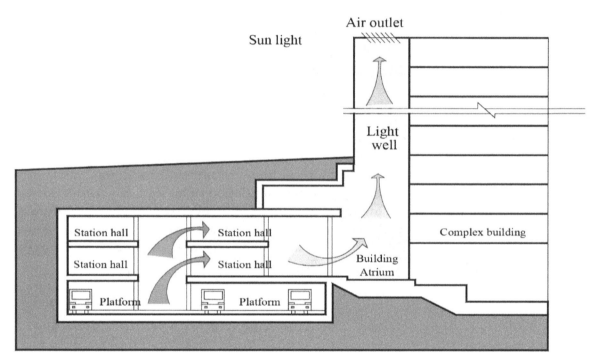

Figure 20. Sketch map of lighting through hot pressing the ventilation in a composite space system.

(c) Piston wind power generation for an auxiliary lighting system

According to past research, if a large number of fans are installed in subway tunnels, the blades produce air resistance, resulting in resistance to the train operation. This can lead to more energy consumption. However, if a few small micro-sized wind power generators are laid out in the subway station, the machines will generate renewable energy for the platform, station hall layer, and artificial lighting for the complex connection space. This is a useful way of using wind energy in subway tunnels. According to the survey distributed for this research, the average wind speed at a height of 1.5 m above the platform is 1.5 m/s, and the average wind speed near the upper end of the metro vehicle can reach 7.5 m/s. Previous research has shown that large, lightweight twist blades are the best choice for subway tunnels. In micro-sized wind turbines, the starting wind speed for vertical axis force generator equipment is 1 m/s, and the rated wind speed is 11 m/s. The safe velocity is 45~60 m/s, and the rated power is 200 W, 300 W, or 400 W. If the subway's daily schedule runs from 6:00 a.m. to 10:00 p.m., the system will be in operation for 16 hours a day. If the 400 W fan is selected, two units can be installed at the corner of each platform floor (for a total of eight units). This would result in 51.2 KW·h of electricity sent per day for artificial lighting demands. According to the general industrial and commercial electricity fee collection standards in Beijing, the annual payback period for each of these typhoon machines would be two to three years, with an average value of 1 Rmb/degree (as shown in Table 12).

Table 12. Economic benefit calculation for piston wind power generation for an auxiliary lighting system.

Rated Power	Impeller Diameter	Lump Sum Investment	Additional Investment	Service Life	Annual Maintenance Cost	Daily Power Output	Annual Power Output	Payback Period of Investment
200	0.47	2000	1200	10	100	3.2	1100	2.99
300	0.66	2100	1500	10	150	4.8	1750	2.24
400	0.66	2400	2000	10	300	6.4	2300	2.16

Note	1.	A one-time investment would be the total cost of the system's installation.
	2.	Additional investments would be used to replace two battery charges for 10 years of life.

4. Conclusions

This study focused on the various tubular space forms in subway station building complexes; the goal was to identify those that would have a moderating effect on passive potential space. The result of this investigation, research, and analysis was an improvement of the indoor environment in terms of comfort and energy consumption. This work followed four key pathways.

(1) It attempted to establish a scientific and logical method for verifying the value of tubular space by establishing causal relationships among the parameterized building space information factors, occupancy satisfaction elements, physical environment comfort aspects, and climate conditions. This research adopted an analytical hierarchical process methodology for classifying each quantized building information factor, and then compared this information to the actual fieldwork physical environment test and occupant satisfaction vote data, in order to discern the key advantages and weaknesses in tubular space design in subway station building complexes.

(2) Based on the actual field investigation, a database of physical environment performance data and users' subjective satisfaction information was established. The results showed that 59% and 57.4% time in in middle tubular space was out of thermal and humidity comfort zone, and the uncomfortable temperature and humidity at the tube entrance measuring point reached 100%. Air quality was the worst in middle tube space, 65% of the time is unhealthy which exposing significant environmental problems. By analyzing correlations therein, the comfort and health problems found in different locations that were related to tubular spaces, as tube as the potential for passive utilization, were able to be discussed.

(3) According to the investigation of the objective physical environment and subjective comfort feelings, it can be concluded that the potential for passive utilization of tubular space includes two aspects: improvement in comfort level itself and utilization of climate between natural or artificial. Based on the measured data, an integrated design method for tube path spaces exhibiting high levels of performance and low amounts of energy consumption in subway station building complexes was put forward.

(4) The research described three typical composite tubular space designs, including wind tunnels/solar chimneys (atriums), which are composite tubular spaces with air filtration systems; composite space systems that provide lighting by hot-pressing the ventilation; and piston wind power generation for use in auxiliary lighting systems. These provide methods and ideas for future research and design.

Author Contributions: Conceptualization, J.L.; methodology, J.L.; software, S.L.; validation, Q.W.; formal analysis, J.L.; investigation, J.L., S.T., Y.J.; data curation, J.L.; writing—original draft preparation, J.L.; writing—review and editing, S.L.; visualization, Q.W.; project administration, J.L.; funding acquisition, J.L.

Acknowledgments: Bejing Jiaotong University; Beijing MTR Construction Administration Corporation.

References

1. Wey, W.; Zhang, H.; Chang, Y. Alternative transit-oriented development evaluation in sustainable built environment planning. *Habitat Int.* **2016**, *55*, 109–123. [CrossRef]

2. Yang, J.; Chen, J.; Le, X.; Zhang, Q. Density-oriented versus development-oriented transit investment: Decoding metro station location selection in Shenzhen. *Transp. Policy* **2016**, *51*, 93–102. [CrossRef]

3. Papa, E.; Bertolini, L. Accessibility and Transit-Oriented Development in European metropolitan areas. *J. Transp. Geogr.* **2015**, *47*, 70–83. [CrossRef]

4. Peng, Y.; Li, Z.; Choi, K. Transit-oriented development in an urban rail transportation corridor. *Transp. Res. Part B Methodol.* **2017**, *103*, 269–290. [CrossRef]

5. Yang, Y.; Zhang, P.; Ni, S. Assessment of the Impacts of Urban Rail Transit on Metropolitan Regions Using System Dynamics Model. *Transp. Res. Procedia* **2014**, *4*, 521–534. [CrossRef]

6. Liu, J. *Rail Transit Complex Design Based on TOD*; Beijing Jiaotong University: Beijing, China, 2013.

7. Hu, Y.; Zhao, C. Brief Discussion on Architectural Design of Green Subway Station. *Build. Sci.* **2014**, *6*, 132–138.

8. Song, Y.; Wang, J.; Zhu, N. Pondering over the Passive Design Strategy for Native Green Buildings of China. *Archit. J.* **2013**, *7*, 94–99.

9. Wang, Z.; Chen, F.; Shi, Z. Prediction on Medium and long term energy consumption of urban rail transit network in Beijing. *China Railw. Sci.* **2013**, *34*, 133–136.

10. Chen, J.; Gao, G.; Wang, X.; Wang, X. Calculation method of whole life-cycle energy consumption for urban rail transit. *J. Traffic Transp. Eng.* **2014**, *14*, 89–97.

11. Gao, G.; Guan, W.; Li, J.; Dong, H.; Zou, X.; Chen, W. Experimental investigation of an active–passive integration energy absorber for railway vehicles. *Thin-Walled Struct.* **2017**, *117*, 89–97. [CrossRef]

12. Marzouk, M.; Abdelaty, A. Monitoring thermal comfort in subways using building information modeling. *Energy Build.* **2014**, *84*, 252–257. [CrossRef]

13. Li, X.; Wang, Z.; Ma, C.; Liu, L.; Liu, X. Energy Consumption Test and Analysis of Large Public Buildings Based on Gray Box Model. *Procedia Eng.* **2016**, 146–150. [CrossRef]

14. He, J.; Yan, Z.F.; Liu, H.J. On the Current Energy Consumption and Countermeasures of Large Public Buildings in China. *Adv. Mater. Res.* **2012**, 840–842. [CrossRef]

15. Lam, J.C.; Yang, L.; Liu, J. Development of passive design zones in China using bioclimatic approach. *Energy Convers Manag.* **2006**, *47*, 746–762. [CrossRef]

16. Badescu, V.; Laaser, N.; Crutescu, R.; Crutescu, M.; Dobrovicescu, A.; Tsatsaronis, G. Modeling, validation and time-dependent simulation of the first large passive building in Romania. *Renew. Energy* **2011**, *36*, 142–157. [CrossRef]

17. Sadineni, S.B.; Madala, S.; Boehm, R.F. Passive building energy savings: a review of building envelope components. *Renew. Sustain Energy Rev.* **2011**, *15*, 3617–3631. [CrossRef]

18. Feist, W.; Schnieders, J.; Dorer, V.; Haas, A. Re-inventing air heating: convenient and comfortable within the frame of the Passive House concept. *Energy Build.* **2005**, *37*, 1186–1203. [CrossRef]

19. Zhang, H.; Li, J.; Dong, L.; Chen, H. Integration of sustainability in Net-zero House: Experiences in Solar Decathlon China. *Energy Procedia* **2014**, *57*, 1931–1940. [CrossRef]

20. Yiing, C.F.; Yaacob, N.M.; Hussein, H. Achieving Sustainable Development: Accessibility of Green Buildings in Malaysia. *Procedia Soc. Behav. Sci.* **2013**, *101*, 120–129. [CrossRef]

21. Russell-Smith, S.V.; Lepech, M.D.; Fruchter, R.; Littman, A. Impact of progressive sustainable target value assessment on building design decisions. *Build. Environ.* **2015**, *85*, 52–60. [CrossRef]

22. Olgyay, V. *Design with Climate*, New ed.; John Wiley & Sons Inc.: Hoboken, NJ, USA, 1992.

23. Li, J. *Passive Adjustment Performance of Intermediary Space in Buildings*; Tsinghua University: Beijing, China, 2016.

24. Jiang, Y. *The Study on Ventilation Design of Deep Plan Buildings with Vertical Sapce*; Shenyang Architecture University: Shenyang, China, 2011; Volume 12, p. 15.

25. Brown, G.Z.; Dekey, M. *Sun, Wind & Light: Architecture Design Strategies*; John Wiley & Sons Inc.: Hoboken, NJ, USA, 2006.

26. Pieter, D.W.; Marinus, V.D.V. Providing computational support for the selection of energy saving building components. *Energy Build.* **2004**, *36*, 749–758.

27. Kenji, M.; Keishi, D.; Li, M. Tokyo Tokyu Toyoko line Shibuya Station. *Archit. J.* **2009**, *4*, 40–45.
28. Katavoutas, G.; Assimakopoulos, M.N.; Asimakopoulos, D.N. On the determination of the thermal comfort conditions of a metropolitan city underground railway. *Sci. Total Environ.* **2016**, *566*, 877–887. [CrossRef] [PubMed]
29. Li, Y.; Geng, S.B.; Zhang, X.S.; Zhang, H. Study of thermal comfort in underground construction based on field measurements and questionnaires in China. *Build. Environ.* **2017**, *116*, 45–54. [CrossRef]
30. Kim, M.J.; Richard, D.B.; Kim, J.T.; Yoo, C.K. Indoor air quality control for improving passenger health in subway platforms using an outdoor air quality dependent ventilation system. *Build. Environ.* **2015**, *92*, 407–417. [CrossRef]
31. *Chip Sullivan.Garden and Climate*; China Building Industry Press: Beijing, China, 2005.
32. Wu, C.; Wang, Z.; Lu, Z. Strategy of urban development from the view of Shibuya Hikarie. *Archit. Tech.* **2015**, *11*, 40–47.
33. Zhang, F.; Zhou, X. From the utilization of underound space to the conformity of underground space and the city: a case study of the two renovations of Les Hasses. *Mod. Urban Res.* **2014**, *29*, 29–38.
34. Li, J. Multi-criteria Approach to Impact Evaluation of Passive Adjustment Performance of Intermediary Space in Buildings. *Archit. J.* **2016**, *2*, 50–55.
35. Song, Y.; Li, J.; Wang, J.; Hao, S.; Zhu, N.; Lin, Z. Multi-criteria approach to passive space design in buildings: Impact of courtyard spaces on public buildings in cold climates. *Build. Environ.* **2015**, *89*, 295–307. [CrossRef]
36. Li, B. *The Research on Climatic-Active Design Strategy of Building Skin in Hot-Summer and Cold-Winter Zone*; Tsinghua University: Beijing, China, 2004; pp. 127–281.
37. Qiu, J. *Evaluation Science: Theory Method Practice*; Science Press: Beijing, China, 2010. (In Chinese)
38. Haas, R.; Meixner, O. An Illustrated Guide to the Analytic Hierarchy Process. Available online: https://mi.boku.ac.at/ahp/ahptutorial.pdf (accessed on 22 February 2019).
39. Li, J.; Song, Y.; lv, S.; Wang, Q. Impact evaluation of indoor environmental performance of animate space in buildings. *Build. Environ.* **2015**, *94*, 353–370. [CrossRef]
40. McMulla, R. *Environmental Science in Building*; Macmillan: Basingstoke, UK, 2007.
41. Fanger, P.O. Fundamentals of thermal comfort. *Adv. Sol. Energy Technol.* **1988**, *4*, 3056–3061.
42. Leadership in Energy and Environmental Design (LEED). The U.S. Green Building Council (USGBC): Washington, DC, USA, 2001. Available online: http://www.usgbc.org/ (accessed on 22 February 2019).
43. MOHURD. *Assessment Standard for Green Building of China*; GB/T50378-2014; China Building Industry Press: Beijing, China, 2014.
44. Kang, J.; Zhang, M. Semantic differential analysis of the soundscape in urban open public spaces. *Build. Environ.* **2010**, *45*, 150–157. [CrossRef]
45. Sun, Z.; Wei, D.; Xie, J. Research on subway tunnel wind power system. *Renew. Energy Resour.* **2016**, *9*, 1333–1341.
46. Hummelgaard, J.; Juhl, P.; Sabjornsson, K.O.; Clausen, G.; Toftum, J.; Langkilde, G. Indoor air quality and occupant satisfaction in five mechanically and four naturally ventilated open-plan office buildings. *Build. Environ.* **2007**, *42*, 4051–4058. [CrossRef]
47. Assimakopoulos, M.N.; Katavoutas, G. Thermal comfort conditions at the platforms of the Athens Metro. *Procedia Eng.* **2017**, *180*, 925–931. [CrossRef]
48. Moreno, T.; Reche, C.; Minguillón, M.C.; Capdevila, M.; de Miguel, E.; Querol, X. The effect of ventilation protocols on airborne particulate matter in subway systems. *Sci. Total Environ.* **2017**, *584*, 1317–1323. [CrossRef] [PubMed]
49. Burnett, J.; Pang, A.Y. Design and performance of pedestrian subway lighting systems. *Tunn. Undergr. Space Technol.* **2004**, *19*, 619–628. [CrossRef]
50. Han, L.; Feng, L.; Yuan, Y. Numerical Analysis of Cooling Effect of Tunnel Ventilation System in Subway Station. *Refrig. Air Cond.* **2016**, *2*, 1–4.
51. Yan, L. *The Effect of Piston Wind on Subway Environment and Energy Saving Character*; Beijing University of Technology: Beijing, China, 2015.
52. *Standard for Urban Rail Transit Lighting*; GB/T16275-2008; Ministry of Housing and Urban-Rural Development of the People's Republic of China: Beijing, China, 2008.
53. *Indoor Air Quality Standard GB/T18883-2002*; Ministry of Housing and Urban-Rural Development of the People's Republic of China: Beijing, China, 2002.

54. *Environemt Air Quality Standard GB3095-2012*; Ministry of Housing and Urban-Rural Development of the People's Republic of China: Beijing, China, 2012.

55. ASHRAE (American Society of Heating, Refrigerating, and Air Conditioning Engineers). *ASHRAE Standard 62.1—2013.Ventilation for Acceptable Indoor Air Quality*; ASHRAE: Atlanta, GA, USA, 2013.

56. Ren, M.; Guo, C.; Guo, Q.; Yang, Y.X.; Kang, G.Q.; Luo, H.L. Numerical Analysis and Effectively Using of Piston-effect in Subway. *J. Shanhaijiaotong Univ.* **2008**, *8*, 1376–1380, 1391.

57. Hu, Y. An Analysis of the Transitional Space of Traffic Hubs Case Studies of Traffic Hubs in Japan. *Archit. J.* **2014**, *6*, 109–113.

58. Wang, S.; Zhu, X.; Jiang, Y. Measurement and analysis of wall heat flow in thermal environment of Beijing Subway. *J. Undergr. Work. Tunn.* **1997**, *3*, 32–37.

6

Towards Sustainable Neighborhoods in Europe: Mitigating 12 Environmental Impacts by Successively Applying 8 Scenarios

Modeste Kameni Nematchoua [1,2,3,4,*], **Matthieu Sevin** [2] **and Sigrid Reiter** [2]

[1] Beneficiary of an AXA Research Fund Postdoctoral Grant, Research Leaders Fellowships, AXA SA 25 avenue Matignon, 75008 Paris, France

[2] LEMA, UEE, ArGEnCo Department, University of Liège, 4000 Liège, Belgium; matthieu.severin@uliege.be (M.S.); sigrid.reiter@uliege.be (S.R.)

[3] Department of Architectural Engineering, 104 Engineering Unit A, Pennsylvania State University, State College, PA 16802-1416, USA

[4] The University of Sydney, Indoor Environmental Quality Lab, School of Architecture, Design and Planning, Sydney, NSW 2006, Australia

* Correspondence: mkameni@uliege.be.

Abstract: The purpose of this research is to determine the most impactful and important source of environmental change at the neighborhood level. The study of multiple scenarios allows us to determine the influence of several parameters on the results of the life cycle analysis of the neighborhood. We are looking at quantifying the impact of orientation, storm water management, density, mobility and the use of renewable energies on the environmental balance sheet of a neighborhood, based on eleven environmental indicators. An eco-neighborhood, located in Belgium, has been selected as the modeling site. The results show that the management of mobility is the parameter that can reduce the impact the most, in terms of greenhouse effect, odor, damage to biodiversity and health. With the adaptation of photovoltaic panels on the site, the production exceeds the consumption all through the year, except for the months of December and January, when the installation covers 45% and 75% of the consumption, respectively. Increasing the built density of the neighborhood by roof stacking allows the different environmental impacts, calculated per inhabitant, to be homogeneously minimized.

Keywords: life cycle assessment; sustainable neighborhood; Belgium; urban scale; roof stacking

1. Introduction

From the 1970s, a general awareness had been created with regard to environmental problems. The first oil crises, the end of the Thirty Glorious Years and the emergence of mass unemployment were highlights that questioned the idealistic aspect of a "model society", which had been in force since the end of the Second World War in Western countries. People were beginning to notice the incompatibility between the well-being of the productivity system, which is infinite growth, and the survival of the ecosystem as we knew it then. Thus, many people had come to believe that it would be beneficial for everyone to change the way our society operated [1]. Between 1960 and 1971, two large non-governmental organizations had emerged to fight for the protection of nature—Greenpeace and the World Wildlife Fund (WWF). The following year, that is, in 1972, the Club of Rome published its report under the tutelage of the globally respected Massachusetts Institute of Technology. They warned that the frenetic development of major industrialized nations would lead to the depletion of the world's reserves of non-renewable resources. Many environmental disasters that were publicized a lot more played an important role in this awareness [2]. The oil spills, Chernobyl, hurricane Katrina, or the heat wave of 2003, which caused the death of 15,000 people in France, made their mark. At the beginning of the twenty-first century, a leap was made regarding the international awareness of environmental problems. The Communiqués of the United Nations Intergovernmental Panel on Climate Change (IPCC) were significant in making this happen [3].

These communiqués claim that humans are 90% responsible for the worsening of the greenhouse effect and that this could lead to a rise in water levels of more than 40 cm [4,5]. They point to the economic, environmental and social risks that global warming could create [6]. Even the Catholic Church has reacted by defining a new sin, the sin of pollution. Targets are set at the European level to address environmental issues. The 2030 Package fixed by all the member countries of the European Union revealed at least a 40% cut in greenhouse gas emissions (from 1990 levels); at least a 32% share in renewable energy; and at least a 32.5% improvement in energy efficiency. This objective can be reached if all these countries work in collaboration.

In industrialized countries, the construction sector is responsible for 42% of the final energy consumption [7,8], 35% of greenhouse gas emissions [9] and 50% of greenhouse gas emissions from extractions, from all the materials combined [6]. In addition, the urban sprawl is increasing land use, and between 1980 and 2000, the built space in Europe has increased by 20% [10]. Buildings are responsible for different types of soil consumption: a so-called primary consumption, that is to say, their building footprint; and also a secondary consumption, that is, the extraction, production, transportation and end-of-life treatment of construction products [11]. This type of impact is minimally considered in most studies, if at all considered. However, the life cycle analysis also studies the environment around the built area [12]. The culmination of all thermal and energy regulations is the European Zero Energy Building (nZEB) goal [13]. It aims to ensure that all new buildings have a neutral annual energy consumption; that is, they produce as much energy locally as they consume over the course of a year. This concept can be extended to the neighborhood scale to target the zero-energy level at the community scale [14].

A life cycle assessment is a method, an engineering tool, initially developed for industry. It aims to quantify the environmental impact generated by a product, a system or an activity. This requires an analysis of material consumptions, energy and emissions in the environment, throughout the life cycle [15]. An environmental impact is considered to be any potential effect on the natural environment, human health or the depletion of natural resources [1,16]. Thus, LCA is an objective process that allows for the establishment of various means to ensure increased respect for the environment [17]. Nowadays, a life cycle assessment (LCA) is the most reliable method for assessing environmental impacts associated with buildings and materials.

In 2002, Guinee et al. [18] stated that one of the first (unpublished) LCA studies on the analysis of aluminum cans was conducted by the Midwest Research Institute (MRI) for the Coca-Cola Company. Buyle et al. [19] showed that the first LCA in the construction sector was performed in the 1970s. In the early 1980s, life-cycle analysis widened its interest to the field of construction. Different studies used different methods, approaches and terminologies. There was a clear lack of scientific discussion and consultation on this subject [19]. In the 1990s, we saw many more multi-criteria approaches, such as environmental audits and assessments that studied the entire life cycle of products. These methods were beginning to get standardized; conferences were organized, and many more scientific publications were produced on the subject. From 1994, the International Organization for Standardization (ISO) was also involved in the field of life cycle analysis and in 1997 published, for the first time, its ISO 14040 and 14044 standards on the harmonization of procedures [20,21].

From the beginning of the twenty-first century, interest in LCA and reflections on the complete life cycle are increasing. Many more scientific studies are being published. Today, LCA is recognized as the most successful and objective multi-criteria assessment tool for environmental impacts, on the entire building scale [22,23]. In Belgium and several European countries, most of the studies on LCA in the construction sector concentrate on the building level [23–25].

In the literature, it is important to note that many studies on LCA have several limitations: some use a single scale when analyzing the reduction of the environmental footprint in the construction sector; others use only one indicator (energy demand) to conduct a study within a building; while others focus on a single stage of the life cycle (the occupation stage). To deepen this study, we carried out this work by pushing the reflection further. Thus, we will no longer work on the scale of a single building, such as many studies, but on the neighborhood scale. We will not study a single indicator, but more than ten. We will not focus on one step, but we will study the whole life cycle.

In Belgium, the first thermal regulation was born in Wallonia in 1985. The EPB Directive—European Directive "Energy Performance of Building Directive" (EPBD) (European Parliament, 2002) —has been applied in Belgium since 2008 and was regularly reinforced in the years that followed. It is important to note that European regulations have the merit of reducing the energy consumption of buildings during their occupation phase. However, they focus only on this phase. All other phases of the life cycle—the extraction of raw materials, production, choice of building materials, their transportation and even their recycling at the end of their life cycle—are not taken into account. Furthermore, the there is a lack of integration and following of different requirements, which is responsible for an asymmetry of compliance in the member countries of the EU.

Thus, our study has a goal to go beyond the occupation phase and take into account the complete life cycle. In addition, these regulations are interested in only one aspect, which is the consumption of energy. We want to study the set of different environmental impacts that are significant and known. Finally, the current studies on which the energy standards are based are often carried out only at the scale of the building. We want to broaden the reflection at the neighborhood level, as it is clear to us that the environmental issues of tomorrow will be resolved at the urban scale. We believe that this type of approach is the logical continuation of the current regulations and that it is important to take the plunge. It is thanks to this type of analysis, from the cradle to the grave, that one can judge the real and lasting aspect of a construction. Indeed, we can expect that the environmental cost of energy will decrease as well as the consumption in the building sector. As a result, the relative share of non-occupancy phases in the overall environmental balance will continue to increase.

This research proposes a more efficient method for analyzing the life cycle at the scale of a neighborhood, and compares the results obtained with those of other existing research. The design and analysis of several scenarios allow us to assess several important characteristics of the LCA applied to an eco-neighborhood.

2. A Review on Current Researches Regarding Life Cycle Analysis in the Building Sector

The life cycle assessment (LCA) method is a clearly validated scientific method and is even standardized at the European level [20,21]. The LCA allows one to carry out different types of comparative studies [23], for example: (i) comparison of two entire systems or part of their life cycle; (ii) comparison between different phases of the cycle life; (iii) comparison of two different versions of the same system; and (iv) comparison of a system with a reference. The method also makes it possible to quantify an environmental impact on the complete life cycle of a product or only on one stage of the cycle without necessarily making a comparison. Thus, it is a tool that can not only serve as a decision aid, but also allows targeting of the phases of the life cycle of a product, which would need to be reworked with attention paid to the environment. The normative framework of the LCA [20,21] defines four different steps to follow: (1) the definition of the objective of the study; (2) the "Life Cycle Inventory" (LCI); (3) the "Life Cycle Impact Assessment" (LCIA); and (4) the interpretation of results—all these phases are organized independently and iteratively.

Specific standards were established for the LCA of the building sector by the European Committee for Standardization (CEN) in 2011: EN 15978 [26] and EN 15643-2 [27]. These standards are increasingly used to define and/or reduce the impacts of buildings on the environment. It is currently the only scientifically sound approach to carry out an environmental assessment at the building scale. It allows a quantitative study of the construction over their entire life cycle. However, its use at the urban or neighborhood scale is recent [28,29]. Despite the novelty of the LCA application at the neighborhood level, it is considered the most reliable method. It is a challenge and a fascinating research topic to test the application of the LCA method to an eco-neighborhood, especially since no other study to our knowledge has focused on the comparison of so many parameters and environmental indicators at the community level. Note also that many sustainable building certification schemes are based on the LCAs of the building materials [30], such as BREEAM, DGNB and Valid in Belgium.

2.1. Building Scale

Several studies in different countries studied LCA at the building level.

In 2011, Rossi et al. [25] compared the LCA of the same building located in three cities distributed in three different European countries and climates: Brussels (Belgium), Coimbra (Portugal) and Luleå (Sweden). A difference of less than 17.4% was obtained on comparing the operational energy and carbon. Stephan et al. [31] analyzed the life cycle energy use in a passive building in Belgium and, then, carried out a comparison with other building types. The results showed that new techniques of construction had to be applied for improving the house energy efficiency. The passive house embodied energy accounts for 55% of the total energy on the 100-year life cycle. Cabeza et al. [32] gave a review of the life cycle assessment (LCA), life cycle energy analysis (LCEA) and life cycle cost analysis (LCCA) in numerous kinds of buildings, located in different countries with varied climates. The results showed that few of the LCA and LCEA studies were carried out in traditional buildings. In their research, Kellenberger and Althaus [33] carried out the LAC of many house components (roof, wall, etc.), with the purpose of evaluating the performance of the materials. The transportation of the building materials and other parameters were also studied. For deepening the knowledge of the environmental characteristics of the building materials and energy, Bribián et al. [34] applied three environmental impact categories for comparing the most used material in the new designs. The results showed that the impact of a material can be significantly reduced by applying the new methods of eco-innovation. Vilches et al. [35] showed that a majority of the LCA was based on energy demand compared at every stage of the life cycle. This research focused on the environmental impact of buildings system retrieval. A strong review on the life cycle energy analyses of buildings from 73 cases, applied in 13 countries and taking into consideration office and residential buildings, was shown by Ramesh et al. [36]. Rashid and Yusoff [37] assessed the phase and material that significantly affected the environment. In the research carried out by Chau and Leung [38], the results showed that the use of different functional units did not allow easy comparison of the studies found in the literature.

2.2. Neighborhood Scale

We will now broaden the field of study and move to the neighborhood scale as a whole. Now that much progress has been made on the energy consumption of new buildings, other issues emerge [39]. We emphasized the importance of focusing on phases other than occupation, whose relative impacts increase with a decrease in consumption [40]. It is also necessary to tackle the thermal renovation in a more serious way. Beyond the building scale, the concepts of a zero-carbon city, a city without CO_2 or a post-carbon city are emerging around the world. Cities are now welcoming 50% of humanity. However, energy is not the only environmental problem. Cities are aware of the need to preserve biodiversity and green spaces [41]. To achieve these different objectives, new tools and methods are needed, to be able to measure, at an urban scale, the consequences of architectural and urban choices on the environment [41]. Many methodologies exist to quantify the environmental impacts at the city scale, but according to Anderson et al. [42], LCA is again the dominant method at the urban scale. Indeed, after the study of the different existing methods, Loiseau et al. [43] showed that LCA provides an appropriate framework and is the only method to avoid transferring environmental loads from one phase of the life cycle to another, from one environmental impact to another, or from one territory to another.

There is currently a need at the neighborhood level to integrate reflections on bioclimatic design, shared facilities, urban density or mobility issues, in order to achieve better environmental performance. Olivier-Solà et al. [44] explained that it is highly likely that the environmental and energy issues we are currently dealing with at the building level will soon be transferred to the urban scale. Thus, neighborhood-level LCA is starting to get into practice. Some works and publications concerning this method have been written but they remain rare and heterogeneous [45]. Some studies carried out by Ecole des Mines ParisTech within the energy and process center aim to scientifically develop the LCA method at the neighborhood level. The goal is even more ambitious because this work aims to make the method a tool for decision support, from the design phase [41].

2.3. Goal

We want to study various parameters that impact the environmental balance of a neighborhood. We considered several environmental indicators that we detailed. We wished to identify the most important parameters that have the greatest impact on the environmental quality of a neighborhood. This may include, for example, orientation, the presence or absence of permeable soils, renewable energy sources or integration with public transit systems. Even if the general influence of some of these parameters was known, we wished to quantify precisely the environmental impacts and compare their importance with that of the other studied parameters. For this, we conducted the environmental analysis of a neighborhood and we varied the different design parameters to quantify their impacts. Thus, we were able to provide recommendations regarding certain design choices and their potential environmental impacts.

3. Methodology

This study is constituted of many important sections, such as (a) the survey; (b) modeling; (c) application of new scenarios; and (d) analysis.

3.1. Location

This study was carried out in a neighborhood located in Liege city in Belgium. This city is dominated by a continental climate in a temperate zone. During the year, we can note four seasons: winter, autumn, summer and spring. The neighborhood evaluated in this study is located nearby the University of Liege. This site is home to an extension of Liege University, and heavily dominated by green space, which is shown in Figure 1. Several categories of buildings are found in this new neighborhood. Notably, there are apartments with two, three and even four facades. Most of these surfaces were developed for social housing. Figures 2 and 3 show the location of this neighborhood.

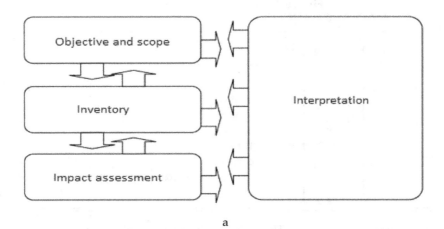

a

Figure 1. *Cont.*

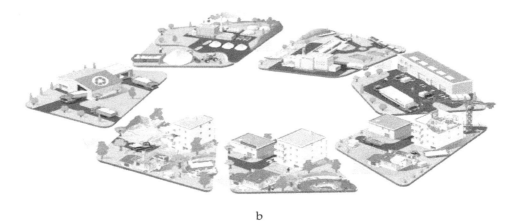

b

Figure 1. (**a**) Life cycle assessment (LCA) stage according to ISO 14044 [21]. (**b**) 3D modeling of some habitats in the studied site.

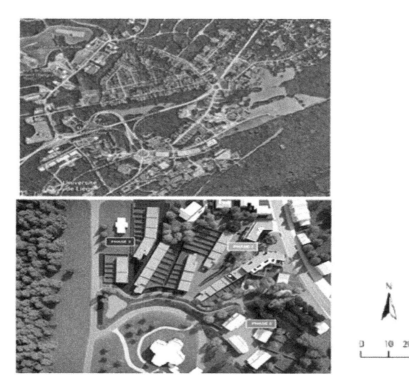

Figure 2. The study area location.

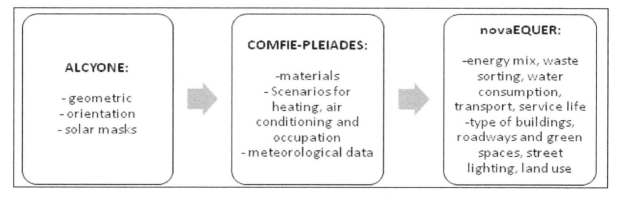

Figure 3. Chaining analysis software. The neighborhood studied was newly built by the Belgian government by adopting the concept of sustainable development.

3.2. Structure Analysis

In this city, several residences were considerate with respect to the energy demand suggested by international standards [46,47]. A total of 50% of the studied residences are semi-detached. This eco-neighborhood was built on a 3.51 ha plot. It has 17,000 m^2 of totally green space, and a total of 219 people in the residences. The life cycle assessment of the neighborhood is fixed at 80 years in the case of this study. We have taken some environmental data from the ECOINVENT database.

3.3. Simulation Tool

In this study, we used Pleiades software, version (4.19). It is divided into six modules: Library, Modeler, BIM, Editor, Results and LCA. Indeed, these tools were applied in several publications [48–56]. The analysis chain was as shown in Figure 3. Other different details regarding these simulation tools are also given in [57–60].

Under the base of the modeler tool of this software, it is easy to model all the buildings with their main characteristics. It is also possible to make the first simulations. All the results are automatically saved in the "result module", which will be requested to evaluate the ACV of the neighborhood. The analysis of an LCA is not easy, because we must associate any constituent of the neighborhood in the software (buildings, roads, garden, water, people, climate, waste, energy mix, etc.). The environmental impact of all the main elements of the site is automatically added to form the global neighborhood.

3.4. Scenarios

Some methods applied in this research were found in [54–60]. Globally, in this study, numerous scenarios were established, such as (1) building orientations; (2) water management; (3) mobility; (4) density; and (5) photovoltaic solar installation. It was very important to know the impacts of all these scenarios for improving the future planning of the new neighborhoods.

3.5. Modeling

We began the modeling of our study area by studying the project's characteristics data and the graphic modeling of the buildings on the site. A note was made of the geometrical parameters attributed to each of the walls of the buildings and their thermal properties, and the zoning and scenarios of use were also defined. Once all the parameters were defined, the dynamic thermal simulation calculations were started. All the characteristics of the buildings are described in Table 1.

It was necessary to model some elements of our buildings, such as the walls, joinery, surface conditions and thermal bridges. With regard to the walls, we not only reveal the materials and elements of construction, their thickness, and their characteristics, but also the possible thermal bridges. At this stage, we have modeled the actual walls of the project with their precise characteristics. It is also necessary to obtain information on the surface state of the different walls, in order to manage their behavior with respect to radiation.

Tables 2 and 3 show the characteristics of the heat transmissions of the frame and glazing, as well as the thermal bridges.

Table 1. Wall composition.

Element	Component	E (cm)	ρ*e (kg/m²)	λ (w/m.k)	R (m²·K/W)
Coated exterior wall	Exterior coating	1.5	26.0	1.150	0.01
	Expanded polystyrene	32.0	8.0	0.032	10.0
	Limestone silico block	15.0	270.0	0.136	1.10
	Ceiling	1.3	11.0	0.325	0.04
Barded outer wall	Cement fiber cladding	2.0	36.0	0.950	0.02
	Air blade	1.2	0.0	0.080	0.15
	polyurethane	24.0	7.0	0.025	9.60
	Limestone silico block	15.0	27.0	0.136	1.10
	Ceiling	1.3	11.0	0.325	0.04
High floor	PDM sealing	-	-	-	-
	Polyurethane	40.0	12	0.025	16.00
	Concrete slab	25.0	325	1.389	0.18
	Ceiling	1.3	11	0.325	0.04
Intermediate floor	Chappe + coating	8.0	144	0.700	0.11
	Polyurethane	1.0	0	0.030	0.33
	Aerated concrete	8.0	48	0.210	0.38
	Concrete slab	25.0	325	1.389	0.18
	Ceiling	1.3	11	0.325	0.04
Low floor	Chappe + coating	8.0	144	0.700	0.11
	Polyurethane	25.0	8	0.025	10.00
	Concrete slab	25.0	575	1.750	0.14
	Ceiling	1.3	11	0.325	0.04
Internal wall	Limestone silico block	15.0	270	0.136	1.1
	Expanded polystyrene	4.0	1	0.032	1.25
	Limestone silico block	15.0	270	0.136	1.10
	Ceiling	1.3	11	0.325	0.04

Thickness (e), the mass per unit area (ρ*e), thermal conductivity (λ) and thermal resistance (R).

Table 2. Haracteristics of the joinery.

Material	U_{frame} (W/m²·K)	$U_{glazing}$ (W/m²·K)	Sw	Ti
Low emission double glazed	2.1	1.695	0.549	0.68
Insulating door	1	-	0.04	-

Solar factors (Sw) and light transmission factors (Ti).

The hourly temperature data, global and diffuse horizontal radiation, wind speeds, relative humidity, atmospheric pressure and precipitation of the studied sites, over the last forty years, were downloaded from American satellites by the Meteonorm software, and subsequently converted so as to implement them in the Pleiades software. We have modeled the walls, floors, slabs, roofs, openings and solar masks (Figure 4). The geometries of the buildings and the actual openings have been scrupulously valued. The significance of modeling the neighborhood in three dimensions is that we now take into account the orientation and the solar masks that different buildings make on each other. In this manner, we will be able to study the impact of a change in the orientation of the mass plan or the increase in height of certain buildings.

Table 3. Characteristics of the thermal bridges.

Input Data	Ψ (W/m²·K)
Windows support	0.15
Door step	0.15
Low floor	0.2
Outgoing angle	0.08
Incoming angle	0.03

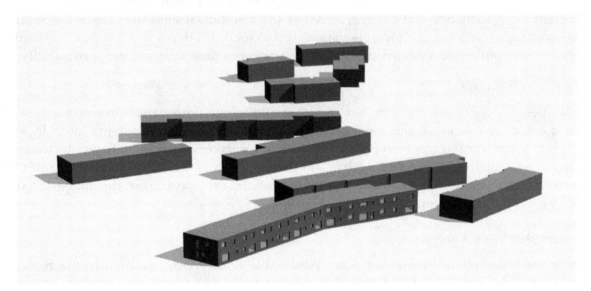

Figure 4. View of the 3D model of the neighborhood as presented in the Pleiades software.

3.6. Other Input Data

Several important results were obtained after simulation, such as (i) the detailed characteristics of all the simulated residences; and (ii) the different needs related to the consumption of water and energy. The lifespan of the different building materials was set at 80 years, such as those of the buildings.

There were different impacts resulting from the renovation phase. This different energy data were analyzed under the reference of the Belgian energy mix integrated in the software. According to the report of the International Panel of Climate Change in 2016, the Belgian energy mix is set at 4% coal, 27% natural gas, 17% renewable and 52% nuclear. It was important to notice that the consumption related to heating and domestic hot water (DHW) were calculated using the most recent data.

The supply system was a natural gas condensing boiler, having a 92% lower heating value (PCI) efficiency. The water consumption was estimated at 100 L/occupant/day. In the case of waste disposal, the new waste sorting policy was applied for this purpose (less waste.wallonie.be), which was set at 90% for glass waste and 75% for the paper and cardboard. This percentage of waste was applied as recycled, and not buried. With regard to the different Belgian statistics, it is found that 40% of the 1500 g of daily household waste per occupant are directly sent for incineration with an estimated yield of 85%, with the distance between the site and the landfill being 10 km, 100 km to the incineration plant and 50 km to the recycling site.

3.7. Orientation Scenario

We studied the different orientation effects of buildings at the neighborhood level. Initially, all the buildings were installed so as to more easily orient the different facades towards the south and the north. We have called this "scenario o". Subsequently, we tried this test on several other orientations by guiding the mass plane to successive rotations of 45°, 90° and even 180°. Subsequently, we calculated the standard deviation of all the buildings studied affected by each of its orientations.

We chose the most unfavorable orientation to perform the LCA analysis of the neighborhood. Subsequently, we rigorously compared the different results of the new LCA in the neighborhood with that of the central neighborhood in order to evaluate the real effects of orientation on the LCA of a neighborhood.

3.8. Water-Use Scenario

The main objective of this new scenario is to collect all the rainwater and the discharged directly into a sanitary sewer network. If it were possible by pipeline to recover more than 95% of the rainwater in the different valleys, ditches, and cisterns, then it would be totally useless to evaluate the permeability of the different types of flooring. It is thus paramount to focus on two scenarios: one is based on the different rainwater collection systems and the other one is oriented towards the permeability of soils.

3.8.1. Rainwater Scenario

In the specific case of this scenario, we modeled all the rainwater tanks. In summary, rainwater was used to clean the interior and exterior of buildings, cleaning instruments, etc. In addition, reclaimed rainwater was fed from a separate network of reservoirs, ditches, valleys and water bodies. Garden water was collected by several ditches and turned towards the water. Rainwater from the roof was directly poured into the tanks. We assumed that all the rainwater from this place of study was controlled by a separate network.

3.8.2. Permeable Floors Analysis

In this scenario, we implemented more permeable floor coverings than in the basic option. In this manner, aisles, squares and car parks are constructed with unrepaired concrete pavements and concrete–grass slabs. Thus, the total impermeability of the site goes from 66% in the initial state to 58% once the permeable coatings are implemented. This small difference between the average permeability of the two scenarios is explained by the high proportion of green spaces in our neighborhood that do not see the modified permeability between the two scenarios. The large proportion of green areas of the site explains its relatively high permeability from the initial state. In this scenario, we consider that no other rainwater harvesting system be implemented. All of the water does not infiltrate directly into the soil; it is sent to the wastewater network.

3.9. Urban Mobility Impact Analysis

Let us now look at the impact of mobility on the neighborhood's environmental record.

In our basic scenario we considered a significant use of the car for daily commuting. We will compare this scenario with a second one, where the site is considered urban, perfectly integrated with public transport networks and at a short distance from the shops of primary needs. Let us recapitulate the mobility hypotheses: (i) Initial scenario: 80% of the occupants commute daily; the 20 km distance from home to work is carried out daily by car; and the 5 km distance from home to the shops is done weekly by car. (ii) "Urban Site" scenario: 100% of the occupants make the trip, daily; the 2.5 km distance from home to work is done daily by bus; the 300 m distance from home to the shops is carried out weekly by bike or on foot.

3.10. Urban Density Impact Analysis

The purpose of these scenarios is to analyze the different effects of density on the life cycle of a neighborhood.

3.10.1. Vertical Scenario

We introduce another floor to each building. The configuration of the neighborhood remains the same. Overall, we are increasing the number of occupants as well as the construction area of our new neighborhood. The new district will have 100 more inhabitants than the old district.

3.10.2. Horizontal Scenario

The objective is to assess the impact of horizontal densification. For this, we decided to add some residential buildings on the used area (see Figure 5). In total, four buildings were added. The total population is identical as in the reference scenario, which allowed us to keep the same configuration as that of the vertical density scenario. To achieve this goal, we had to occupy a small part of the public parking space. All the buildings added had the same characteristics as those existing; this in order to better compare these two methods and choose the best.

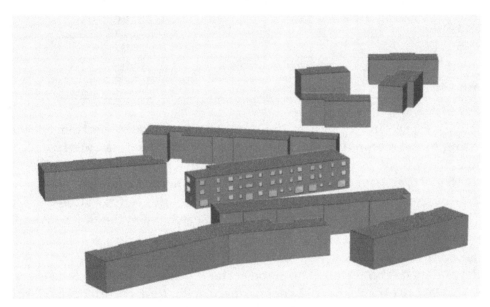

Figure 5. View of the 3D model of the neighborhood in the "density +" (horizontal density) configuration as presented in the Pleiades software (openings only appear on the selected building).

3.11. Urban Renewable Energies Use Impact

In the initial scenario, all the electricity used came from the Belgian electricity grid and the production impacts were taken into account. In this new configuration, we will have a photovoltaic system on all the roofs on the site, and we consider having a panel area equivalent to two-thirds of the roof area of each building. It must be noted that our homes use electricity only for light and to power household appliances. The installed installation will consist of mono crystalline photovoltaic solar panels. The sensors will be placed using a support on the roof terrace. They will be oriented south and inclined at 35°, the optimal inclination in Belgium. We then performed the thermal simulation of each building and completed the final LCA of the neighborhood.

4. Results

In this study, we obtained a heating requirement of 15.4 kWh/m^2·year. The main requirement for meeting passive standards is to have a heating requirement of less than 15 kWh/m^2·year. We notice that some buildings do not respect the passive standard. This may be due to their wrong orientation. The average results of the LCA under building scales are shown in Table 4. These results showed that in the studied buildings, after 80 years, the average greenhouse gas was expected to be 2586.1 tCO$_2$ eq, while the cumulative energy demand would be 73,935.2 GJ.

Table 4. Average LCA results for the buildings in terms of calculated impacts.

Components	Value over 80 Years	Value/Inhabitant per Year	Value/m^2 per Year
Greenhouse gas (Tco$_2$ eq.)	2586.1	1.115	0.035
Acidification (kg SO$_2$ eq.)	10,198.5	4.396	0.139
Cumulative Energy Demand (GJ)	73,935.2	31.869	1.010
Waste water (m^3)	135,936.1	58.593	1.858
Waste product (t)	2173.5	0.937	0.030
Depletion of abiotic resource (kg antimony eq.)	23,845.2	10.278	0.326
Eutrophication (kg PO$_4$ eq.)	4423.5	1.907	0.060
Photochemical ozone product (kg ethylene eq.)	705.1	0.304	0.010
Biodiversity damage (PDF·m^2·year)	123,317.3	53.154	1.685
Radioactive waste (dm^3)	76.5	0.033	0.001
health damage (DALYS)	2.9	0.001	0.000
Odor (Mm^{3}air)	74,614.9	32.162	1.020

The radioactive waste would increase to 76.5 dm^3. The waste water was from 58.593 m^3/inhabitant per year. According to the research of Marique and Reiter [24], heating energy was estimated to be between 190 and 200 kW h/m^2, in the case of the conventional neighborhood. In this case, the heating energy is around 16 kW h/m^2 as requested by several international standards. Moreover, according to Lotteau et al. [29], the greenhouse gas is between 11 and 124 kgCO$_2$/m^2; in this research, it is around 35 kgCO$_2$/m^2. This means that the results found in this research are in the range given in the literature.

The average odor concentration was 32.162 Mm^{3}air/inhabitant per year. Table 5 shows some simulation results of the LCA on the neighborhood scale. It was seen that the total greenhouse gas was expected to be 21,733.64 tCO$_2$ eq after 100 years, whereas the total cumulative energy demand was 532,385.49 GJ. The health damage was 22.29 DALYS (disability-adjusted life years), and the potential for degradation related to land use obtained was 28,630.00 m^2/year.

DALYs are the sum of the YLDs and YLLs, per disease category or outcome, and per age and sex class:

$$DALY = YLD + YLL \tag{1}$$

where YLD (the morbidity component of the DALYs) = number of cases * disease duration * disability weight; YLL (the mortality component of the DALYs) = number of deaths * life expectancy at age of death.

It was interesting to notice that the total eutrophication assessed was 42,794.03 kg PO$_4$ eq. The analysis of the most important sources of impact (greenhouse gas and cumulative energy demand) and their distributions according to the different stages of the life cycle is given in Figures 6 and 7. In Figure 6, we notice the strong predominance of the occupation phase, which concentrates on 93% greenhouse gas production. In this phase, mobility is strongly in the majority with 46% emissions. Heating and domestic hot water accounts for 24% of emissions, while waste treatment accounts for 15% of the emissions during the use phase. Only 1% of greenhouse gas comes from a "public space". It is interesting to note that emissions from the household waste management are equivalent to those from the production of hot sanitary water (15% of the use phase emissions), whereas the emissions from heating accounted for only two-thirds of the emissions, from the production of hot sanitary water.

Table 5. LCA results at the neighborhood scale (initial case).

Components	Buildings	Use	Renewal	Demolition	Total
Greenhouse gas (100 year) (tCO$_2$ eq.)	1340.73	20,242.92	97.60	52.38	21,733.64
Acidification (kg SO$_2$-eq.)	4126.27	78,756.54	1032.04	377.64	84,292.49
Cumulative Energy Demand (GJ)	17,663.39	50,9331.29	4081.31	1309.50	532,385.49
Waste water (m^3)	79,708.77	72,431.40	1759.22	975.23	806,760.62
Waste product (t)	826.20	7281.24	371.63	5332.21	13,811.28
Depletion of abiotic resource (kg antimony eq.)	7267.52	165,099.14	1548.55	564.07	174,479.29
Eutrophication (kg PO$_4$ eq.)	807.14	41,545.40	366.27	75.22	42,794.03
Photochemical ozone product (kg ethylene eq.)	336.16	5256.74	55.91	11.04	5659.85
Biodiversity damage (PDF·m^2·year)	22,614.47	848,177.80	18,298.61	1004.49	890,095.38
Radioactive waste (dm^3)	106.39	446.00	1.48	0.21	554.08
Health damage (DALYS)	1.06	20.85	0.31	0.08	22.29
Odor (Mm^{3}air)	11,152.42	2,085,882.72	2698.31	1866.63	2,101,600.08

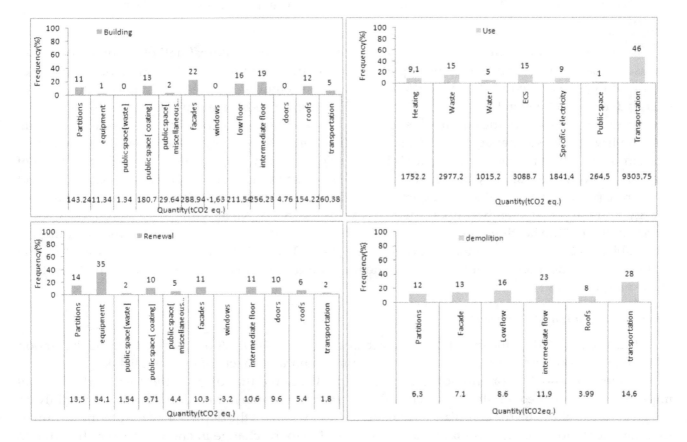

Figure 6. Detailed results of the calculation of the "greenhouse effect" impact at the neighborhood level (initial case).

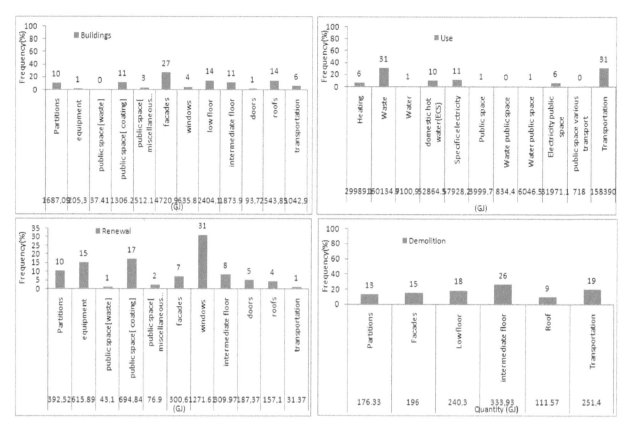

Figure 7. Detailed results of the "cumulative energy demand" impact calculation at the neighborhood scale (initial case).

Emissions due to the mobility of the inhabitants accounted for almost half of the emissions from the use phase. These characteristics were perhaps due to the fact that the high thermal performance of our buildings greatly reduced their heating consumption. Figure 7 showed the impact of cumulative energy demand defined in [61] .As in the previous Figure 6, it was seen in this figure that the use phase was predominant (96% of the cumulative total energy demand). This could be due to the accounting for mobility and waste management. The cumulative demand for energy due to waste management was almost identical to that due to the mobility of residents and was equivalent to almost one-third of the demand of the occupancy phase. In addition, this was because the cumulative energy demand from transportation and waste management during the use phase was 60% of the total cumulative energy demand of the neighborhood, over its entire life cycle. Meanwhile, the cumulative energy demand due to "the heating and domestic hot water" was only half of that required for the transport of inhabitants or the management of household waste. These different results show a very strong participation of the mobility component and the household waste management component in the LCA at the neighborhood level.

4.1. Orientation Impact Assessment

This section studied the orientation impact assessment of the LCA outcomes at the neighborhood level. Figure 8 showed the comparison of the environmental impacts of the established scenarios to "0° orientation" and "90° orientation", in percentage. We noted that once all the neighborhood-level impacts were accounted for, the influence of the orientation became minimal. Indeed, it was mainly on the greenhouse effect, on the cumulative demand of energy, and on the depletion of the abiotic resources that the orientation had an important effect. This was due to the change in energy consumption due to heating. However, we observed only a relative increase of less than 1% of these impacts. Moreover, this evolution only affected the phase of use. On the other hand, we observed a 1% increase in greenhouse gas emissions, as well as the depletion of abiotic resources and cumulative energy demand during the use phase in the case of a rotation to 90°. This is could be due to the increase in gas consumption caused by the increase in heating needs.

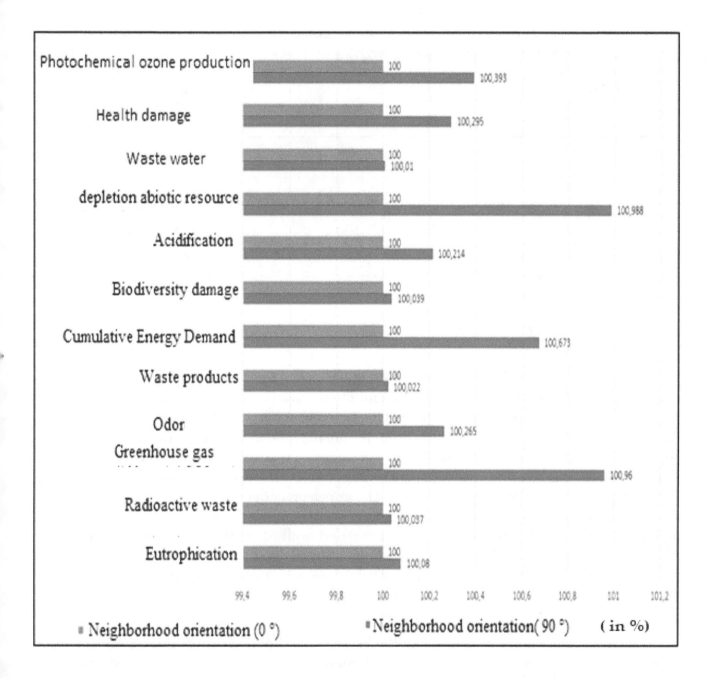

Figure 8. Comparison of the environmental impacts of the "0° orientation" and "90° orientation" scenarios (functional unit: entire neighborhood), in percentage.

Comparing the scores of the environmental indicators only for the heating items during the use phase, as also for both orientations, we notice an 11% increase in the greenhouse effect, as well as a cumulative energy demand and depletion of abiotic resources for the 90° orientation. Thus, the orientation has an impact on heating consumption and on environmental indicators relating only to these [62–65]. However, at the neighborhood level, this orientation has little impact on the overall results of the LCA. However, even if the orientation has little influence on the LCA results at the neighborhood level, at the building level it can be decisive, especially for obtaining the passive label.

The different quantities of environmental impacts are shown in the Table 6.

Table 6. Details of orientation scenario per square meter. (1) Greenhouse gas; (2) acidification; (3) cumulative energy demand; (4) waste water; (5) waste products; (6) depletion of abiotic resource; (7) eutrophication; (8) photochemical ozone production; (9) biodiversity damage; (10) radioactivity waste; (11) health damage; (12) odor.

Environmental Impact	1	2	3	4	5	6	7	8	9	10	11	12
Scenario 0: Orientation 0° (m^2/year)	0.041	0.160	1.011	1.532	0.026	0.331	0.081	0.0107	1.69	0.001	0	3.99
Scenario 1: Orientation 90°	0.042	0.160	1.018	1.532	0.026	0.335	0.081	0.01	1.69	0.001	0.0004	3.99

4.2. Water Management Impact Assessment

In Figure 9a, we note that setting up rainwater harvesting systems has a strong impact on certain environmental indicators. Indeed, collecting all the rainwater can reduce eutrophication by 32%. This significant decrease is due to the fact that the runoff water is entirely recovered on the site by the valleys and infiltration basins. Thus, the nutrients are not strained, but retained on the site. On the other hand, it was noticed that drinking water consumption is also strongly impacted. Indeed, with a well-sized tank, it is possible to use only rainwater to feed the washing machines and flushes with water. This will save drinking water up to 6000 L per person per year, which implies a 14% reduction in water consumption of the neighborhood on a scale of its total life cycle—a 7% decrease in waste produced over the entire life cycle of the neighborhood. Indeed, on the use phase, 15% less waste is produced. This is the runoff water that is no longer directed to the treatment plants, and therefore no longer needs to be treated. Moreover, we have observed a decrease of about 4% in damage to biodiversity, damage to health and acidification.

The analysis in Figure 9b shows that the impact of soil permeability on the total LCA of the neighborhood is lower. In fact, the concerned indicators are still eutrophication and waste production. In this case, the use of permeable soils reduces the impact of eutrophication by 5% and the production of waste by 1% over the entire life cycle of the neighborhood. In fact, the amount of water that infiltrates into the ground, thanks to the permeable pavements, is less than the quantity that can be recovered by the recovery systems presented in the previous scenario. Figure 10 shows the comparison of the three scenarios (initial scenario, and with and without permeable floor coverings). The analysis of this figure shows that it is more efficient to install recovery systems like cisterns, valleys or infiltration basins at the neighborhood level. However, implementing permeable floor coverings on areas that cannot benefit from recovery systems will have a positive impact on the amount of wastewater to be treated and on eutrophication.

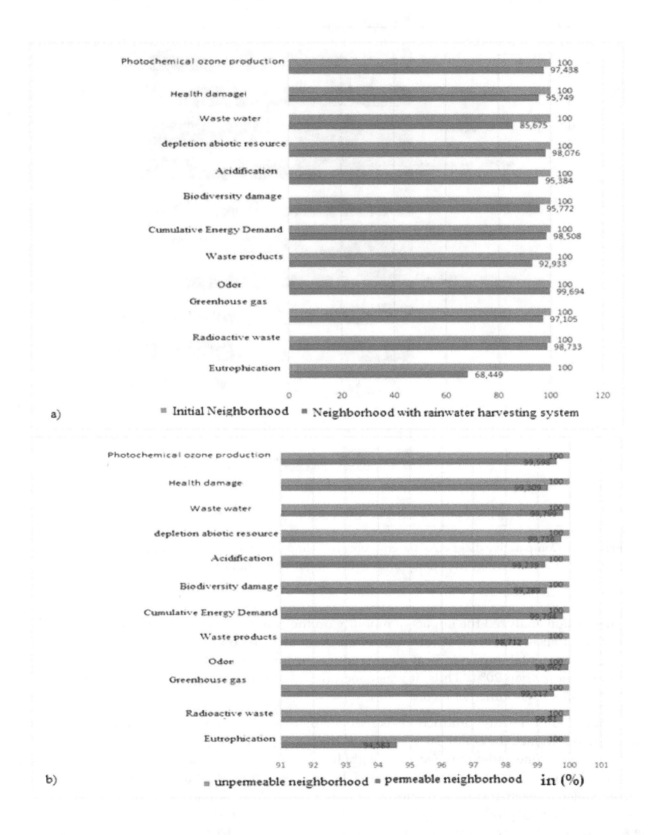

Figure 9. Comparative diagram of the environmental impacts of scenarios with and without rainwater harvesting systems (**a**), and with and without permeable floor coverings (**b**) (functional unit: entire neighborhood), in percentage.

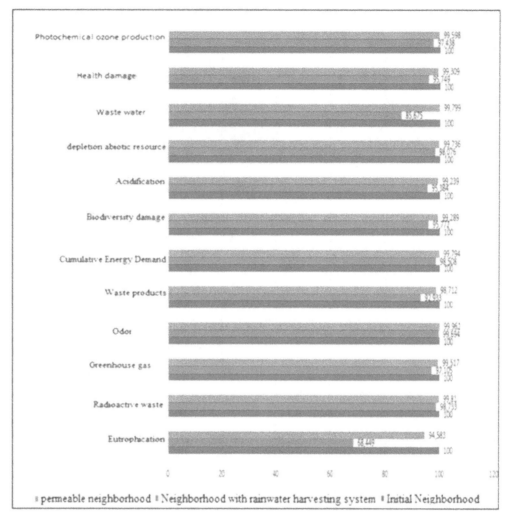

Figure 10. Comparative diagram of the environmental impacts of the initial scenarios, with and without permeable floor coverings (functional unit: entire neighborhood), in percentage.

4.3. Mobility Flow Impact Assessment

This section analyzed the impact of mobility on the neighborhood's environmental record.

Figure 11 shows an analysis of the environmental impact on the mobility scenarios. It is seen that all environmental impact indicators are reduced from 6% to 50%. Seven indicators out of 12 are reduced by more than 20%. Thus, it was concluded that mobility has a significant impact on the neighborhood's environmental record.

"Photochemical ozone production" is reduced by more than 50% over the entire life cycle of the neighborhood. In fact, the combustion of fuels is the main source of nitrogen oxide production, which transforms into ozone under the effect of sunlight [64]. In our urban site scenario, 54% of photochemical ozone production in the use phase is avoided, by reducing the use of automobiles. Indeed, 95% of transport-related ozone production during the operational phase is avoided in this scenario. Another photochemical ozone production station is waste management. The previous figure (Figure 11) shows the same observation with the "greenhouse gases". Indeed, a decrease of 40% of the emissions is observed on the total life cycle of the neighborhood, thanks to a decrease of 93% transport emissions during the use phase. On the other hand, it is interesting to note that "acidification" has also been strongly impacted by the suppression of automobile use. We have observed a 35% decrease in this impact indicator over the entire life cycle of the neighborhood. It is the same for "depletion of abiotic resources" and "damage to health", which saw their score reduced by 34% and 32%, respectively.

Indeed, much less fuel and fossil resources are consumed and the pollution responsible for many health problems is also greatly reduced.

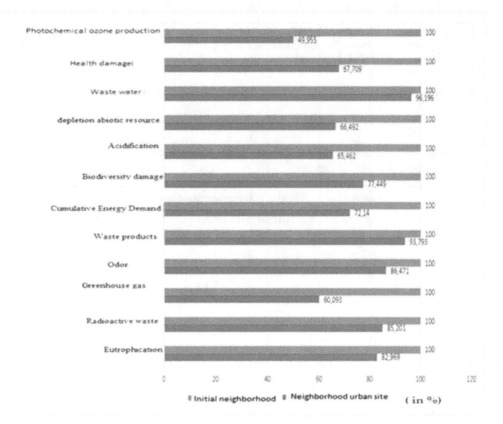

Figure 11. Comparative diagram of the environmental impacts of mobility scenarios (functional unit: entire neighborhood). For example, mobility and the use of personal vehicles to carry out daily commuting distances have a huge impact on the neighborhood's environmental record. Climate impact indicators are the most affected. It is possible to reduce them by half. The cumulative demand for energy, acidification, depletion of biotic resources and damage to health can be reduced by a third, thanks to a mobility scenario.

Decreasing the use of cars can create huge savings in energy. Public transport uses the energy contained in fuels in a more efficient and rational manner. Thus, the cumulative energy demand is reduced by 28%. It is also shown that there has been a 23% decrease in damage to biodiversity, 17% in eutrophication, 15% in radioactive waste, 13% in odors and 6% in waste produced.

Some data are showed on Table 7.

Table 7. Water management scenario.

Environmental Impact	1	2	3	4	5	6	7	8	9	10	11	12
Water management scenario (m²/year)	0.032	0.123	0.855	1.285	0.023	0.269	0.049	0.008	1.43	0	0	3.665

(1) Greenhouse gas; (2) acidification; (3) cumulative energy demand; (4) waste water; (5) waste products; (6) depletion of abiotic resource; (7) eutrophication; (8) photochemical ozone production; (9) biodiversity damage; (10) radioactivity waste; (11) health damage; (12) odor.

Detailed responses on the urban mobility are showed in Table 8.

Table 8. Urban mobility scenario.

Environmental Impact	1	2	3	4	5	6	7	8	9	10	11	12
Urban mobility scenario (m²/year)	0.041	0.160	1.011	1.532	0.026	0.331	0.082	0.01	1.69	0.001	0	3.99

(1) Greenhouse gas; (2) acidification; (3) cumulative energy demand; (4) waste water; (5) waste products; (6) depletion of abiotic resource; (7) eutrophication; (8) photochemical ozone production; (9) biodiversity damage; (10) radioactivity waste; (11) health damage; (12) odor.

4.4. Density Impact Assessment

Table 9 estimates the heating requirements of the various buildings in the study area.

Table 9. Heating requirements of the different neighborhood buildings in the basic and high configuration of a floor.

Name	Heating Requirements (kWh/m².year)	
Buildings	Initial Situation	First Floor
A3	15	14
B2	12	12
B3	14	13
D1	19	20
D2	20	20
D3	20	21
D4	18	19
C1	12	11
C2	13	12
C3	13	11
Mean	15.6	15.3

Analysis of this data showed that the heating requirements with an additional floor dropped slightly. We thought that the additional shading created should act as solar masks, which would reduce solar gain and increase heating needs. However, it seemed that the increase in compactness caused by the rise of the buildings was more impacting. Figure 12 shows the comparative diagram of the environmental impacts of the scenarios.

In Figure 12a, the results are expressed on the basis of a functional unit encompassing the entire neighborhood. This is because the indicator scores had all increased in fairly similar proportions, from about 25% to 30%. Indeed, the share of the indicators related to the buildings was modified, but not that related to the district, which remained unchanged. This functional unit did not allow us to draw any interesting conclusions. This is why we are going to translate the results of the study into the "Occupant" functional unit, to be able to compare per capita impacts in both configurations.

As shown in Figure 12b, if we compare the environmental indicators by reporting them to the number of inhabitants, we notice that the high-rise one-story has a better environmental performance. The odor indicator is reduced by 26% and eutrophication by 19%. The other ten indicators are reduced between 11% and 15%. Indeed, even the site welcomes more occupants and the consumption by these added to the initial consumption, all impacts from the site itself and public spaces remain unchanged. Thus, the built surface is more profitable.

In the case of an increase in density built by adding buildings to the site (Figure 13a), the results were not as favorable as in the previous case. In fact, apart from odors, radioactive waste and eutrophication, the scores of which decreased by 21%, 3% and 10%, respectively, the other indicators had increased. They all earned between 1% and 5%. Indeed, we did not benefit here from a gain in compactness and we did not pool the networks. In addition, the construction of new buildings was greener in materials and energy than the rise of a floor. The analysis of Figure 13b showed that densifying the neighborhood vertically was more remarkable environmentally. The impact on the total

LCA of the district was much more pronounced than during horizontal densification, for which the assessment was mixed.

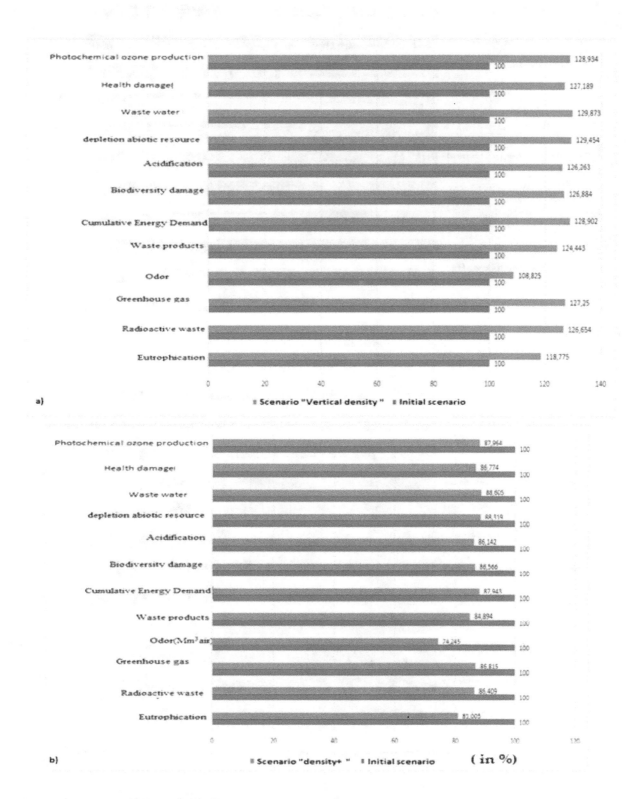

Figure 12. Comparative diagram of the environmental impacts of: (**a**) "Initial" and "Vertical Density" (functional unit: occupant); and (**b**) "Initial" and "Density +" scenarios (functional unit: entire neighborhood).

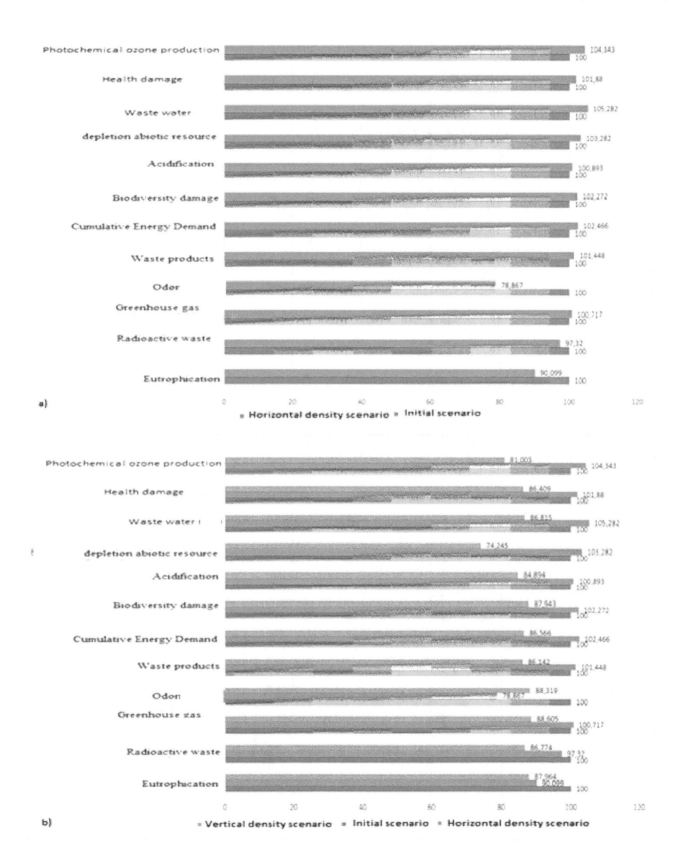

Figure 13. Comparative diagram of the environmental impacts of the "Initial" and "Horizontal Density" scenarios (**a**); and "Initial", "Horizontal Density" and "Vertical Density" (functional unit: occupant) (**b**).

Some results are showed on the Table 10.

Table 10. Vertical and horizontal density scenarios.

Environmental Impact	1	2	3	4	5	6	7	8	9	10	11	12
Vertical density scenario (m^2/year)	0.033	0.130	0.877	1.503	0.025	0.278	0.074	0.008	1.499	0	0	3.688
Horizontal density scenario (m^2/year)	0.041	0.163	1.118	2.604	0.032	0.356	0.088	0.010	1.897	0.001	0	3.931

(1) Greenhouse gas; (2) acidification; (3) cumulative energy demand; (4) waste water; (5) waste products; (6) depletion of abiotic resource; (7) eutrophication; (8) photochemical ozone production; (9) biodiversity damage; (10) radioactivity waste; (11) health damage; (12) odor.

4.5. Impact of Renewable Energy Uses

Taking into account the dynamic thermal simulation, the consumption and electricity production were calculated. For all buildings, production exceeded consumption throughout the year, except for the months of December and January, where the installation covered 45% and 75% of the consumption, respectively. In fact, the buildings consumed, on an average, 12 kWh/m^2 of electricity per year. These results were consistent with the Belgian averages for dwellings that did not heat up with electricity. Photovoltaic panels produced an average of 26 kWh/m^2 over the year. Thus, except for the months of January and December, no electrical energy was drawn from the Belgian network. The effects on the LCA of the neighborhood are presented in Figure 14.

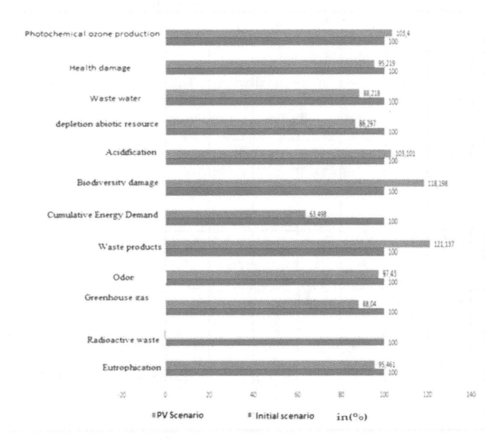

Figure 14. Comparative diagram of the environmental impacts of the "Initial" and "PV" scenarios (functional unit: entire neighborhood).

Of all the configurations studied, the one comprising the addition of photovoltaic panels is the one that produces the most heterogeneous results on the neighborhood's LCA. Indeed, some indicators are greatly reduced, while others see their score increase considerably.

The most affected impact is the production of radioactive waste. Over the entire life cycle, the production of radioactive waste is reduced by 102%. Indeed, even if this production of waste increases during the construction (9%) and renovation (1893%) phases, because of the impact of the manufacture of panels, the use phase makes up for this delay. The enormous increase in the usage phase score is explained by the fact that the panels are changed every 20 years and that in the previous scenario, the production of radioactive waste of this phase was insignificant. That being said, the production of radioactive waste during the use phase decreases by 127%. This is explained by the fact that production is higher than consumption. As a result, not only is the construction and maintenance of the system offset, but the production of radioactive waste from the use phase is also eliminated. Moreover, it allows other homes to benefit from the clean energy produced. Thus, our neighborhood reduces the production of radioactive waste from other neighborhoods, which gives a negative score for this indicator.

The second-most impacted indicator is the cumulative demand for energy. The total energy needed by the neighborhood to operate over its entire life cycle is reduced by 37%. Once again, the construction and renovation phases are negatively impacted. The construction phase saw its energy consumption increase by 75% and the renovation phase by 978%, due to the manufacture of the panels. However, the occupation phase saw its demand decrease by 47%.

The depletion of abiotic resources and the greenhouse effect also decreased by 14% and 12%, respectively, over the entire life cycle. The evolution of the indicators once again followed the same pattern: a significant increase in the construction and renovation phases. However, once again these increases are offset by a reduction in the environmental impact of the use phase, the most impactful phase of the life cycle. We observed a 25% drop in greenhouse gas emissions over this phase and a 26% decrease in the depletion of the abiotic resources.

Conversely, some indicators see their score increase. This is the case of the production of waste. The renovation phase saw its waste production increase by 742%. In fact, 4400 m^2 of the panel area had to be replaced thrice over the neighborhood's life cycle and in addition included their initial installation. The 15% decrease in waste production during the use phase did not make up for this increase. As a result, the neighborhood's total waste generation over its entire life cycle was up by 21% (Figure 14). Some results are showed on the Table 11.

Table 11. Impact of renewable energy.

Environmental Impact	1	2	3	4	5	6	7	8	9	10	11	12
PV scenario (m^2/year)	0.039	0.164	0.722	1.375	0.029	0.317	0.055	0.011	1.946	0	0	3.977

(1) Greenhouse gas; (2) acidification; (3) cumulative energy demand; (4) waste water; (5) waste products; (6) depletion of abiotic resource; (7) eutrophication; (8) photochemical ozone production; (9) biodiversity damage; (10) radioactivity waste; (11) health damage; (12) odor.

Finally, paradoxically, the damage done to biodiversity is also increasing. It is again the manufacture and the replacement of the panels which is in question. The impact of the construction phase increases by 229% and that of the renovation phase by 849%. The 6% drop in impact during the use phase does not compensate for these losses. Thus, over the cycle, the damage to biodiversity increases by 18%.

4.6. Global Analysis of All the Scenarios

In order to classify the different scenarios and define the design parameters to take into account their priority, we calculated the sum of the variations, as a percentage of all the indicators compared to the initial scenario. We chose to apply no weighting but will remove the indicator "odors", which distorts the results by its important variations. Table 12 shows some obtained results.

Table 12. Changes in environmental indicators for all scenarios considered compared to the initial scenario (functional unit: occupant).

	Eutrophication (kg PO$_4$ eq.)	Radioactive Waste (dm^3)	Greenhouse Gas (100 years) (tCO$_2$ eq.)	Odor (Mm^3 air)	Product Waste (t)	Cumulative Energy Demand (GJ)	Biodiversity Damage (PDF·m^2·year)	Acidification (kg SO$_2$ eq.)	Depletion Abiotic Resource (kg Antimony eq.)	Waste Water (m^3)	Health Damage (DALYS)	Photochemical Ozone Product (kg Ethylene eq.)
Initial neighborhood	100	100	100	100	100	100	100	100	100	100	100	100
90° orientation neighborhood	99.9	99.9	99.0	99.7	99.9	99.3	99.9	99.8	99.1	99.9	99.7	99.6
rainwater harvesting neighborhood	68.4	98.7	97.1	99.7	92.9	98.5	95.7	95.4	98.1	85.7	95.7	97.4
Permeable neighborhood	94.5	99.8	99.5	99.9	98.7	99.8	99.2	99.2	99.7	99.8	99.3	99.6
Urban site neighborhood	82.9	85.2	60.1	86.5	93.8	72.1	77.4	65.5	66.5	96.2	67.7	49.9
Vertical density neighborhood	87.9	86.7	88.6	88.3	86.1	86.6	87.9	84.8	74.2	86.8	86.4	81
PV Scenario	95.5	−1.5	88.0	97.43	121.1	63.5	118.2	103.1	86.3	88.2	95.2	103.4
horizontal density scenario	90.1	97.3	100.7	78.8	101.4	102.5	102.3	100.9	103.3	105.3	101.9	104.3

It was noted by analyzing this table that mobility has an impact of 282% of the cumulative decrease on all indicators: vertical density (163%), renewable energies (138%), rainwater harvesting (76%), soil permeability (11%), orientation (4%) and horizontal density (−10%).

5. Discussion

Overall, it was seen in this study that the installation of photovoltaic panels has a mixed record. Indeed, the installation was heavily oversized. Several results found in this research are similar to those assessed by Lotteau et al. [65,66]. Indeed, those asserted by Lotteau et al. [65] and the divergence of methodology among different researchers with regard to LCA prevented an easy comparison of results at the neighborhood level. However, in the known research, several aspects common to the LCA were studied, such as (i) the operational energy consumption analysis of buildings; (ii) the quantitative analysis of the construction materials; and (iii) the transport requirement analysis and so on. The process ranged from statistical data collection from neighborhoods to detailed simulations based on physical modelling. Mobility management was the most significant element. Indeed, it was the parameter that allowed a reduction in most of the impacts in terms of greenhouse effects, odors, damage to biodiversity and health, acidification, depletion of abiotic resources and photochemical ozone production. The day-to-day use of individual transportation by local residents has a huge impact on the neighborhood's LCA. Eliminating the use of personal vehicles for the benefit of public transport makes it possible to limit the greenhouse effect four times more than to generate all the electricity of the district, thanks to the photovoltaic panels. Thus, mobility management must be one of the issues to be addressed as a matter of priority in any urban reflection. Designing a neighborhood that is sustainable and environmentally friendly, while being disconnected from public transport, is not always the ideal solution. In the past, Mohamad Monkiz et al. [67] also found that mobility management was one of the most important aspects in the LCA study. The criterion of vertical density was also a fundamental element. Increasing the built density of the neighbourhood by elevation of the buildings was environmentally very beneficial. This made it possible to pool many flows, to increase the energy and environmental efficiency of the neighbourhood, and, thus, homogeneously minimize the different environmental impacts. These results were almost similar to those of André Stephan et al. [16], who found that by replacing an area, part suburb, with apartment buildings, allowed to decrease the total energy consumption by 19.6%.

An eco-district must therefore have a certain density. One of the criteria for a sustainable neighbourhood covers this aspect and imposes a density of 30 to 40 dwellings per hectare [68,69]. It was found in this study that the implementation of renewable energy production systems showed a significant environmental balance, as was seen in several research results [70]. This method was useful for limiting the production of radioactive waste and for the cumulative demand for energy. However, the manufacture of photovoltaic panel systems has a negative impact on the LCA in terms of damage to biodiversity and waste produced. Thus, their large-scale implementation does not necessarily seem to be a priority, at least not until their manufacturing and recycling processes are cleaner. On the other hand, integrating rainwater harvesting systems into the neighbourhood has been shown to have a strong impact on the results of an LCA, especially in terms of eutrophication and water use. Intelligent rainwater management should be a priority when designing a neighbourhood. Finally, soil permeability and orientation are parameters that can also improve the environmental record of a neighbourhood, but to a lesser extent. As for the choice of applying the concept of horizontal density to the neighbourhood, by adding more buildings, it can be counter-productive. In the studied neighbourhood, it is seen that the annual energy savings and avoided GHG emissions were less significant than those recorded in one neighbourhood of New York City (7.3 GJ and 0.4 metric tonnes). The main results of this research may be of interest to construction companies, public officials and decision makers for applying the environmental criteria to the planning process of new and existing neighborhoods.

6. Limitations

All scientific research has some limitations. In the case of this study, it was seen that

- it is difficult to compare the results of the life cycle assessment at the neighborhood scale because the type and form of neighborhood vary from country to country;
- this study is based on the analysis of the LCA of a sustainable neighborhood. It would have been better to study the case of a more conventional neighborhood more suited to the new climate;
- the functional unit adopted in this study is the square meter per living area, whereas it would have been better to also assess per capita;
- certain hypotheses fixed in this study and depending on the morphology of the neighborhood studied (such as mobility, transport, etc.) are not applicable to all the other neighborhoods.

7. Conclusions

Despite the complexity and limitations of the LCA method, this tool has proven to fit the needs of this study perfectly. Even as the majority of the LCA study at the building level has been focused on a very limited number of indicators and often only one parameter, we have been determined in studying more than ten indicators and eight scenarios. This wide range of studied parameters has allowed us to make several interesting observations. First is the need to broaden the environmental thinking on the urban scale. The predominance of the impacts due to mobility and waste management in the overall environmental assessment of the district attests to this. We have shown that these typical problems of urban development are to be treated as a priority, given their considerable influence on the LCA of an already energy performing neighbourhood. Thus, once these urban issues are taken into account, the parameters influencing the scale of the building become insignificant. This is the case with guidance, which, as we have observed, has very little impact on a neighbourhood LCA. Given the internal design parameters of the neighbourhood, it is noted that some are more environmentally impacting than others. The density or management of rainwater parameters need to be carefully studied and prioritized, as they have a strong impact on the neighborhood's environmental performance. We have shown that it is highly preferable to densify the neighbourhood vertically rather than horizontally and that rainwater harvesting systems are more efficient than permeable soils. The installation of photovoltaic panels proved to be mitigated from the point of view of sustainability. This study focused on a theme that seemed most urgent in this line of study. However, many other parameters remain to be studied in order to provide designers with the complete lines of conduct. Thus, this study remains open and will be completed at the scale of a great metropolis and a country.

Author Contributions: Conceptualization, M.K.N. and S.R.; methodology, M.S.; software, M.S.; validation, M.K.N. and M.S.; formal analysis, M.K.N.; investigation, M.S.; resources, S.R.; data curation, M.S.; writing—Original draft preparation, M.K.N.; writing—Review and editing, M.K.N.; visualization, M.K.N.; supervision, S.R.; project administration, S.R. All authors have read and agreed to the published version of the manuscript.

Acknowledgments: The authors would like to acknowledge and thank the AXA Company for their support of this study, as well as the LEMA laboratory team who helped conduct this study.

References

1.	Simonen, K. Life Cycle Assessment. Pocket Architecture: Technical Design Series; Routledge: London, UK; New York, NY, USA, 2014.
2.	Libération. La Prise de Conscience Environnementale (2009). Available online: http://www.liberation.fr/terre/2009/11/30/de-1970-a-2009-histoire-d-une-prise-de-conscience_596573 (accessed on 3 March 2019).
3.	IPCC. *Climate Change: The Physical Science Basis. Contribution of Working Group I to the Fourth Assessment Report of the Intergovernmental Panel on Climate Change*; Cambridge University Press: Cambridge, UK, 2007.

4.   IPCC. *Climate Change: Impact, Adaptation and Vulnerability. Summary for Policymakers, Working Group II Contribution to the Fifth Assessment Report of the Climate*; Cambridge University Press: Cambridge, UK, 2014.

5.   Nematchoua, M.K.; Ricciardi, P.; Orosa, J.A.; Buratti, C. A detailed study of climate change and some vulnerabilities in Indian Ocean; A case of Madagascar island. *Sustain. Cities Soc.* **2018**, *41*, 886–898. [CrossRef]

6.   European Commission. Roadmap to a Resource Efficient Europe. 2011. Available online: https://www.eea.europa.eu/policy-documents/com-2011-571-roadmap-to (accessed on 10 January 2020).

7.   International Energy Agency. Energy Related Environmental Impact of Buildings, Technical Synthesis Report Annex 31: International Energy Agency Buildings and Community Systems. 2005. Available online: http://www.ecbcs.org/docs/annex_31_tsr_web.pdf (accessed on 20 April 2019).

8.   International Energy Agency. World Energy Outlook. 2011. Available online: https://www.iea.org/topics/world-energy-outlook (accessed on 10 January 2020).

9.   Metz, B.; Davidson, O.; Bosch, P.; Dave, R.; Meyer, L. *Contribution of Working Group III to the Fourth Assessment Report of the Intergovernmental Panel on Climate Change (2007)*; Cambridge University Press: Cambridge, UK, 2009.

10.   European Environment Agency. *Urban Sprawl in Europe, the Ignored Challenge*; Europeans Environmental Agency: Copenhagen, Denmark, 2006; Available online: https://www.eea.europa.eu/publications/eea_report_2006_10 (accessed on 22 November 2019).

11.   Allacker, K.; Sala, S. Land use impact assessment in the construction sector: An analysis of LCIA models and case study application. *Life Cycle Assess.* **2014**, *19*, 1799–1809. [CrossRef]

12.   Trigaux, D.; Oosterbosch, B.; De Troyer, F.; Allacker, K. A design tool to assess the heating energy demand and the associated financial and environmental impact in neighbourhoods. *Energy Build.* **2017**, *152*, 516–523. [CrossRef]

13.   European Commission. Available online: https://ec.europa.eu/energy/topics/energy-efficiency/energy-efficient-buildings/nearly-zero-energy-buildings_en (accessed on 25 May 2020).

14.   Marique, A.-F.; Reiter, S. A simplified framework to assess the feasibility of zero-energy at the neighbourhood/community scale. *Energy Build.* **2014**, *82*, 114–122. [CrossRef]

15.   Allacker, K. Sustainable Building—The Development of an Evaluation Method. Ph.D. Thesis, Katholieke Universiteit Leuven, Leuven, Belgium, 2010.

16.   Stephan, A.; Crawford, R.H.; de Myttenaere, K. Multi-scale life cycle energy analysis of a low-density suburban neighbourhood in Melbourne, Australia. *Build. Environ.* **2013**, *68*, 35–49. [CrossRef]

17.   Gervasio, H.; Santos, P.; da Silva, L.-S.; Vassart, O.; Hettinger, A.-L.; Huet, V. *Large Valorisation on Sustainability of Steel Structures, Background Document*; European Commission: Brussels, Belgium, 2014; ISBN 978-80-01-05439-0.

18.   Guinee, J.-B.; Udo De Haes, H.-A.; Huppes, G. Quantitative life cycle assessment of products: 1:goal definition and inventory. *J. Clean. Prod.* **1993**, *1*, 313. [CrossRef]

19.   Buyle, M.; Braet, J.; Audenaert, A. Life cycle assessment in the construction sector: A review. *Renew. Sustain. Energy Rev.* **2013**, *26*, 379–388. [CrossRef]

20.   ISO (International Standardization Organization). International Standard ISO Environmental Management—Life Cycle Assessment—Principles and Framework. 2006. Available online: https://www.iso.org/standard/38498.html (accessed on 10 January 2020).

21.   ISO (International Standardization Organization). International Standard ISO Environmental Management—Life Cycle Assessment—Requirements and Guidelines. 2006. Available online: https://sii.isolutions.iso.org/obp/ui#iso:std:iso:14044:ed-1:v1:en (accessed on 10 January 2020).

22.   Nematchoua, M.K.; Orosa José, A.; Reiter, S. Life cycle assessment of two sustainable and old neighbourhoods affected by climate change in one city in Belgium: A review. *Environ. Impact Assess. Rew.* **2019**, *78*, 106282. [CrossRef]

23.   Nematchoua, M.K.; Teller, J.; Reiter, S. Statistical life cycle assessment of residential buildings in a temperate climate of northern part of Europe. *J. Clean. Prod.* **2019**, *229*, 621–631. [CrossRef]

24.   Rossi, B.; Marique, A.-F.; Glaumann, M.; Reiter, S. Life-cycle assessment of residential buildings in three different European locations, basic tool. *Build. Environ.* **2012**, *51*, 395–401. [CrossRef]

25.   Rossi, B.; Marique, A.-F.; Reiter, S. Life-cycle assessment of residential buildings in three different European locations, case study. *Build. Environ.* **2012**, *51*, 402–407. [CrossRef]

26. CEN. Sustainability Assessment of Construction Works—Assessment of Environmental Performance of Buildings—Calculation Method; NSAI. 2011. Available online: https://standards.globalspec.com/std/1406797/EN%2015978 (accessed on 6 April 2019).

27. CEN Sustainability of Construction Works—Assessment of Buildings—Part 2: Framework for the Assessment of Environmental Performance; NASI. 2011. Available online: https://infostore.saiglobal.com/en-us/Standards/EN-15643-2-2011-336673_SAIG_CEN_CEN_772658/ (accessed on 6 April 2020).

28. Peuportier, B.; Popovici, E.; Troccmé, M. Analyse du cycle de vie à l'échelle du quartier, bilan et perspectives du projet ADEQUA. Build. Environ. 2013, 3, 17.

29. Lotteau, M.; Loubet, P.; Pousse, M.; Dufrasnes, E.; Sonnemann, G. Critical review of life cycle assessment (LCA) for the built environment at the neighborhood scale. Build. Environ. 2015, 93, 165–178. [CrossRef]

30. Wolf, M.-A.; Pant, R.; Chomkhamsri, K.; Sala, S.; Pennington, D. The International Reference Life Cycle Data System (ILCD) Handbook; European Commission, Joint Research Centre, Institute for Environment and Sustainability; Publications Office of the European Union: Luxembourg, 2012.

31. Stephan, A.; Crawford, R.H.; de Myttenaere, K. A comprehensive assessment of the life cycle energy demand of passive houses. Appl. Energy 2013, 112, 23–34. [CrossRef]

32. Cabeza, L.F.; Rincón, L.; Vilariño, V.; Pérez, G.; Castell, A. Life cycle assessment (LCA) and life cycle energy analysis (LCEA) of buildings and the building sector: A review. Renew. Sustain. Energy Rev. 2014, 29, 394–416. [CrossRef]

33. Kellenberger, D.; Althaus, H.-J. Relevance of simplifications in LCA of building components. Build. Environ. 2009, 44, 818–825. [CrossRef]

34. Bribián, I.Z.; Capilla, A.V.; Usón, A.A. Life cycle assessment of building materials: Comparative analysis of energy an environmental impacts and evaluation of the eco-efficiency improvement potential. Build. Environ. 2011, 46, 1133–1140. [CrossRef]

35. Vilches, A.; Garcia-Martinez, A.; Sanchez-Montanes, B. Life cycle assessment (LCA) of building refurbishment: A literature review. Build. Environ. 2017, 135, 286–301. [CrossRef]

36. Ramesh, T.; Prakash, R.; Shukla, K.K. Life cycle energy analysis of buildings: An overview. Build. Environ. 2010, 42, 1592–1600. [CrossRef]

37. Rashid, A.F.A.; Yusoff, S. A review of life cycle assessment method for building industry. Renew. Sustain. Energy Rev. 2015, 45, 244–248. [CrossRef]

38. Chau, C.K.; Leung, T.M.; Ng, W.Y. A review on Life Cycle Assessment, Life Cycle Energy Assessment and Life Cycle Carbon Emissions Assessment on buildings. Appl. Energy 2015, 143, 395–413. [CrossRef]

39. Colombert, M.; De Chastenet, C.; Diab, Y.; Gobin, C.; Herfray, G.; Jarrin, T.; Trocmé, M. Analyse de cycle de vie à l'échelle du quartier: Un outil d'aide à la décision? Le cas de la ZAC Claude Bernard à Paris (France). Environ. Urbain Urban Environ. 2011, 5, c1–c21.

40. Anderson, J.E.; Wulfhorst, G.; Lang, W. Energy analysis of the built environment—A review and outlook. Renew. Sustain. Energy Rev. 2015, 44, 149–158. [CrossRef]

41. Loiseau, E.; Junqua, G.; Roux, P.; Bellon-Maurel, V. Environmental assessment of a territory: An overview of existing tools and methods. Environ. Manag. 2012, 112, 213–225. [CrossRef] [PubMed]

42. Oliver-Sol, J.; Josa, A.; Arena, A.P.; Gabarrell, X.; Rieradevall, J. The GWP-Chart: An environmental tool for guiding urban planning processes. Application to concrete sidewalks. Cities 2011, 28, 245–250. [CrossRef]

43. Albertí, J.; Balaguera, A.; Brodhag, C.; Fullana-I-Palmer, P. Towards lifecycle sustainability assessment of cities. A review of Background knowledge. Sci. Total Environ. 2017, 609, 1049–1063. [CrossRef]

44. Blengini, G.-A. Life cycle of buildings, demolition and recycling potential: A case study in Turin. Build. Environ. 2009, 44, 319–330. [CrossRef]

45. Blengini, G.-A.; Di Carlo, T. The changing role of life cycle phases, subsystems and materials in the LCA of low energy buildings. Energy Build. 2009, 42, 869–880. [CrossRef]

46. Nematchoua, M.K.; Reiter, S. Analysis, reduction and comparison of the life cycle environmental costs of an eco-neighborhood in Belgium. Sustain. Cities Soc. 2019, 48, 101558. [CrossRef]

47. Ecoinvent LCI Database. Available online: https://simapro.com/databases/ecoinvent/?gclid=CjwKCAjwsdfZBRAkEiwAh2z65sg-fOlOpNksILo (accessed on 17 December 2019).

48. Nematchoua, M.K. Simulation of the photochemical ozone production coming from neighbourhood: A case applied in 150 countries. Health Environ. 2020, 1, 38–47. [CrossRef]

49. Tsoka, S. Optimizing indoor climate conditions in a sports building located in Continental Europe. 6th International Building Physics Conference, IBPC 2015. *Energy Procedia* **2015**, *78*, 2802–2807.

50. Salomon, T.; Mikolasek, R.; Peuportier, B. Outil de simulation thermique du bâtiment, COMFIE. In *Journée SFT-IBPSA, Outils de Simulation Thermo-Aéraulique du Bâtiment*; La Rochelle, France. 2005. Available online: https://www.researchgate.net/publication/225075851_Outil_de_simulation_thermique_du_batiment_Comfie (accessed on 8 November 2019).

51. Kinnan, O.; Sinnott, D.; Turner, W.J.N. Evaluation of passive ventilation provision in domestic housing retrofit. *Build. Environ.* **2016**, *106*, 205–218. [CrossRef]

52. Jolliet, O.; Saadé, M.; Grettaz, P.; Shaked, S. *Analyse du Cycle de Vie: Comprendre et Réaliser un Ecobilan*, 2e édition mise à Jour et Augmentée ed.; Polytechniques et Universitaire, Collection Gérer L'environnement: Lausanne, Switzerland, 2010; ISBN 978-2-88074-886-9.

53. Roux, C.; Schalbart, P.; Peuportier, B. Analyse de cycle de vie conséquentielle appliquée à l'étude d'une maison individuelle. In Proceedings of the Conférence IBPSA 2016, Champs-sur-Marne, France, 23–24 May 2016.

54. Roux, C. Analyse de Cycle de vie Conséquentielle Appliquée aux Ensembles Bâtis. Construction Durable. Ph.D. Thesis, PSL Research University, Paris, France, 2016.

55. Kemajou, A.; Mba, L. Matériaux de construction et confort thermique en zone chaude Application au cas des régions climatiques camerounaises. *Rev. Energ. Renouv.* **2011**, *14*, 239–248.

56. Bacot, P.; Neuveu, A.; Sicard, J. Analyse modale des phenomenes thermiques en régime variable dans le batiment. *Rev. Génerale Therm.* **1985**, *28*, 111–123.

57. Blay, D. Comportement et performance thermique d'un habitat bioclimatique à serre accolée. *Batim. Energ.* **1986**. Available online: https://www.researchgate.net/publication/281916706_Des_eco-techniques_a_l\T1\textquoterighteco-conception_des_batiments (accessed on 10 December 2019).

58. Scientific Assessment Working Group of IPCC. *Radiative Forcing of Climate Change*; World Meteorological Organization and United Nation Environment Programme. 1994. Available online: https://www.worldcat.org/search?q=au%3AIntergovernmental+Panel+on+Climate+Change.+Scientific+Assessment+Working+Group.&qt=hot_author (accessed on 10 December 2019).

59. Reiter, S. Life Cycle Assessment of Buildings—A Review. In Proceedings of the Arcelor-Mittal International Network in Steel Construction, Sustainable Workshop and Third Plenary Meeting, Bruxelles, Belgium, 7 July 2010.

60. Riera Perez, M.G.; Rey, E. A multi-criteria approach to compare urban renewal scenarios for an existing neighborhood. Case study in Lausanne (Switzerland). *Build. Environ.* **2013**, *65*, 58–70. [CrossRef]

61. Azizi, M.M. *Sustainable Neighborhood Criteria: Temporal and Spatial Changes, 2013–2014*; School of Urban Planning, College of Fine Arts, University of Tehran: Tehran, Iran, 2014.

62. Nematchoua, M.K.; Andrianaharison, Y.; Kalameu, O.; Somayeh, A.; Ruchi, C.; Reiter, S. Impact of climate change on demands for heating and cooling energy inhospals: An in-depth case study of six islands located in the Indian Ocean region. *Sustain. Cities Soc.* **2019**, *44*, 629–645. [CrossRef]

63. VDI. *Cumulative Energy Demand—Terms, Definitions, Methods of Calculation (in Germany)*; Benth: Berlin, Germany, 1997.

64. Lotteau, M.; Yepez-Salmon, G.; Salmon, N. Environmental Assessment of Sustainable Neighborhood Projects through NEST, a Decision Support Tool for Early Stage Urban Planning. *Procedia Eng.* **2015**, *115*, 69–76. [CrossRef]

65. Khasreen, M.M.; Banfill, P.F.G.; Menzies, G.F. Life-Cycle Assessment and the Environmental Impact of Buildings: A Review. *Sustainability* **2009**, *1*, 674–701. [CrossRef]

66. Börjesson, P.; Gustavsson, L. Greenhouse gas balances in building construction: Wood versus concrete from life-cylce and forest land-use perspectives. *Energy Policy* **2000**, *28*, 575–588. [CrossRef]

67. Dezfooly, R.G. Sustainable Criteria Evaluation of Neighbourhoods Through Residents' Perceived Needs. *Int. J. Archit. Urban Dev.* **2013**, *3*, 2.

68. Reap, J.; Roman, F.; Duncan, S.; Bras, B. A survey of unresolved problems in life cycle assessment: Part 2: Impact assessment and interpretation. *Int. J. Life Cycle Assess.* **2008**, *13*, 374–388. [CrossRef]

69. Nematchoua, M.K.; Reiter, S. Life cycle assessment of an eco-neighborhood: Influence of a sustainable urban mobility and photovoltaic panels. In Proceedings of the International Conference on Innovative Applied Energy (IAPE'19), Oxford, UK, 14–15 March 2019; ISBN 978-1-912532-05-6.

70. Spatari, S.; Yu, Z.; Montalto, F.A. Life cycle implications of urban green infrastructure. *Environ. Pollut.* **2011**, *159*, 2174–2179. [CrossRef] [PubMed]

7

Cooking/Window Opening and Associated Increases of Indoor PM$_{2.5}$ and NO$_2$ Concentrations of Children's Houses

Yu-Chuan Yen [1], Chun-Yuh Yang [1], Kristina Dawn Mena [2], Yu-Ting Cheng [1] and Pei-Shih Chen [1,3,4,5,*]

[1] Department of Public Health, College of Health Science, Kaohsiung Medical University, Kaohsiung City 807, Taiwan; ponder10003@gmail.com (Y.-C.Y.); chunyuh@kmu.edu.tw (C.-Y.Y.); llssdog@hotmail.com (Y.-T.C.)

[2] Epidemiology, Human Genetics, and Environmental Sciences, School of Public Health, University of Texas Health Science Center at Houston, Houston, TX 77046, USA; Kristina.D.Mena@uth.tmc.edu

[3] Institute of Environmental Engineering, College of Engineering, National Sun Yat-Sen University, Kaohsiung City 807, Taiwan

[4] Department of Medical Research, Kaohsiung Medical University Hospital, Kaohsiung City 807, Taiwan

[5] Research Center for Environmental Medicine, Kaohsiung Medical University, Kaohsiung City 807, Taiwan

* Correspondence: pschen@kmu.edu.tw.

Abstract: High concentrations of air pollutants and increased morbidity and mortality rates are found in industrial areas, especially for the susceptible group, children; however, most studies use atmospheric dispersion modeling to estimate household air pollutants. Therefore, the aim of this study was to assess the indoor air quality, e.g., CO, CO$_2$, NO$_2$, SO$_2$, O$_3$, particulate matter with aerodynamic diameter less than 2.5 μm (PM$_{2.5}$), and their influence factors in children's homes in an industrial city. Children in the "general school", "traffic school", and "industrial school" were randomly and proportionally selected. Air pollutants were sampled for 24 h in the living rooms and on the balcony of their houses and questionnaires of time–microenvironment–activity-diary were recorded. The indoor CO concentration of the traffic area was significantly higher than that of the industrial area and the general area. In regard to the effects of window opening, household NO$_2$ and PM$_{2.5}$ concentrations during window opening periods were significantly higher than of the reference periods. For the influence of cooking, indoor CO$_2$, NO$_2$, and PM$_{2.5}$ levels during the cooking periods were significantly higher than that of the reference periods. The indoor air quality of children in industrial cities were affected by residential areas and household activities.

Keywords: indoor air quality; children's house; industrial city; window opening; cooking

1. Introduction

According to the Environmental White Paper of Taiwan Environmental Protection Agency (Taiwan EPA), the annual average concentrations of ambient CO, NO$_2$, SO$_2$, and O$_3$ in 2008 were 0.47 ppm, 16.90 ppb, 4.35 ppb, and 29.09 ppb, respectively. The Kaohsiung–Pingtung area was the worse polluted area in Taiwan and accounted for 5.93% of station-days of the Pollutant Standards Index (PSI) > 100. Especially, Kaohsiung is a heavy industrial city. In industrial areas, high concentrations of air pollutants and increased morbidity and mortality rates are found, depending on the types of industrial activities and exposure concentrations in residential areas [1,2]. Children are more susceptible to the health effects of air pollution than adults due to not having full development of their pulmonary metabolic capacity [3]. Long-term exposure of air pollution may affect children's lung development [4]. Previously, most of the studies revealed that ambient pollution such as particulate matter with aerodynamic diameter less than 10 and 2.5 μm (PM$_{10}$ and PM$_{2.5}$), sulfur dioxide (SO$_2$), nitrogen dioxide (NO_2), volatile organic compounds (VOCs), etc. in industrial areas may increase the risk of respiratory symptoms, and attacks of asthma in children [1,5,6]. Therefore, indoor air quality of children's homes may be very important to children's health, especially in industrial cities, since children spend most of their time at home [7].

Indoor air quality may be affected by indoor human activities such as cooking, smoking, cleaning, etc. and the infiltration of outdoor pollutants produced from the traffic or industrial sources [8–11]. For example, SO_2, NO_x, $PM_{2.5}$, and carbon monoxide (CO), the major conventional air pollutant in steel plants, oil refineries, and vehicular exhaust emissions [12–14], may enter a house through cracks and windows [15,16]. In addition, if the indoor air is not well ventilated, the air pollutants may accumulate in the indoor environment, and then seriously affects the health of the inhabitants [17].

Previously, atmospheric dispersion modeling was used to estimate the household concentrations of indoor air pollutants in industrial areas [1,18,19]. Only a few studies actually measured individual exposure [20] and household concentrations [21–23], and these studies only focused on PM mass concentrations, elemental composition, and VOCs concentrations. However, other air pollutants e.g., CO, carbon dioxide (CO_2), NO_2, SO_2, and ozone (O_3) in households in industrial cities also need to be considered. Therefore, the main aim of this study was to assess the indoor air quality including CO, CO_2, NO_2, SO_2, O_3, and $PM_{2.5}$, temperature and relative humidity, and their influence factors (e.g., window opening and cooking) in children's homes in an industrial city—Kaohsiung City. To our knowledge, this is the first study to assess the indoor air quality including CO, CO_2, NO_2, SO_2, and O_3 in children's homes in an industrial city. In addition, the second aim was to evaluate potential determinants of indoor air pollutants levels of occupants' activities, including cooking and window opening, etc. It is also the first study to reveal the differences of air pollutants between cooking periods/window opening periods and reference periods through a time–microenvironment–activity-diary via a questionnaire in one-hour time segments.

2. Materials and Methods

2.1. Study Area

Kaohsiung City (22°38′ N, 120°17′ E), located in southern Taiwan and with the population density of 9962.6/km^2 in 2010, is the largest industrialized harbor city in Taiwan with intense traffic and heavy industries including the largest steel plant (the China Steel Corporation, which also ranked the 19th steel mill in the world in 2005), the largest oil refinery (the CPC Corporation), the largest international shipbuilding (it ranked 6th in the world in 2005) in Taiwan, and many petrochemical industries.

2.2. Study Design

In April 2010, we selected three elementary schools in Kaohsiung City. One elementary school had a general air quality monitoring station of Taiwan EPA on the roof of the 4th floor, so we called this school a "general school". Another elementary school was 0.33 km from Taiwan EPA's traffic air quality monitoring station and was regarded as a "traffic school". The "industrial school" was an elementary school located near the Xiaogang Industrial Zone in Kaohsiung City and about 0.30 km from Taiwan EPA's air monitoring station. The study population was limited to children who attended these schools. The number of students in the "general school", "traffic school", and "industrial school" were 1669, 987, and 960, respectively. After obtaining the assented of the child and the permission of the parents, we recorded the subjects who agreed to home visits for environmental sampling. Children were randomly and proportionally selected from each school to participate in this study. Finally, the home visits of 32, 16, and 12 participants in the "general school", "traffic school", and "industrial school", respectively, were completed between April 2010 and October 2010.

2.3. Air Sampling

Indoor air pollutants including CO, CO_2, NO_2, SO_2, O_3, $PM_{2.5}$, temperature, and relative humidity were measured by real-time monitoring equipment for 24 h in the living rooms. We also measured the atmospheric CO, CO_2, NO_2, SO_2, O_3, and $PM_{2.5}$ on the balcony as outdoor concentrations. All instruments were placed on the bench at a height of approximately 1 m above the ground. The PM was measured by a real-time optical scattering instrument (DUSTTRAK™ DRX Aerosol Monitor Models 8533, TSI Incorporated, Shoreview, MN, USA) and the measurements were taken every 1 s by the flow rate of 3.0 L/min with detectable concentration from 0.001 to 150 mg/m³. The CO, CO_2, NO_2, SO_2, indoor temperature, and relative humidity were also recorded (KD-airboxx, KD Engineering, Blaine, WA, USA) every 15 s with the measuring range of 0 to 500 ppm, 0 to 10,000 ppm, 0 to 20 ppm, 0 to 20 ppm, 0 to 50 °C, and 5% to 95%, respectively. The accuracy of CO, CO_2, NO_2, and SO_2 were ±3% of reading or 2 ppm (whichever was greater), ±5% of reading or 60 ppm (whichever was greater), 0.25 ppm, and 0.25 ppm, respectively. The resolution of CO, CO_2, NO_2, and SO_2 were 0.1 ppm, 1 ppm, 0.01 ppm, and 0.01 ppm, respectively. In terms of O_3, it was detected by a real-time monitoring (Model 202 Ozone monitor™, 2B Technologies Inc, Boulder, CO, USA) every 5 min with the measuring range of 0 to 250 ppm.

All real-time monitors were manufacturer-calibrated for the study in the beginning of this study and every six months. Before every field sampling, the DUSTTRAK™ DRX Aerosol Monitor Models 8533 was calibrated using emery oil aerosol and nominally adjusted to the respirable mass of standard ISO 12103-1, A1 test dust, (Arizona Dust); and the KD-airboxx, the Model 202 Ozone monitor™ were calibrated using zero gas and span gas. In addition, the zero calibrators of instruments were carried out, and the flow rate of sampling pump also was adjusted by Gilian Gilibrator-2NIOSH Primary Standard Air Flow Calibrator (Sensidyne, St. Petersburg, FL, USA) before every household sampling.

2.4. Household Characteristics

In addition, household characteristics including the number of occupants, air-conditioning use, smoking, incense burning, etc. were also recorded in the questionnaires. In addition to household characteristics, data on potential determinants of indoor air pollutants levels of occupants' activities, including cooking and window opening, etc. were obtained through a time–microenvironment–activity-diary via a questionnaire in one-hour time segments. We also actually evaluated the effects of window opening and cooking on indoor air pollutants. The window opening periods were defined from a time–microenvironment–activity-diary and two one-hour periods before and after window opening periods were defined as the reference periods. In terms of cooking, the cooking periods were the periods recorded by participants as cooking from a time–microenvironment–activity-diary and the reference periods were defined as the one-hour periods before the cooking periods.

2.5. Ethics

This study was approved by the Institutional Review Board of the Kaohsiung Medical University Chung-Ho Memorial Hospital (the protocol number was KMU-IRB-990045). Informed written consent was obtained from each child (the phonetic version of the consent form that the children read and signed) and their legal guardians.

2.6. Statistical Analyses

Statistical analyses in this study were performed using SAS version 9.3 (SAS Institute of Taiwan Ltd, Taipei, Taiwan). Descriptive statistics were used to describe the 24-hour of average of exposure data (indoor/outdoor air pollutant concentrations, temperature, and relative humidity). The concentrations of air pollutants were not normally distributed (data not shown), therefore we analyzed our data by nonparametric statistics, also known as distribution-free statistics. A paired Student's t-test was used to assess the difference in the average concentration of air pollutants between indoor and outdoor, between window opening periods and reference periods, and between cooking periods and reference periods. With the objective of evaluating significant differences among the areas (general, traffic, and industry) for all air pollutants variables, data were analyzed using one-way analysis of variance (ANOVA) with Scheffe multiple comparison test. The generalized estimating equations (GEE) is a general statistical method in a longitudinal study with small samples for adjusting time interference, in which each time point is an independent event. Finally, the relationships between the 24-hour average concentrations of indoor air pollutants (dependent variable) and household characteristics (independent variable) were analyzed using GEE, adjusting for other household characteristics, and time interference. A p-value of less than 0.05 was considered significant.

3. Results

Table 1 shows the descriptive statistics of 24-h average indoor and outdoor air pollutants, temperature, and relative humidity in 60 houses. When indoor air pollutants were paired with outdoors within the same home, we found that the 24-hour average concentrations of indoor CO, CO_2, and NO_2 were significantly higher than the 24-hour average of outdoors concentrations, whereas, outdoor O_3 and $PM_{2.5}$ concentrations were significantly higher than indoor concentrations (all $p < 0.01$). The average distance between homes of subjects and their school were 0.86 km, 0.94 km, and 1.46 km in general, traffic, and industrial areas, respectively, as well as, the average distance between homes of subjects and the nearest air monitoring station were 1.07 km, 0.97 km, and 1.46 km in general, traffic, and industrial areas, respectively.

Table 1. Descriptive statistics of 24-h average indoor and outdoor air pollutants, temperature, and relative humidity in 60 houses.

		Mean	Median	Standard Deviation	Minimum	Maximum	p-Value [#]
CO (ppm)	indoor	3.47	0.83	4.29	0.00	12.27	0.004 [‡]
	outdoor	0.60	0.38	0.55	0.00	1.98	
CO_2 (ppm)	indoor	655.43	479.55	321.60	413.82	1320.00	<0.001 [‡]
	outdoor	322.22	319.92	17.23	285.83	353.90	
NO_2 (ppb)	indoor	185.30	177.97	41.52	127.28	251.41	0.008 [‡]
	outdoor	107.54	118.22	36.83	39.90	149.80	
SO_2 (ppm)	indoor	0.00	0.00	0.00	0.00	0.00	0.193
	outdoor	0.01	0.00	0.02	0.00	0.06	
O_3 (ppb)	indoor	11.04	8.50	8.93	1.06	32.29	0.006 [‡]
	outdoor	13.46	9.20	12.34	0.24	45.50	
$PM_{2.5}$ ($\mu g/m^3$)	indoor	60.00	40.00	50.00	10.00	210.00	0.001 [‡]
	outdoor	110.00	90.00	90.00	30.00	410.00	
Temperature (°C)	indoor	31.00	31.00	1.76	26.00	34.00	-
Relative humidity (%)	indoor	72.00	72.00	4.98	62.00	84.00	-

[#] Paired Student's t-test, [‡] $p < 0.01$.

In comparison with household air pollutants of three areas, Table S1 shows descriptive statistics of 24-h average concentration of indoor air pollutants in the houses of traffic, industry, and general areas. Figure 1 shows the 24-h average concentration of indoor air pollutants (A) CO, (B) CO_2, (C) NO_2, and (D) O_3 in the houses of traffic, industry, and general areas. We found the 24-hour average concentration of indoor CO concentration of the traffic area was significantly higher than that of the industrial area, and the general area with all $p < 0.01$ (Figure 1, Table S1). In addition, the 24-hour average concentration of indoor CO_2 level of the general area was significantly lower than that of the traffic area, and industrial area (all $p < 0.01$) (Figure 1, Table S1). Finally, both the 24-hour average concentration of household NO_2 and O_3 concentrations of the industrial area were significantly lower than that of the traffic area, and general area (all $p < 0.01$) (Figure 1, Table S1). Moreover, there was no statistical significant difference of the 24-hour average concentration of indoor SO_2 and $PM_{2.5}$ between the three areas (Table S1).

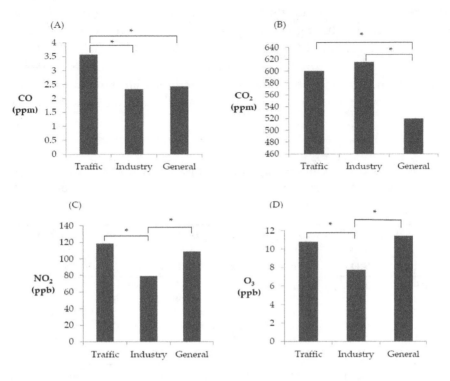

Figure 1. The 24-h average concentration of indoor air pollutants (**A**) CO, (**B**) CO_2, (**C**) NO_2, and (**D**) O_3 in the houses of traffic, industry, and general areas. * Scheffe test $p < 0.01$.

Table 2 shows the percentage of household characteristics including window opening, residents >4 people, cooking, etc. in traffic, industry, and general areas. We found compared with traffic area and industrial area, the general area had a higher percentage of window opening, cooking, and air-conditioning use; moreover, a lower percentage of residents > 4 people, smoker, incense burning, mosquito coil burning, and essential oil using.

Table 2. The percentage (%) of household characteristics in traffic, industry, and general areas.

	Area		
	Traffic	**Industry**	**General**
Window opening	68.75	66.67	87.50
Occupants (>4 people)	40.40	57.01	34.41
Cooking	87.50	83.33	95.83
Air-conditioning use	62.50	83.33	95.83
Making tea	31.25	30.00	0
Smoker	63.64	40.00	26.09
Incense burning	72.73	50.00	29.17
Mosquito coil burning	37.50	22.22	12.50
Essential oil using	31.25	33.33	25.00

The following Table 3 shows the ratios of air pollutants during window opening periods to the reference periods and the differences in air pollutants between window opening periods and reference periods. The median ratios of pollutants during window opening periods to the reference periods for NO_2 and $PM_{2.5}$ were 1.56 and 1.13, respectively with the maximum values up to 5.23 and 1.85 respectively (Table 3). The NO_2 and $PM_{2.5}$ levels during window opening periods were significantly higher than that of the reference periods, and the maximum increased values were 53.25 ppb and 44 μg/m³, respectively. Table 4 shows the ratios of air pollutants during cooking periods to reference periods and the differences in air pollutants between window opening periods and reference periods. The median ratios of pollutants during cooking periods to the reference periods for CO, CO_2, NO_2, and $PM_{2.5}$ were 0.93, 1.06, 1.11, and 1.09, respectively. The concentrations of CO_2, NO_2, and $PM_{2.5}$ during the cooking periods were significantly higher than those of reference periods with increased concentrations of 26.17 ppm, 5.40 ppb, and 5 μg/m³, respectively. However, the CO level during cooking periods was significantly lower than that of the reference periods with the decreased concentration of 0.25 ppm.

Table 3. The ratios of air pollutants during window opening periods to reference periods and the differences in air pollutants between window opening periods and reference periods.

	Ratios (Window Opening Periods/Reference Periods [§])				Differences (Window Opening Periods − Reference Periods [§])				p-Value [#]
	Median	S.D.	Min.	Max.	Median	S.D.	Min.	Max.	
CO (ppm)	0.98	1.34	0.57	4.44	0.00	1.31	−2.42	3.67	0.53
CO_2 (ppm)	1.05	0.18	0.73	1.43	29	128	−141	296	0.21
NO_2 (ppb)	1.56	1.30	0.94	5.23	18.71	16.05	−9.40	53.25	<0.01 [‡]
SO_2 (ppm)	0.00	0.92	0.00	3.27	0.00	0.02	0.00	0.08	0.21
O_3 (ppb)	1.18	0.59	0.56	2.19	0.81	4.44	−11.91	10.05	0.52
$PM_{2.5}$ (μg/m³)	1.13	0.31	0.69	1.85	7	16.20	−6	44	0.04 [†]

[#] Paired Student's t-test, [†] $p < 0.05$, [‡] $p < 0.01$. [§] Reference periods were two one-hour periods before and after window opening periods.

Table 4. The ratios of air pollutants during cooking periods to reference periods and the differences in air pollutants between during cooking periods and reference periods.

	Ratios (Cooking Periods/Reference Periods [§])				Differences (Cooking Periods − Reference Periods [§])				p-Value [#]
	Median	S.D.	Min.	Max.	Median	S.D.	Min.	Max.	
CO (ppm)	0.93	0.22	0.46	1.51	−0.25	0.84	−3.53	0.61	<0.01 [‡]
CO_2 (ppm)	1.06	0.14	0.85	1.58	26.17	90.21	−111.67	342.5	<0.01 [‡]
NO_2 (ppb)	1.11	0.98	0.51	5.43	5.40	29.71	−71.17	101.75	<0.01 [‡]
O_3 (ppb)	1.08	0.69	0.46	4.36	0.27	8.89	−35.14	17.08	0.94
$PM_{2.5}$ (μg/m³)	1.09	0.30	0.60	2.56	5	14	−45	56	0.04 [†]

[#] Paired Student's t-test, [†] $p < 0.05$, [‡] $p < 0.01$. [§] Reference periods were the one-hour period before cooking periods.

Table 5 shows the association between air pollutants concentrations (24-h average concentration of air pollutants in each house as dependent variable), and household characteristics by using the generalized estimating equations model. This study revealed that CO concentrations were positively associated with the number of occupants, cleaning, smoking, incense burning, mosquito coil burning, and negatively correlated to cooking with a statistical significance. Indoor CO_2 concentrations were positively associated with the number of occupants, air-conditioning use, smoking, incense burning, and negatively correlated to mosquito coil burning with a statistical significance. In addition, significantly higher NO_2 levels were found in the homes with smokers than homes without smokers. There were significantly positive associations between indoor SO_2 concentrations and smoking and incense burning. In terms of O_3, indoor O_3 concentrations were positively associated with the window opening and negatively correlated to the number of occupants, incense burning, and essential oil use with a statistical significance. For $PM_{2.5}$, it was positively associated with cleaning and incense burning with a statistical significance.

Table 5. Association between air pollutants concentrations (24-h average concentration of air pollutants in each house as dependent variable), and household characteristics: generalized estimating equations.

	CO (ppm)	CO_2 (ppm)	NO_2 (ppb)	SO_2 (ppm)	O_3 (ppb)	$PM_{2.5}$ ($\mu g/m^3$)
Window opening (Yes vs. No)	0.32	84.84	−0.61	0.44	24.34 ‡	−0.021
Occupants	0.52 ‡	51.62 ‡	3.02	−0.008	−3.49 ‡	0.004
Cleaning (Yes vs. No)	4.73 †	−317.49	1.39	0.43	−6.24	0.047 †
Cooking (Yes vs. No)	−3.89 †	228.02	−28.01	−0.21	1.79	0.065
Fan using (Yes vs. No)	1.42	−32.97	10.58	−0.0003	−2.07	0.002
Air- conditioning use (Yes vs. No)	−1.22	246.99 ‡	87.87	0.25	21.59	0.008
Making tea (Yes vs. No)	37.04	-	−0.45	−0.13	14.21	−0.050
Smoking (Yes vs. No)	17.21 †	1988.44 ‡	547.36 ‡	2.98 ‡	1.69	0.173
Incense burning (Yes vs. No)	18.21 †	2927.87 †	193.11	3.66 ‡	−108.9 ‡	0.416 ‡
Mosquito coil burning (Yes vs. No)	41.55 ‡	−892.64 †	673.52	2.67	2.29	-
Essential oil use (Yes vs. No)	12.76	269.25	74.89	−0.66	−89.29 ‡	−0.022

Generalized estimating equations (GEE) † $p < 0.05$, ‡ $p < 0.01$.

4. Discussion

Our results showed that the outdoor concentrations of O_3 and $PM_{2.5}$ were significantly higher than indoor concentrations. The Kaohsiung City is a city with intense traffic and heavy industries, and previous studies believed SO_2, NO_x, $PM_{2.5}$, and CO were the major conventional air pollutant in steel plants, oil refineries, and vehicular exhaust emissions [12–14,24]. In addition, outdoor O_3 might be formed by the photochemical reaction of nitrogen oxides absorbing sunlight, and VOCs [25,26]. According to the PSI database from 2010 to 2012 of Taiwan EPA, only O_3 and total suspended particulate (TSP) would exceed the standard [27]. This may be the reason why outdoor $PM_{2.5}$ and O_3 concentrations were higher than indoor concentrations. In our study, outdoor median $PM_{2.5}$ levels (90 $\mu g/m^3$) were higher than both the National Ambient Air Quality Standards of UAS and Taiwan EPA with the 24-hour standard for $PM_{2.5}$ of 35 $\mu g/m^3$. In addition, the median value of indoor $PM_{2.5}$ concentrations (40 $\mu g/m^3$) was also higher than the criteria of indoor air quality (IAQ) standards of Taiwan EPA (35 $\mu g/m^3$/24 h). In our study, indoor CO, CO_2, and NO_2 levels were significantly higher than outdoor levels. The number of occupants and human activities such as cooking, smoking, etc. might be the factors affecting indoor pollutants whereas liquefied petroleum gas (LPG), not electric stoves, was the main cooking way in Kaohsiung City [28,29]. In addition, most of the houses were just by the roads and very close to the mobile sources in Kaohsiung City, which was thought of as a traffic-intensive city with the number of cars and motorcycles of approximately 430,000 and 1,230,000, respectively, in 2010 [30]. Thus, the main combustion products of vehicular engines such as CO, NO_x, etc. entering the houses through cracks and windows might be the reason why the indoor concentrations of CO, CO_2, and NO_2 were higher than outdoor concentrations [15,16].

In comparison with traffic, industrial, and general areas, the highest household CO concentration was found in the traffic area among the three areas. According to the previous study, the greatest source of CO (more than 90%) in cities was motor vehicles [24]. The high traffic flow in the traffic area might be the reason for the observation. For CO_2, our study indicated that the lowest household CO_2 level was in the general area among the three areas. The main source of CO_2 was from human respiration [24,31]. The number of residents might be one possible reason since the number of residents > 4 people in the traffic area, the industrial area, and general area were 40.40%, 57.01%, and 34.41%, respectively. We also found both household NO_2 and O_3 concentrations of the industrial area were lowest among the three areas, which was not consistent with the observations of previous studies that ambient NO_2 was related to industrial activities [24], and outdoor O_3 might be formed by the photochemical reaction of nitrogen oxides absorbing sunlight, and VOCs [25,26]. We believed these may be related to Taiwan EPA' s policies and efforts to control air pollution from stationary sources after that the "Stationary Pollution Source Air Pollutant Emissions Standards" was passed in 1992, and the "Air Pollution Control Act Enforcement Rules" was also implemented in 2003.

In regard to the effects on the window opening, our study displayed that household NO_2 and $PM_{2.5}$ concentrations during window opening periods were significantly higher than that during reference periods. NO_X and PM were related to traffic emissions [24,32], and most of the houses in Taiwan were adjacent to roads, so window opening might increase indoor NO_2 and $PM_{2.5}$. For the influence of cooking, there were many simulated experiments exploring the air pollutant emissions of cooking-related fuel combustion [29,33–36], and they demonstrated that CO, CO_2, NO_X, and $PM_{2.5}$ would be emitted by the process of the experiments. Although CO also was produced by cooking, it was revealed that combustion of high-grade fuels (such as natural gas, and LPG which contained propane, butane, etc.), the main fuel-burning stoves use in Taiwan households usually produce much less CO than combustion of low-grade fuels [29,33]. In the previous study, Delp et al. revealed the residential cooking exhaust hoods could not completely capture the pollutants and their efficiency was highly variable [37]. Our results showed that indoor CO_2, NO_2, and $PM_{2.5}$ levels during cooking periods were significantly higher than during reference periods, but the indoor CO level during cooking periods was lower than during reference periods, possibly indicating that the emission rate of CO_2, NO_2, and $PM_{2.5}$ might be higher than the capture rate of the exhaust hood and the emission rate of CO might be lower than the pollutants capture rate of the exhaust hood.

In terms of influence factors, we found there were significantly positive correlations between the number of occupants and CO and CO_2 concentrations. Our study was consistent with the observations of the previous study that CO_2 was produced by human respiration [24,31]. In addition to the combustion, the indoor CO also was related to the status of residents; the previous studies revealed either a smoking person or person with inflammatory diseases exhaled higher CO levels than control group [38,39]. We also found smoking was significantly positively associated with household CO in our study. According to previous studies, smoking, incense burning, and mosquito coil burning were significantly positively associated with CO, CO_2, SO_2, NO_X, and PM [40–42], and these results were consistent with our observation. The cleaning behavior would increase indoor $PM_{2.5}$ and CO levels; it was consistent with the previous study that indoor $PM_{2.5}$ and PM_5 levels could be elevated by the cleaning behavior of dry dust, and vacuuming [43]. In addition, commercial cleansers and disinfectants contain VOCs [44], and El Fadel et al. found VOCs concentration was positively correlated with CO concentration [45]. We also revealed that air-conditioning use was positively associated with indoor CO_2 concentrations with a statistical significance, which was consistent with a previous observation that CO_2 levels were higher in mechanically ventilated buildings than in naturally ventilated buildings [46]. There was a significantly negative association between essential oil use and O_3 concentration. The commercially available essential oils contain many VOCs (e.g., D-limonene, α- pinene, etc.) [47], in addition, a study displayed that indoor VOCs level had increased significantly after burning essential oils [48]. O_3 was one of the indoor oxidants [49,50], and Waring et al. demonstrated that 68% of all O_3 reactions were with D-limonene, and 26% of all O_3 reactions are with α-pinene [50]. This might be the reason why the essential oil use could decrease the O_3 level. Finally, by questionnaire, it was found that window opening was significantly correlated with increased O_3 concentration, which was not consistent with the results from the time–microenvironment–activity-diary that only NO_2 and $PM_{2.5}$ levels during the window opening periods were significantly higher than that of reference periods. We believed O_3 was a major component of photochemical pollution, so it is more relevant to outdoor sources than indoor sources. Thus, compared with the households which closed the windows, the households which opened the windows had a significantly higher 24-hour average concentration of O_3. When comparing the window opening periods with the reference periods (two one-hour periods before and after window opening periods), there was no significant variation in atmospheric O_3 concentration in a short time (within three hours). For $PM_{2.5}$ and NO_X levels, there was no significant difference between households which closed and opened the windows, the possible reason might be that $PM_{2.5}$ and NO_X could come from both indoor (cooking) and outdoor (traffic) sources.

5. Conclusions

This study explored the concentration of indoor air pollutants in different areas including traffic, industrial, and general areas within an industrial city. Moreover, this study also revealed household NO_2 and $PM_{2.5}$ concentrations during window opening periods were significantly higher than that of the reference periods with increased concentrations of 18.71 ppb, and 7 $\mu g/m^3$, respectively. For the influence of cooking, indoor CO_2, NO_2, and $PM_{2.5}$ levels during the cooking periods were significantly higher than that of the reference periods with increased concentrations of 26.17 ppm, 5.40 ppb, and 5 $\mu g/m^3$, respectively.

Author Contributions: Conceptualization, C.-Y.Y.; formal analysis, Y.-C.Y.; writing—original draft preparation, Y.-T.C. and Y.-T.C.; writing—review and editing, P.-S.C. and K.D.M.

References

1. Idavain, J.; Julge, K.; Rebane, T.; Lang, A.; Orru, H. Respiratory symptoms, asthma and levels of fractional exhaled nitric oxide in schoolchildren in the industrial areas of Estonia. *Sci. Total. Environ.* **2019**, *650*, 65–72. [CrossRef] [PubMed]

2. Martuzzi, M.; Pasetto, R.; Martin-Olmedo, P. Industrially Contaminated Sites and Health. *J. Environ. Public Health* **2014**, *2014*, 198574. [CrossRef] [PubMed]

3. Kurt, O.K.; Zhang, J.; Pinkerton, K.E. Pulmonary health effects of air pollution. *Curr. Opin. Pulm. Med.* **2016**, *22*, 138–143. [CrossRef] [PubMed]

4. Rice, M.B.; Rifas-Shiman, S.L.; Litonjua, A.A.; Oken, E.; Gillman, M.W.; Kloog, I.; Luttmann-Gibson, H.; Zanobetti, A.; Coull, B.A.; Schwartz, J.; et al. Lifetime Exposure to Ambient Pollution and Lung Function in Children. *Am. J. Respir. Crit. Care Med.* **2016**, *193*, 881–888. [CrossRef] [PubMed]

5. Deng, Q.; Lu, C.; Norbäck, D.; Bornehag, C.G.; Zhang, Y.; Liu, W.; Yuan, H.; Sundell, J. Early life exposure to ambient air pollution and childhood asthma in China. *Environ. Res.* **2015**, *143*, 83–92. [CrossRef] [PubMed]

6. Marchetti, P.; Marcon, A.; Pesce, G.; Paolo, G.; Guarda, L.; Pironi, V.; Fracasso, M.E.; Ricci, P.; de Marco, R. Children living near chipboard and wood industries are at an increased risk of hospitalization for respiratory diseases: A prospective study. *Int. J. Hyg. Environ. Health* **2014**, *217*, 95–101. [CrossRef] [PubMed]

7. Silvers, A.; Florence, B.T.; Rourke, D.L.; Lorimor, R.J. How Children Spend Their Time: A Sample Survey for Use in Exposure and Risk Assessments. *Risk Anal.* **1994**, *14*, 931–944. [CrossRef]

8. Cannistraro, G.; Cannistraro, A.; Cannistraro, M.; Galvagno, A. Analysis of air pollution in the urban center of four cities sicilian. *Int. J. Heat Technol.* **2016**, *34*, S219–S225.

9. Cannistraro, M.; Cannistraro, G.; Chao, J.; Ponterio, L. New Technique Monitoring and Transmission Environmental Data with Mobile. *Instrum. Meas. Metrol.* **2018**, *18*, 549–562. [CrossRef]

10. Cannistraro, M.; Chao, J.; Ponterio, L. Experimental Study of Air Pollution in the Urban Centre of the City of Messina. *Model. Meas. Control C* **2018**, *79*, 133–139. [CrossRef]

11. Leung, D.Y.C. Outdoor-indoor air pollution in urban environment: Challenges and opportunity. *Front. Environ. Sci.* **2015**. [CrossRef]

12. Brand, A.; McLean, K.E.; Henderson, S.B.; Fournier, M.; Liu, L.; Kosatsky, T.; Smargiassi, A. Respiratory hospital admissions in young children living near metal smelters, pulp mills and oil refineries in two Canadian provinces. *Environ. Int.* **2016**, *94*, 24–32. [CrossRef] [PubMed]

13. Tian, L.; Zeng, Q.; Dong, W.; Guo, Q.; Wu, Z.; Pan, X.; Li, G.; Liu, Y. Addressing the source contribution of PM2.5on mortality: An evaluation study of its impacts on excess mortality in China. *Environ. Res. Lett.* **2017**, *12*, 104016. [CrossRef]

14. Topacoglu, H.; Katsakoglou, S.; Ipekci, A. Effect of exhaust emissions on carbon monoxide levels in employees working at indoor car wash facilities. *Hippokratia* **2014**, *18*, 37–39. [PubMed]

15. Srithawirat, T.; Latif, M.T.; Sulaiman, F.R. *Indoor PM10 and Its Heavy Metal Composition at a Roadside Residential Environment*; Atmosfera: Phitsanulok, Thailand, 2016; Volume 29, pp. 311–322.

16. United States Environmental Protection Agency. Available online: https://www.epa.gov/report-environment/indoor-air-quality (accessed on 20 August 2019).

17. World Health Organization. Household (Indoor) Air Pollution. 2014. Available online: https://www.who.int/indoorair/en/ (accessed on 20 August 2019).

18. Han, I.; Guo, Y.; Afshar, M.; Symanski, E.; Stock, T.H. Comparison of trace elements in size-fractionated particles in two communities with contrasting socioeconomic status in Houston, TX. *Environ. Monit. Assess.* **2017**, *189*, 67. [CrossRef] [PubMed]

19. Hart, J.E.; Garshick, E.; Dockery, D.W.; Smith, T.J.; Ryan, L.; Laden, F. Long-term ambient multipollutant exposures and mortality. *Am. J. Respir. Crit. Care Med.* **2011**, *183*, 73–78. [CrossRef] [PubMed]

20. Milà, C.; Salmon, M.; Sanchez, M.; Ambrós, A.; Bhogadi, S.; Sreekanth, V.; Nieuwenhuijsen, M.; Kinra, S.; Marshall, J.D.; Tonne, C. When, Where, and What? Characterizing Personal PM2.5 Exposure in Periurban India by Integrating GPS, Wearable Camera, and Ambient and Personal Monitoring Data. *Environ. Sci. Technol.* **2018**, *52*, 13481–13490. [CrossRef]

21. Pekey, B.; Bozkurt, Z.B.; Pekey, H.; Doğan, G.; Zararsız, A.; Efe, N.; Tuncel, G.; Zararsiz, A. Indoor/outdoor concentrations and elemental composition of PM10/PM2.5 in urban/industrial areas of Kocaeli City, Turkey. *Indoor Air* **2010**, *20*, 112–125. [CrossRef]

22. Pekey, H.; Arslanba, D. The Relationship Between Indoor, Outdoor and Personal VOC Concentrations in Homes, Offices and Schools in the Metropolitan Region of Kocaeli, Turkey. *Water Air Soil Pollut.* **2008**, *191*, 113–129. [CrossRef]

23. Liu, W.; Zhang, J.; Zhang, L.; Turpin, B.; Weisel, C.; Morandi, M.; Stock, T.; Colome, S.; Korn, L. Estimating contributions of indoor and outdoor sources to indoor carbonyl concentrations in three urban areas of the United States. *Atmos. Environ.* **2006**, *40*, 2202–2214. [CrossRef]

24. Najjar, Y. Gaseous Pollutants Formation and Their Harmful Effects on Health and Environment. *Innov. Energy Policies* **2011**, *1*. [CrossRef]

25. Du, X.; Liu, J. Relationship between outdoor and indoor ozone pollution concentration. *Trans. Tianjin Univ.* **2009**, *15*, 330–335. [CrossRef]

26. Jianhui, B.; Singh, N.; Chauhan, S.; Singh, K.; Saud, T.; Saxena, M.; Soni, D.; Mandal, T.K.; Bassin, J.K.; Gupta, P.K. Study on surface O_3 chemistry and photochemistry by UV energy conservation. *Atmos. Pollut. Res.* **2010**, *1*, 118–127. [CrossRef]

27. Taiwan Environmental Protection Agency. Available online: https://taqm.epa.gov.tw/taqm/tw/PSIOver100MonthlyReport.aspx (accessed on 20 August 2019).

28. Persily, A.; De Jonge, L. Carbon dioxide generation rates for building occupants. *Indoor Air* **2017**, *27*, 868–879. [CrossRef] [PubMed]

29. Penney, D.; Benignus, V.; Kephalopoulos, S.; Kotzias, D.; Kleinman, M.; Verrier, A. *WHO Guidelines for Indoor Air Quality: Selected Pollutants*; WHO Regional Office for Europe: Denmark, 2010.

30. Taiwan Environmental Protection Agency. Available online: https://erdb.epa.gov.tw/DataRepository/Statistics/StatSceAreapop.aspx (accessed on 20 August 2019).

31. Satish, U.; Mendell, M.J.; Shekhar, K.; Hotchi, T.; Sullivan, D.; Streufert, S.; Fisk, W.J. Is CO_2 an Indoor Pollutant? Direct Effects of Low-to-Moderate CO_2 Concentrations on Human Decision-Making Performance. *Environ. Health Perspect.* **2012**, *120*, 1671–1677. [CrossRef]

32. Chao, H.-R.; Hsu, J.-W.; Ku, H.-Y.; Wang, S.-L.; Huang, H.-B.; Liou, S.-H.; Tsou, T.-C. Inflammatory Response and PM2.5 Exposure of Urban Traffic Conductors. *Aerosol. Air Qual. Res.* **2018**, *18*, 2633–2642. [CrossRef]

33. Bilsback, K.R.; Dahlke, J.; Fedak, K.M.; Good, N.; Hecobian, A.; Herckes, P.; L'Orange, C.; Mehaffy, J.; Sullivan, A.; Tryner, J.; et al. A Laboratory Assessment of 120 Air Pollutant Emissions from Biomass and Fossil Fuel Cookstoves. *Environ. Sci. Technol.* **2019**, *53*, 7114–7125. [CrossRef]

34. Shen, G.; Hays, M.D.; Smith, K.R.; Williams, C.; Faircloth, J.W.; Jetter, J.J. Evaluating the Performance of Household Liquefied Petroleum Gas Cookstoves. *Environ. Sci. Technol.* **2018**, *52*, 904–915. [CrossRef]

35. Shen, G.; Gaddam, C.K.; Ebersviller, S.M.; Wal, R.L.V.; Williams, C.; Faircloth, J.W.; Jetter, J.J.; Hays, M.D. A Laboratory Comparison of Emission Factors, Number Size Distributions, and Morphology of Ultrafine Particles from 11 Different Household Cookstove-Fuel Systems. *Environ. Sci. Technol.* **2017**, *51*, 6522–6532. [CrossRef]

36. Smith, K.R.; Uma, R.; Kishore, V.V.N.; Zhang, J.; Joshi, V.; Khalil, M.A.K. Greenhouse Implications of Household Stoves: An Analysis for India. *Annu. Rev. Energy Environ.* **2000**, *25*, 741–763. [CrossRef]

37. Delp, W.W.; Singer, B.C. Performance Assessment of U.S. Residential Cooking Exhaust Hoods. *Environ. Sci. Technol.* **2012**, *46*, 6167–6173. [CrossRef] [PubMed]

38. Deveci, S.E.; Deveci, F.; Açik, Y.; Ozan, A.T. The measurement of exhaled carbon monoxide in healthy smokers and non-smokers. *Respir Med.* **2004**, *98*, 551–556. [CrossRef] [PubMed]

39. Ryter, S.W.; Choi, A.M. Carbon monoxide in exhaled breath testing and therapeutics. *J. Breath Res.* **2013**, *7*, 017111. [CrossRef] [PubMed]

40. Alberts, W.M. Indoor air pollution: NO, NO_2, CO, and CO_2. *J. Allergy Clin. Immunol.* **1994**, *94*, 289–295. [CrossRef] [PubMed]

41. Lee, S.C.; Guo, H.; Kwok, N.H. Emissions of air pollutants from burning of incense by using large environmental chamber. In Proceedings of the Indoor Air 2002: 9th International Conference on Indoor Air Quality and Climate, Monterey, CA, USA, 30 June–5 July 2002; Volume 1.

42. Lee, S.C.; Wang, B. Characteristics of emissions of air pollutants from mosquito coils and candles burning in a large environmental chamber. *Atmos. Environ.* **2006**, *40*, 2128–2138. [CrossRef]

43. Ferro, A.R.; Kopperud, R.J.; Hildemann, L.M. Source Strengths for Indoor Human Activities that Resuspend Particulate Matter. *Environ. Sci. Technol.* **2004**, *38*, 1759–1764. [CrossRef]

44. United States Environmental Protection Agency. Available online: https://www.epa.gov/indoor-air-quality-iaq/volatile-organic-compounds-impact-indoor-air-quality (accessed on 20 August 2019).

45. El Fadel, M.; Alameddine, I.; Kazopoulo, M.; Hamdan, M.; Nasrallah, R. Carbon Monoxide and Volatile Organic Compounds as Indicators of Indoor Air Quality in Underground Parking Facilities. *Indoor Built Environ.* **2001**, *10*, 70–82. [CrossRef]

46. Sribanurekha, V.; Wijerathne, S.N.; Wijepala, L.H.S.; Jayasinghe, C. Effect of Different Ventilation Conditions on Indoor CO_2 Levels. In Proceedings of the International Conference on Disaster Resilience, At Kandalama, Sri Lanaka, 19–21 July 2011.

47. Nematollahi, N.; Kolev, S.D.; Steinemann, A. Volatile chemical emissions from essential oils. *Air Qual. Atmos. Heal.* **2018**, *11*, 949–954. [CrossRef]

48. Chao, C.J.; Wu, P.C.; Chang, H.Y.; Su, H.J. The effects of evaporating essential oils on indoor air quality. In Proceedings of the Indoor Air 2005, 10th International Conference on Indoor Air Quality and Climate, Beijing, China, 4–9 September 2005.

49. Young, C.J.; Zhou, S.; Siegel, J.A.; Kahan, T.F. Illuminating the Dark Side of Indoor Oxidants. *Environ. Sci. Process. Impacts* **2019**, *21*, 1229–1239. [CrossRef]

50. Waring, M.S.; Wells, J.R. Volatile organic compound conversion by ozone, hydroxyl radicals, and nitrate radicals in residential indoor air: Magnitudes and impacts of oxidant sources. *Atmos. Environ.* **2015**, *106*, 382–391. [CrossRef]

8

Indoor Particle Concentrations, Size Distributions and Exposures in Middle Eastern Microenvironments

Tareq Hussein [1,2,*], Ali Alameer [1], Omar Jaghbeir [1], Kolthoum Albeitshaweesh [1],
Mazen Malkawi [3], Brandon E. Boor [4,5], Antti Joonas Koivisto [2], Jakob Löndahl [6], Osama Alrifai [7]
and Afnan Al-Hunaiti [8]

[1]   Department of Physics, The University of Jordan, Amman 11942, Jordan; alameer_hw95@hotmail.com (A.A.);
      omarjaghbeir@gmail.com (O.J.); kolthoum.baitshaweesh@gmail.com (K.A.)
[2]   Institute for Atmospheric and Earth System Research (INAR), University of Helsinki, PL 64, FI-00014 UHEL,
      Helsinki, Finland; joonas.apm@gmail.com
[3]   Regional Office for the Eastern Mediterranean (EMRO), Centre for Environmental Health Action (CEHA),
      World Health Organization (WHO), Amman 11181, Jordan; malkawim@who.int
[4]   Lyles School of Civil Engineering, Purdue University, West Lafayette, IN 47907, USA; bboor@purdue.edu
[5]   Ray W. Herrick Laboratories, Center for High Performance Buildings, Purdue University, West Lafayette,
      IN 47907, USA
[6]   Department of Design Sciences, Lund University, P.O. Box 118, SE-221 00 Lund, Sweden;
      jakob.londahl@design.lth.se
[7]   Validation and Calibration Department, Savypharma, Amman 11140, Jordan; olzr27@gmail.com
[8]   Department of Chemistry, The University of Jordan, Amman 11942, Jordan; a.alhunaiti@ju.edu.jo
[*]   Correspondence: tareq.hussein@helsinki.fi

Abstract: There is limited research on indoor air quality in the Middle East. In this study, concentrations and size distributions of indoor particles were measured in eight Jordanian dwellings during the winter and summer. Supplemental measurements of selected gaseous pollutants were also conducted. Indoor cooking, heating via the combustion of natural gas and kerosene, and tobacco/shisha smoking were associated with significant increases in the concentrations of ultrafine, fine, and coarse particles. Particle number (PN) and particle mass (PM) size distributions varied with the different indoor emission sources and among the eight dwellings. Natural gas cooking and natural gas or kerosene heaters were associated with PN concentrations on the order of 100,000 to 400,000 cm^{-3} and $PM_{2.5}$ concentrations often in the range of 10 to 150 $\mu g/m^3$. Tobacco and shisha (waterpipe or hookah) smoking, the latter of which is common in Jordan, were found to be strong emitters of indoor ultrafine and fine particles in the dwellings. Non-combustion cooking activities emitted comparably less PN and $PM_{2.5}$. Indoor cooking and combustion processes were also found to increase concentrations of carbon monoxide, nitrogen dioxide, and volatile organic compounds. In general, concentrations of indoor particles were lower during the summer compared to the winter. In the absence of indoor activities, indoor PN and $PM_{2.5}$ concentrations were generally below 10,000 cm^{-3} and 30 $\mu g/m^3$, respectively. Collectively, the results suggest that Jordanian indoor environments can be heavily polluted when compared to the surrounding outdoor atmosphere primarily due to the ubiquity of indoor combustion associated with cooking, heating, and smoking.

Keywords: indoor air quality; aerosols; particle size distributions; ultrafine particles; particulate matter (PM); smoking; combustion

1. Introduction

Indoor air pollution has a significant impact on human respiratory and cardiovascular health because people spend the majority of their time in indoor environments, including their homes, offices, and schools [1–9]. The World Health Organization (WHO) has recognized healthy indoor air as a fundamental human right [4]. Comprehensive indoor air quality measurements are needed in many regions of the world to provide reliable data for evaluation of human exposure to particulate and gaseous indoor air pollutants [10].

Indoor air pollutant concentrations depend on the dynamic relationship between pollutant source and loss processes within buildings. Source processes include the transport of outdoor air pollution, which can be high in urban areas [11–13], into the indoor environment via ventilation and infiltration, and indoor emission sources, which include solid fuel combustion, electronic appliances, cleaning, consumer products, occupants, pets, and volatilization of chemicals from building materials and furnishings, among others [10,14–28]. Loss processes include ventilation, exfiltration, deposition to indoor surfaces, filtration and air cleaning, and pollutant transformations in the air (i.e., coagulation, gas-phase reactions). Indoor emission sources can result in substantial increases in indoor air pollutant concentrations, exceeding contributions from the transport of outdoor air pollutants indoors. Air cleaning technologies, such as heating, ventilation, and air conditioning (HVAC) filters and portable air cleaners, can reduce concentrations of various indoor air pollutants.

Evaluation of indoor air pollution and concentrations of particulate and gaseous indoor air pollutants in Middle Eastern dwellings has been given limited attention in the literature. In Jordan, one study investigated the effects of indoor air pollutants on the health of Jordanian women [29] and three studies evaluated concentrations of indoor particles in Jordanian indoor environments [30–32]. These studies provided useful insights on the extent of air pollution in selected Jordanian indoor environments and the role of cultural practices on the nature of indoor emission sources. However, these studies did not provide detailed information on the composition of indoor air pollution, including indoor particle number and mass size distributions, concentrations of ultrafine particles (UFPs, diameter < 0.1 µm), and concentrations of various gaseous pollutants.

The objective of this study was to evaluate size-fractionated number and mass concentrations of indoor particles (aerosols) in selected Jordanian residential indoor environments and human inhalation exposures associated with a range of common indoor emission sources prevalent in Jordanian dwellings, such as combustion processes associated with cooking, heating, and smoking. The study was based upon a field campaign conducted over two seasons in which portable aerosol instrumentation covering different particle size ranges was used to measure particle number size distributions spanning 0.01–25 µm during different indoor activities.

2. Materials and Methods

2.1. Residential Indoor Environment Study Sites in Jordan

The residential indoor environments targeted in this study were houses and apartments covering a large geographical area within Amman, the capital city of Jordan (Figure 1). The selection was based upon two main criteria: (1) prevalence of smoking indoors and (2) heating type, such as kerosene heaters, natural gas heaters, and central heating systems. The selected residential indoor environments included two apartments (A), one duplex apartment (D), three ground floor apartments (GFA), and two houses (H). Table 1 lists the characteristics of each study site. All indoor environments were naturally ventilated. The occupants documented their activities and frequency of cooking, heating, and smoking during the measurement campaign.

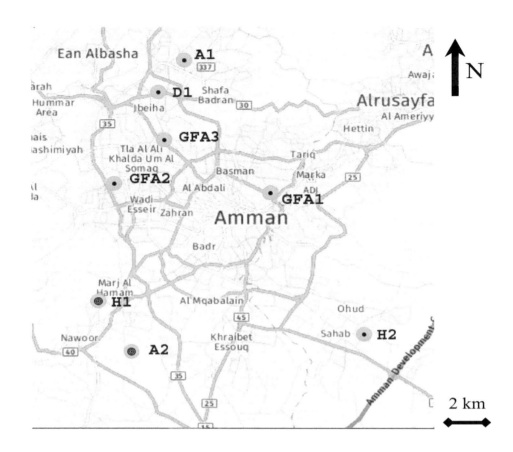

Figure 1. A map showing the Amman metropolitan region with the locations of the selected indoor environment study sites. The type of dwelling is referred to as: (A) apartment, (H) house, (D) duplex apartment, and (GFA) ground floor apartment. Table 1 provides additional details for each dwelling.

Table 1. Characteristics of the selected residential indoor environments. The heating method refers to: kerosene heater (Ker.), natural gas heater (Gas), air conditioning system (AC), electric heaters (El.), and central heating system (Cen.). Cigarette smoking is denoted as (Cig.).

Site ID	Type	Area Type	Kitchen/L. Room	Heating Method					Smoking	
				Ker.	Gas	AC	El.	Cen.	Cig.	Shisha
A1	Apartment (3rd floor)	Suburban	Open	√	√	√				
A2	Apartment (2nd floor)	Rural	Separate			√				
D1	Duplex (2nd and 3rd floors)	Urban Background	Open	√		√				√
GFA1	Ground floor apartment	Urban	Separate	√	√					
GFA2	Ground floor apartment	Urban	Separate				√	√		
GFA3	Ground floor apartment	Urban Background	Open		√				√	√
H1	House	Suburban	Open		√			√		√
H2	House	Rural	Open	√						

2.2. Indoor Aerosol Measurements and Experimental Design

2.2.1. Measurement Campaign

Indoor aerosol measurements were performed during two seasons: winter and summer, as indicated in Table 2. The winter campaign occurred from 23 December 2018 to 12 January 2019. All eight study sites participated in the winter campaign. The summer campaign occurred from 16 May to 22 June 2019. Only GFA2, GFA3, and H2 participated in the summer campaign.

Table 2. Measurement periods and lengths of the two campaigns.

Site ID	Winter Campaign			Summer Campaign		
	Start	End	Length	Start	End	Length
A1	13:15, 23.12.2018	11:50, 25.12.2018	1d 22h 35m	–	–	–
A2	18:20, 04.01.2019	19:50, 05.01.2019	1d 01h 30m	–	–	–
D1	14:10, 28.12.2018	22:10, 30.12.2018	2d 08h 00m	–	–	–
GFA1	15:10, 25.12.2018	14:10, 27.12.2018	1d 23h 00m	–	–	–
GFA2	12:00, 09.01.2019	20:40, 12.01.2019	3d 08h 40m	10:30, 13.06.2019	11:20, 22.06.2019	9d 00h 50m
GFA3	12:30, 31.12.2018	18:30, 02.01.2019	2d 06h 00m	18:50, 16.05.2019	23:40, 23.05.2019	7d 04h 50m
H1	20:20, 02.01.2019	16:30, 04.01.2019	1d 20h 10m	–	–	–
H2	12:30, 06.01.2019	15:30, 09.01.2019	3d 03h 00m	20:50, 24.05.2019	21:30, 29.05.2019	5d 00h 40m

2.2.2. Aerosol Instrumentation

Aerosol instrumentation included portable devices to monitor size-fractionated particle concentrations. Supplemental measurements of selected gaseous pollutants were also conducted. The aerosol measurements included particle number and mass concentrations within standard size fractions: submicron particle number concentrations, micron particle number concentrations, PM_{10}, and $PM_{2.5}$. Table 3 provides an overview of the portable aerosol instrumentation deployed at each study site. The use of portable aerosol instruments has increased in recent years, with a number of studies evaluating their performance in the laboratory, the field, or through side-by-side comparisons with more advanced instruments [33–46]. The instruments were positioned to sample side-by-side without the use of inlet extensions. The instruments were situated on a table approximately 60 cm above the floor inside the living room of each dwelling. The sample time was set to 1 min for all instruments, either by default or through time-averaging of higher sample frequency data.

Table 3. List of the portable air quality instruments and the measured parameters.

Instrument	Model	Aerosol Size Fraction	Metric	Performance Ref.
Laser Photometer	TSI DustTrak DRX 8534	PM_{10}, $PM_{2.5}$, and PM_1	Mass	Wang et al. [33]
Personal Aerosol Monitor	TSI SidePak AM520	$PM_{2.5}$	Mass	Jiang et al. [34]
Optical Particle Counter	TSI AeroTrak 9306-V2	D_p 0.3–25 µm (6 bins)	Number	Wang et al. [33]
Condensation Particle Counter	TSI CPC 3007	D_p 0.01–2 µm	Number	Matson et el. [35]
Condensation Particle Counter	TSI P-Trak 8525	D_p 0.02–2 µm	Number	Matson et el. [35]
Gas monitor	AeroQual S500	O_3, HCHO, CO, NO_2, SO_2, TVOC	ppm	Lin et al. [36]

Two condensation particle counters (CPCs) with different lower size cutoffs (TSI 3007-2: cutoff size 10 nm; TSI P-Trak 8525: cutoff size 20 nm) were used to measure total submicron particle number concentrations. The maximum detectable concentration (20% accuracy) was 10^5 cm^{-3} and 5×10^5 cm^{-3} for the CPC 3007 and the P-Trak, respectively. The sample flow rate for both CPCs was 0.1 lpm (inlet flow rate of 0.7 lpm). A handheld optical particle counter (AeroTrak 9306-V2, TSI, MI, USA) was used to monitor particle number concentrations within 6 channels (user-defined) in the diameter range of 0.3–25 µm. The cutoffs for these channels were defined as 0.3, 0.5, 1, 2.5, 10, and 25 µm. The sample flow rate was 2.83 lpm. A handheld laser photometer (DustTrak DRX 8534, TSI, MI, USA) monitored particle mass (PM) concentrations (PM_1, $PM_{2.5}$, respirable (PM_4), PM_{10}, and total) in the diameter range of 0.1–15 µm (maximum concentration of 150 mg/m^3). The sample flow rate for the DustTrak was 3 lpm. A personal aerosol monitor (SidePak AM520, TSI, MI, USA) with a $PM_{2.5}$ inlet was used for additional measurements of $PM_{2.5}$ concentrations. The SidePak is a portable instrument with a small form factor equipped with a light-scattering laser photometer. The CPCs were calibrated in the laboratory [40], whereas the AeroTrak, DustTrak, and SidePak were factory calibrated. Additionally, a portable gas monitor (S500, AeroQual, New Zealand) estimated the concentrations of gaseous pollutants by installing factory calibrated plug-and-play gas sensor heads. The sensor heads included ozone (O_3), formaldehyde (HCHO), carbon monoxide (CO), nitrogen dioxide (NO_2), sulfur dioxide (SO_2), and total volatile organic compounds (TVOCs).

Each instrument was started at different times during the campaigns; and thus, they did not record concentrations at the same time stamp. Therefore, we interpolated the concentrations of each instrument into a coherent time grid so that we evaluated the number of concentrations in each size fraction with the same time stamp. The built-in temperature and relative humidity sensors used in the aerosol instruments cannot be confirmed to be accurate for ambient observations because these sensors were installed inside the instruments and can be affected by instrument-specific conditions, such as heat dissipation from the pumps and electronics. Therefore, those observations were not considered here.

2.3. Processing of Size-Fractionated Aerosol Concentration Data

The utilization of portable aerosol instruments with different particle diameter ranges and cutoff diameters enables derivations of size-fractionated particle number and mass concentrations [47]: Super-micron (1–10 μm) particle number and mass concentrations, submicron (0.01–1 μm) particle number concentrations, $PM_{2.5}$ mass concentrations, PM_{10} mass concentrations, and PM_{10-1} mass concentrations. Additionally, we derived the particle number size distribution $\left(n_N^0 = \frac{dN}{dlog(D_p)}\right)$ within eight diameter bins:

- 0.01–0.02 μm via the difference between the CPC 3007 and the P-Trak.
- 0.02–0.3 μm via the difference between the P-Trak and the first two channels of the AeroTrak.
- 0.3–0.5 μm, 0.5–1 μm, 1–2.5 μm, 2.5–5 μm, 5–10 μm, and 10–25 μm via the AeroTrak.

The particle mass size distribution was estimated from the particle number size distribution by assuming spherical particles:

$$n_M^0 = \frac{dM}{dlog(D_p)} = \frac{dN}{dlog(D_p)}\frac{\pi}{6}D_p^3\rho_p = n_N^0\frac{\pi}{6}D_p^3\rho_p \tag{1}$$

where n_M^0 is the particle mass size distribution, dM is the particle mass concentration within a certain diameter bin normalized to the width of the diameter range $\left(dlog(D_p)\right)$ of that diameter bin, dN is the particle number concentration within that diameter bin (also normalized with respect to $dlog(D_p)$ to obtain the particle number size distribution, n_N^0), D_p is the particle diameter, and ρ_p is the particle density, here assumed to be unit density (1 g cm^{-3}). In practice, the particle density is size-dependent and variable for different aerosol populations (i.e., diesel soot vs. organic aerosol); therefore, size-resolved effective density functions should be used. However, there is limited empirical data on the effective densities of aerosols produced by indoor emission sources. Thus, the assumption of 1 g cm^{-3} for the particle density will result in uncertainties (over- or underestimates, depending on the source) in the estimated mass concentrations.

The size-fractionated particle number concentration was calculated as:

$$PN_{D_{p2}-D_{p1}} = \int_{D_{p1}}^{D_{p2}} n_N^0(D_P)\cdot dlog(D_P) \tag{2}$$

where $PN_{D_{p2}-D_{p1}}$ is the calculated size-fractionated particle number concentration within the particle diameter range $D_{p1}-D_{p2}$. Similarly, the size-fractionated particle mass concentration $\left(PM_{D_{p2}-D_{p1}}\right)$ was calculated as:

$$PM_{D_{p2}-D_{p1}} = \int_{D_{p1}}^{D_{p2}} n_M^0(D_P)\cdot dlog(D_P) = \int_{D_{p1}}^{D_{p2}} n_N^0(D_P)\frac{\pi}{6}D_p^3\rho_p\cdot dlog(D_P) \tag{3}$$

$PM_{2.5}$ and PM_{10} can be also calculated by using Equation (3) and integrating over the particle diameter range starting from 10 nm (i.e., the lower cutoff diameter according to our instrument setup) and up to 2.5 μm (for $PM_{2.5}$) or 10 μm (for PM_{10}).

3. Results

3.1. Comparisons between Different Aerosol Instruments—Technical Notes

The co-location of different aerosol instruments covering similar size ranges provides a basis to compare concentration outputs as measured through different techniques. First, the $PM_{2.5}$ and PM_{10} concentrations reported by the DustTrak can be compared to evaluate the contribution of the submicron fraction to the total PM concentration in Jordanian indoor environments. According to the DustTrak measurements, it was observed that most of the PM was in the submicron fraction as the mean $PM_{10}/PM_{2.5}$ ratio was 1.03 ± 0.04 (Figure 2). This was somewhat expected as most of the tested indoor activities in this field study were combustion processes (smoking, heating, and cooking) that produce significant emissions in the fine particle range. However, more sophisticated aerosol instrumentation would be needed to verify this finding, such as an aerodynamic particle sizer (APS) and scanning mobility particle sizer (SMPS).

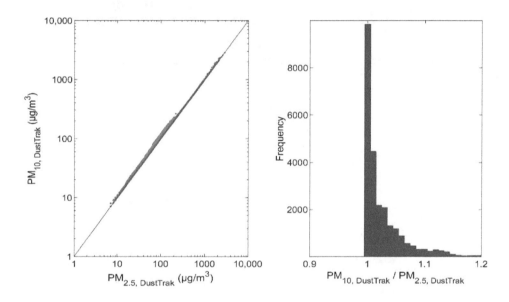

Figure 2. Comparison between the PM_{10} and $PM_{2.5}$ concentrations measured with the DustTrak.

The DustTrak and SidePak both employ a light-scattering laser photometer to estimate PM concentrations. As such, their output can be compared for the same particle diameter range. In general, the $PM_{2.5}$ concentrations measured with the DustTrak were lower than the corresponding values measured with the SidePak (Figure 3). This trend was consistent across the measured concentration range from approximately 10 to >1000 µg/m³. The mean SidePak/DustTrak $PM_{2.5}$ concentration ratio was 2.15 ± 0.48. These differences can be attributed to technical matters related to the internal setup of the instruments and their factory calibrations. For example, the SidePak inlet has an impactor plate with a specific aerodynamic diameter cut point (here chosen as $PM_{2.5}$), whereas the DustTrak differentiates the particle size based solely on the optical properties of particles.

Following the methodology outlined in Section 2.3, we converted the measured particle number size distributions (via CPC 3007, P-Trak, and AeroTrak) to particle mass size distributions assuming spherical particles of unit density. From integration of the latter, we calculated the $PM_{2.5}$ and PM_{10} concentrations. The calculated $PM_{2.5}$ and PM_{10} concentrations can be compared with those reported by the DustTrak. The calculated $PM_{2.5}$ concentrations were found to be less than those reported by the DustTrak (Figure 4). More variability was observed for PM_{10}, with the calculated PM_{10} both under- and overestimating the DustTrak-derived values across the measured concentration range. The mean calculated-to-DustTrak $PM_{2.5}$ ratio was 0.63 ± 0.58 and that for PM_{10} was 1.46 ± 1.27.

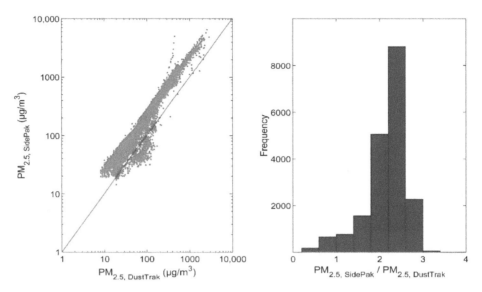

Figure 3. Comparison between the $PM_{2.5}$ concentrations measured with the DustTrak and SidePak.

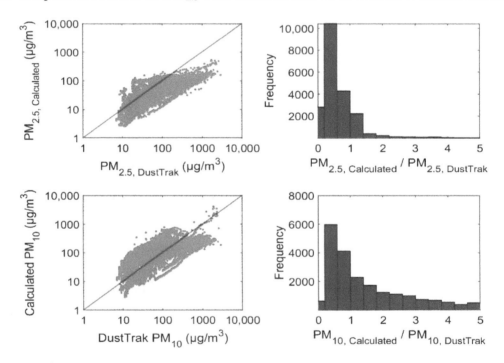

Figure 4. Comparison between the $PM_{2.5}$ and PM_{10} concentrations measured with DustTrak and those calculated using the measured particle number size distributions, assuming spherical particles of unit density.

This brief comparative analysis of the PM concentrations measured by the DustTrak, SidePak, and calculated via measured particle number size distributions illustrates that portable aerosol instruments have limitations and their output is likely to be inconsistent. Relying on a single instrument output may not provide an accurate assessment of PM concentrations. The utilization of an array of portable aerosol instruments can provide lower and upper bounds on PM concentrations in different indoor environments. Calculating PM concentrations from measured particle number size distributions is uncertain in the absence of reliable data on size-resolved particle effective densities for different indoor emission sources.

3.2. *Overview of Indoor Particle Concentrations in Jordanian Dwellings*

3.2.1. Indoor Particle Concentrations during the Winter Season

An overview of the indoor submicron particle number (PN) concentrations and $PM_{2.5}$ and PM_{10} concentrations is presented Tables 4 and 5 (mean ± SD and 95%) and illustrated in Figure 5 for each of the eight Jordanian dwellings investigated in this study. Particle concentration time series are presented in the supplementary material (Figures S1–S8). Indoor particle concentrations (mean ± SD) were also evaluated during the nighttime, when there were no indoor activities reported in the dwellings and the concentrations were observed to be at their lowest levels (Table 6).

Table 4. Indoor particle number and mass concentrations (mean ± SD and 95%) during the winter campaign.

Site ID	CPC 3007 PN ($\times 10^4$/cm^3)		DustTrak PM$_{2.5}$ (μg/m^3)		DustTrak PM$_{10}$ (μg/m^3)		SidePak PM$_{2.5}$ (μg/m^3)	
	Mean ± SD	95%	Mean ± SD	95%	Mean ± SD	95%	Mean ± SD	95%
A1	4.3 ± 6.0	22.6	91 ± 218	612	93 ± 228	628	188 ± 403	1261
A2	1.6 ± 1.7	6.7	44 ± 40	157	47 ± 42	160	–	–
D1	13.3 ± 10.5	30.1	131 ± 202	613	132 ± 202	614	271 ± 448	1446
GFA1	5.4 ± 4.6	22.0	42 ± 26	109	45 ± 30	123	80 ± 38	176
GFA2	3.4 ± 4.0	17.0	29 ± 34	126	29 ± 34	126	–	–
GFA3	6.3 ± 4.8	18.6	433 ± 349	1230	437 ± 350	2140	998 ± 815	2790
H1	11.7 ± 7.4	23.6	138 ± 116	451	141 ± 117	453	325 ± 310	1190
H2	9.7 ± 6.1	25.0	156 ± 190	694	160 ± 190	697	342 ± 477	1690

Table 5. Indoor particle number and mass concentrations (mean ± SD and 95%) during the summer campaign.

Site ID	CPC 3007 PN ($\times 10^4$/cm^3)		DustTrak PM$_{2.5}$ (μg/m^3)		DustTrak PM$_{10}$ (μg/m^3)		SidePak PM$_{2.5}$ (μg/m^3)	
	Mean ± SD	95%	Mean ± SD	95%	Mean ± SD	95%	Mean ± SD	95%
GFA2	1.5 ± 1.4	5.5	30 ± 20	62	31 ± 20	64	58 ± 34	104
GFA3	1.9 ± 1.6	6.3	31 ± 46	179	31 ± 46	180	158 ± 216	819
H2	1.6 ± 0.9	3.8	46 ± 24	101	50 ± 26	107	89 ± 64	305

Table 6. Indoor particle number and mass concentrations (mean ± SD) during the nighttime, when there were no reported indoor activities. The concentrations were calculated for the winter campaign only.

Site ID	CPC 3007 PN ($\times 10^3$/cm^3)	DustTrak PM$_{2.5}$ (μg/m^3)	DustTrak PM$_{10}$ (μg/m^3)	SidePak PM$_{2.5}$ (μg/m^3)
	Mean ± SD	Mean ± SD	Mean ± SD	Mean ± SD
A1	6 ± 3	18 ± 8	18 ± 8	45 ± 19
A2	6 ± 1	10 ± 0	11 ± 1	–
D1	13 ± 2	26 ± 0	26 ± 0	52 ± 3
GFA1	9 ± 1	25 ± 7	26 ± 7	62 ± 15
GFA2	9 ± 3	10 ± 3	10 ± 3	–
GFA3	15 ± 5	67 ± 18	67 ± 18	154 ± 45
H1	10 ± 2	28 ± 6	29 ± 6	59 ± 14
H2	9 ± 2	28 ± 23	29 ± 24	47 ± 28

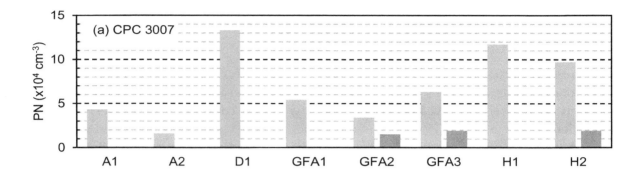

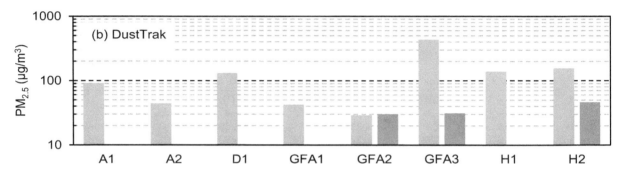

Figure 5. Overall mean indoor particle concentrations during the measurement period in each dwelling: (**a**) submicron particle number (PN) concentrations measured with the condensation particle counter (CPC 3007) and (**b**) $PM_{2.5}$ concentrations measured with the DustTrak. The blue bars represent the winter campaign and the orange bars represent the summer campaign.

Submicron PN concentrations were the lowest in apartment A2, which was equipped with an air conditioning (AC) heating/cooling setting and nonsmoking occupants. For example, the overall mean submicron PN concentrations in A2 was approximately 1.6×10^4 cm^{-3}. The second lowest PN concentrations were observed in the ground floor apartment GFA2, which was equipped with a central heating system (water radiators) and, periodically, electric heaters. Occupants in GFA2 were nonsmokers. The overall mean submicron PN concentration in GFA2 was approximately double that of A2 at 3.2×10^4 cm^{-3}.

The highest submicron PN concentrations were measured in duplex apartment D1, with a mean of 1.3×10^5 cm^{-3}. This apartment had a kerosene heater and one of the occupants smoked shisha (waterpipe or hookah) on a daily basis. The second highest submicron PN concentrations were observed in houses H1 and H2, with overall mean values of 1.2×10^5 cm^{-3} and 9.7×10^4 cm^{-3}, respectively. House H1 was heated by using a natural gas heater and smoking shisha was often conducted by more than one occupant. House H2 was heated with a kerosene heater and cooking activities occurred frequently.

The ground floor apartments, GFA3 and GFA1, showed intermediate submicron PN concentrations among the study sites, with mean concentrations of 6.3×10^4 cm^{-3} and 5.4×10^4 cm^{-3}, respectively. Although occupants in GFA3 heavily smoked tobacco and shisha, the concentrations were lower than those observed in D1 and H1, where shisha was also smoked. The building envelopes of D1 and H1 may be more tightly sealed, with lower infiltration rates compared to GFA3. Furthermore, GFA3 used a natural gas heater and cooking activities were not as frequent. As for GFA1, the heating was a combination of a kerosene heater and a natural gas heater. The cooking activities in GFA1 were minimal and not frequent. Occupants in apartment A1 were nonsmokers. Indoor emission source manipulations were conducted in A1, including various cooking activities and the use of three different types of heating (kerosene heater, natural gas heater, and AC). The overall mean submicron PN concentration in A1 was approximately 4.3×10^4 cm^{-3}.

For $PM_{2.5}$ concentrations, the lowest levels were observed not in A2 (highest submicron PN concentrations), but rather in GFA2, with a mean of approximately 29 $\mu g/m^3$. GFA2 was heated by means of a central heating system and, periodically, with electric heaters. Ground floor apartment GFA1 and apartment A2 exhibited intermediate overall mean $PM_{2.5}$ concentrations among the study sites, with mean values of 42 $\mu g/m^3$ and 44 $\mu g/m^3$, respectively. As previously discussed, the occupants in GFA1 did not conduct frequent cooking activities and heated their dwelling by means of kerosene and natural gas heaters, whereas A2 was heated via an AC. GFA1 was built in the 1970s, whereas A2 was relatively new (less than 10 years old); therefore, A2 is expected to be a more tightly sealed indoor environment compared to GFA1. However, infiltration rate and air leakage (i.e., blower door) measurements were not conducted for the dwellings in this study.

Apartment A1, in which manipulations of various cooking activities and heating methods were conducted, showed an overall mean $PM_{2.5}$ concentration of 91 $\mu g/m^3$. The impact of shisha smoking on $PM_{2.5}$ concentrations in D1 and H1 was clearly evident, with overall mean $PM_{2.5}$ concentrations of 131 $\mu g/m^3$ and 138 $\mu g/m^3$, respectively. The influence of a kerosene heater and intense cooking activities in H2 was also evident, with an overall mean $PM_{2.5}$ concentration of 156 $\mu g/m^3$. The highest $PM_{2.5}$ concentrations were recorded in GFA3 (approximately 433 $\mu g/m^3$), which reflects the frequent shisha and tobacco smoking in this dwelling.

In the absence of indoor activities (Table 6), the submicron PN concentrations were the lowest (approximately 6×10^3 cm^{-3}) in A1 and A2 and the highest in D1 (approximately 1.3×10^4 cm^{-3}) and GFA3 (approximately 1.5×10^4 cm^{-3}). As for the $PM_{2.5}$ concentrations measured with the DustTrak, the lowest concentrations (approximately 10 $\mu g/m^3$) were observed in A2 and GFA2 and the highest concentrations were observed in GFA3 (approximately 67 $\mu g/m^3$). It is important to note that the measured indoor particle concentrations were primarily the result of the transport of outdoor particles indoors via ventilation and infiltration. However, indoor-generated aerosols during the day may still have traces overnight. For example, the dwellings with combustion and smoking activities also had background concentrations higher than other dwellings. Furthermore, differences in background concentrations among dwellings can be due to the geographical location of the dwelling within the city; this might reflect the outdoor aerosol concentrations at a given location [16,48].

3.2.2. Indoor Particle Concentrations: Summer Versus Winter

Indoor aerosol measurements were repeated for three apartments in the summer campaign. We selected a dwelling (H2) that was heated with a kerosene heater and had nonsmoking occupants, a dwelling (GFA2) that was not heated with combustion processes and had nonsmoking occupants, and a dwelling (GFA3) that was heated with a natural gas heater and the occupants were smokers. Although the number of selected indoor environments was fewer in the summer campaign, the measurement period in each dwelling was longer and more extensive than what was measured during the winter campaign.

In general, the observed concentrations during the summer campaign were lower than those observed during the winter campaign (Tables 4 and 5, Figure 5). The overall mean submicron PN concentration during the summer campaign in GFA2 was approximately 1.5×10^4 cm^{-3}, which was about 40% of that during the winter campaign. As for the $PM_{2.5}$ concentrations, the overall mean during the summer campaign was approximately 30 $\mu g/m^3$, which was almost the same as that observed during the winter campaign.

The overall mean submicron PN concentrations in GFA3 and H2 were similar (approximately 1.6–1.9×10^4 cm^{-3}), whereas the corresponding mean $PM_{2.5}$ concentrations were higher in H2 (approximately 46 $\mu g/m^3$) compared to GFA3 (approximately 31 $\mu g/m^3$). The summer/winter ratio for submicron PN concentrations for GFA3 and H2 were 0.3 and 0.2, respectively. The corresponding $PM_{2.5}$ ratios were approximately 0.1 and 0.3. The primary reason for higher particle concentrations during the winter was the use of fossil fuel combustion for heating (i.e., kerosene and natural gas heaters). Furthermore, the dwellings during the summer were more likely to be better ventilated than during the winter, when the dwellings had to conserve energy during heating periods.

3.3. Indoor Particle Number and Mass Size Distributions in Jordanian Dwellings

3.3.1. Indoor Particle Size Distributions in the Absence of Indoor Activities

The mean particle number and mass size distributions for each dwelling in the absence of indoor activities during the winter campaign are presented in Figure S9. Significant differences in the mean particle number and mass size distributions were observed among the eight dwellings. Based on the number size distributions, the submicron PN concentration was the lowest (approximately 6×10^3 cm^{-3}, with a corresponding PM$_{2.5}$ of 5 µg/m^3) in dwellings A1 and A2 and the highest in GFA3 (approximately 1.5×10^4 cm^{-3}, with a corresponding PM$_{2.5}$ of 12 µg/m^3) and D1 (approximately 1.3×10^4 cm^{-3}, with a corresponding PM$_{2.5}$ of 8 µg/m^3). The mean submicron PN concentration was between 9×10^3 cm^{-3} and 10^4 cm^{-3} and the mean PM$_{2.5}$ was 7–9 µg/m^3 in the remainder of the dwellings. It should be noted that GFA3 had the highest submicron PN concentration, whereas H2 had the highest PM$_{2.5}$ concentration (approximately 13 µg/m^3). Differences between the PN and PM concentrations among the eight dwellings is an indicator of variability in the shape and magnitude of the aerosol size distributions, as illustrated in Figure S9.

The coarse PN concentrations were the lowest in A1 (approximately 0.4 cm^{-3}, with a corresponding PM$_{coarse}$ of 0.9 µg/m^3) and D1 (approximately 0.4 cm^{-3}, with a corresponding PM$_{coarse}$ of 1.3 µg/m^3) and the highest was in H2 (approximately 5.2 cm^{-3}, with a corresponding PM$_{coarse}$ of 39.9 µg/m^3) and the second highest was in H1 (approximately 2.5 cm^{-3}, with a corresponding PM$_{coarse}$ of 17.3 µg/m^3). As for A2, GFA1, and GFA3, the coarse PN concentrations were approximately 0.9 cm^{-3} for each of the dwellings, but the corresponding PM$_{coarse}$ was about 6.3, 3.5, and 5.6 µg/m^3, respectively. The similarity in the coarse PN concentrations, compared to the differences observed for the PM$_{coarse}$ concentrations, in these dwellings is an indication of differences in the coarse size fraction of the indoor particle size distributions. This likely reflects differences in indoor emission sources of coarse particles among the dwellings. For example, H2 had the highest coarse PN and PM concentrations which could be explained by the existence of pets (more than two cats), in addition to the geographical location of this dwelling, which was close to an arid area in southeast Amman, where dust events and coarse particle resuspension are common.

3.3.2. Overall Mean Indoor Particle Number and Mass Size Distributions

The overall mean particle number and mass size distributions were calculated for each dwelling for the entire winter measurement campaign (Figures 6 and 7). This includes periods with and without indoor activities. In the following section, we will present and discuss the characteristics of the indoor particle number and mass size distributions during different indoor activities. Each dwelling had a unique set of particle number and mass size distributions that reflected the indoor aerosol emission sources associated with the inhabitants' activities, heating processes, and dwelling conditions. For example, among all dwellings, the lowest UFP concentrations were observed in apartment A2 because combustion processes (i.e., cooking using a natural gas stove) were minimal and the indoor space was heated via AC units. GFA2 had the second lowest UFP concentrations because the heating was via water-based central heating and, occasionally, electric heaters. Furthermore, both A2 and GFA2 were nonsmoking dwellings.

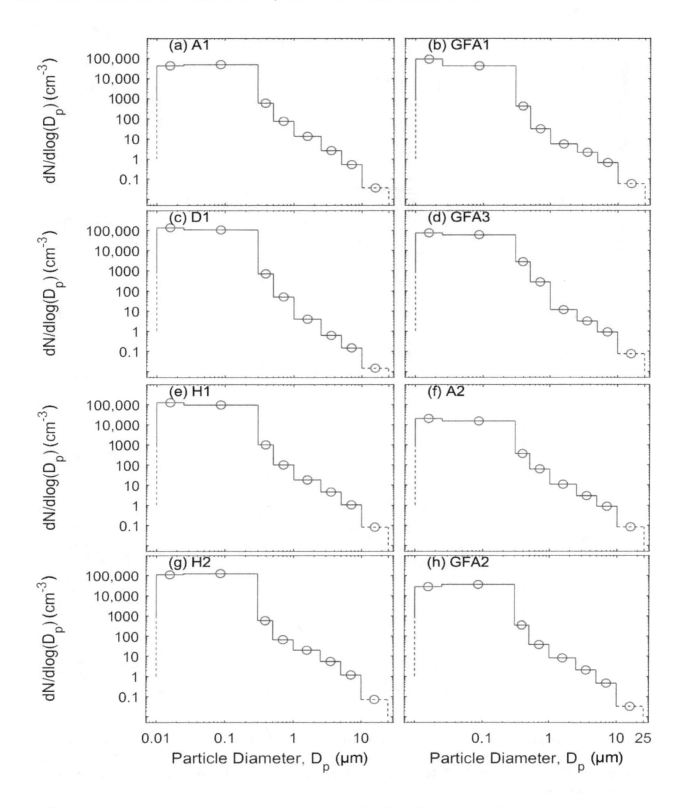

Figure 6. Mean particle number size distributions calculated for the entirety of the winter measurement campaign at each dwelling: (**a**) apartment A1, (**b**) ground floor apartment GFA1, (**c**) duplex D1, (**d**) ground floor apartment GFA3, (**e**) house H1, (**f**) apartment A2, (**g**) house H2, and (**h**) ground floor apartment GFA2.

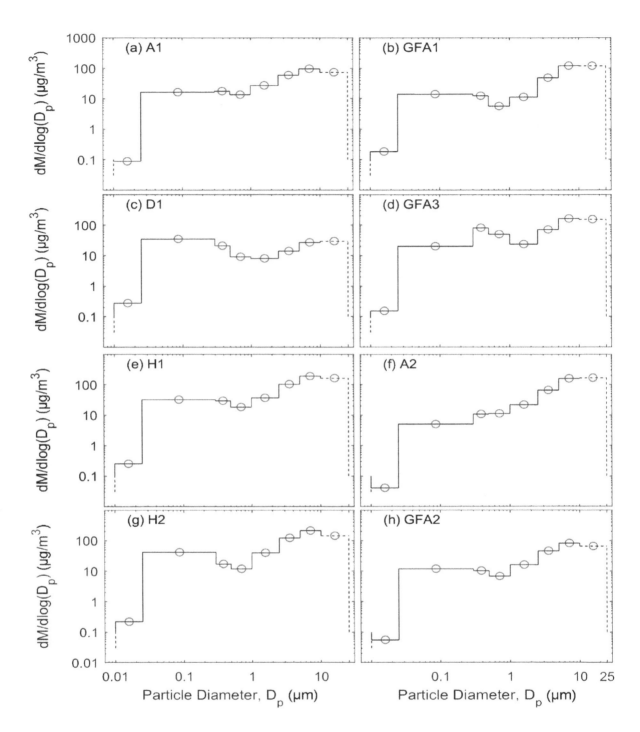

Figure 7. Mean particle mass size distributions calculated for the entirety of the winter measurement campaign at each dwelling: (**a**) apartment A1, (**b**) ground floor apartment GFA1, (**c**) duplex D1, (**d**) ground floor apartment GFA3, (**e**) house H1, (**f**) apartment A2, (**g**) house H2, and (**h**) ground floor apartment GFA2.

Indoor combustion processes had a pronounced impact on submicron particle concentrations, especially UFPs. For example, the impact of using kerosene heaters was evident in A1, D1, GFA1, and H2. Similarly, the impact of using natural gas heaters was evident in A1, GFA1, GFA3, and H1. Shisha smoking was reported in D1, GFA3, and H1, and the impact can be seen in the high concentrations of UFPs that were measured. D1 never obtained a stable background aerosol concentration during the nighttime likely due to traces of the kerosene heater and shisha smoking.

Indoor Particle Concentrations, Size Distributions and Exposures in Middle Eastern Microenvironments
151

3.3.3. The Impact of Indoor Activities on Indoor Particle Size Distributions and Concentrations

As listed in Table 1, the heating processes reported in this study included both combustion (natural gas heater and/or kerosene heater) and non-combustion (central heating, electric, and air conditioning). The cooking activities were reported on stoves using natural gas. The use of microwaves, coffee machines, and toasters were very rare. Table 7 presents a classification of selected activities and the mean PN and PM concentrations during these activities. The location (i.e., dwelling) and duration of the activities are listed in Table S1. Figures S9–S17 in the supplementary material present the mean particle number and mass size distributions during these activities. In this section, the reported PM concentrations were calculated from the particle mass size distributions by assuming spherical particles of unit density, as previously discussed.

Table 7. Classification of indoor activities and corresponding particle number and mass concentrations. Combustion heating is denoted as (Heat.) and the types are natural gas heater (NG) and kerosene heater (K). Cooking on a natural gas stove is denoted as (Stov.) and smoking cigarettes is denoted by (Cig.).

Combustion		Smoking		Non-Combustion		Additional Activity	$PM_{2.5}$ ($\mu g/m^3$)	PM_{10} ($\mu g/m^3$)	PN_1 ($\times 10^3$ cm^{-3})	PN_{10-1} (cm^{-3})
Heat.	Stov.	Shisha	Cig.	Heat.	Other					
√ (NG)							54 ± 26	64 ± 27	214 ± 71	1 ± 0
√ (NG)	√						70 ± 15	81 ± 17	274 ± 38	4 ± 1
√ (NG)	√					Grill burger/sausage	378 ± 101	2094 ± 882	383 ± 82	131 ± 47
√ (NG)	√						9 ± 2	19 ± 3	85 ± 13	1 ± 0
√ (NG)	√						13 ± 7	16 ± 7	68 ± 11	null
√ (NG)	√		√				40 ± 8	189 ± 57	91 ± 18	8 ± 2
√ (NG)	√		√				98 ± 26	158 ± 51	151 ± 37	6 ± 3
√ (NG)	√	√					173 ± 41	424 ± 152	245 ± 53	36 ± 12
√ (NG)	√				√	15 people	65 ± 17	374 ± 91	169 ± 52	13 ± 3
√ (K)	√						130 ± 15	458 ± 110	318 ± 53	27 ± 9
√ (K)	√						82 ± 24	154 ± 60	220 ± 78	7 ± 5
√ (K)	√						78 ± 17	141 ± 36	236 ± 52	5 ± 3
√ (K)	√						43 ± 17	91 ± 60	174 ± 62	5 ± 5
√ (K)	√						99 ± 13	119 ± 14	320 ± 45	1 ± 0
√ (K)	√	√					118 ± 33	139 ± 42	397 ± 60	4 ± 8
√ (K)	√	√					72 ± 24	92 ± 30	330 ± 46	2 ± 1
√ (NG)	√	√ ×2					139 ± 27	288 ± 114	343 ± 72	15 ± 10
√ (NG)	√	√					75 ± 18	226 ± 76	198 ± 47	14 ± 5
√ (NG)	√	√					61 ± 26	168 ± 60	154 ± 39	8 ± 3
√ (NG)	√	√	√				92 ± 33	189 ± 46	123 ± 34	9 ± 6
√ (NG)	√	√ ×2	√				132 ± 31	291 ± 61	242 ± 77	13 ± 5
	√					Cooking soup	40 ± 11	76 ± 17	144 ± 40	3 ± 1
	√					Making chai latte	41 ± 13	49 ± 13	160 ± 44	1 ± 0
	√			√ (C)		Intensive cooking	76 ± 41	191 ± 75	116 ± 29	14 ± 10
	√			√ (C)		Intensive cooking	85 ± 32	181 ± 56	207 ± 78	11 ± 3
	√			√ (C)		Intensive cooking	88 ± 31	201 ± 32	183 ± 91	12 ± 2
	√			√ (C)		Making tea	31 ± 10	52 ± 11	117 ± 43	1 ± 0
	√			√ (C)		Making tea + coffee	16 ± 4	42 ± 10	46 ± 13	1 ± 0
	√			√ (AC)		Intensive cooking	62 ± 19	112 ± 40	74 ± 28	11 ± 5
				√ (AC)		AC operation	10 ± 3	61 ± 28	12 ± 4	3 ± 1
				√ (C)	√	Microwave	17 ± 5	44 ± 11	47 ± 17	1 ± 0
					√	Vacuuming	25 ± 7	181 ± 64	47 ± 15	9 ± 3
					√	Brew coffee	7 ± 2	31 ± 21	11 ± 5	1 ± 1
					√	Brew coffee + toast	14 ± 10	18 ± 11	42 ± 29	null
					√	Toaster	15 ± 6	23 ± 7	44 ± 21	8 ± 2

Cooking Activities without Combustion Processes

Cooking activities were the most commonly reported indoor emission source in all eight dwellings. Periodically, they were reported in the absence of combustion processes (such as a natural gas stove or heating). The non-combustion cooking activities included: microwave (GFA2, Figure S17), brewing coffee (A1, Figure S10), and toasting bread (A1, Figure S10). When compared to the background concentrations (i.e., in the absence of indoor activities), the concentrations during these activities had a minor impact on the indoor air quality in each dwelling.

Brewing coffee had the smallest impact on indoor aerosol concentrations, with a mean calculated $PM_{2.5}$ concentration of approximately 7 $\mu g/m^3$ (submicron PN concentration of 1.1×10^4 cm^{-3}) and mean calculated PM_{10} concentration of approximately 31 $\mu g/m^3$ (coarse PN concentration of 1 cm^{-3}). Using the toaster doubled the $PM_{2.5}$ concentration and increased the submicron PN concentration four-fold. However, it had a negligible impact on the coarse PN and PM concentrations. Using the microwave had a similar impact on concentrations of fine particles as that observed when using a toaster.

Cooking Activities in the Absence of Combustion Heating Processes

Cooking on a stove (natural gas) can be classified as light or intensive. Light cooking activities were reported in dwelling A1 as cooking soup and making chai latte (Figure S10). During these two activities, the mean calculated $PM_{2.5}$ concentration was approximately 40 $\mu g/m^3$. The mean submicron PN concentration was approximately 1.4×10^5 cm^{-3} and 1.6×10^5 cm^{-3} during cooking soup and making chai latte, respectively. The corresponding calculated PM_{10} concentrations were approximately 144 $\mu g/m^3$ and 160 $\mu g/m^3$ and the coarse PN concentrations were approximately 3 cm^{-3} and 1 cm^{-3}, respectively. Here, the differences in the PM_{10} and coarse PN concentrations were unlikely due to the cooking processes, but rather driven by occupancy and occupant movement-induced particle resuspension near the instruments, which was more intense during cooking soup.

Light cooking activities (such as making tea and/or coffee) were also reported in GFA2, which had a central heating system. During the making of tea and coffee, the mean calculated $PM_{2.5}$ concentrations were approximately 16 $\mu g/m^3$ and 31 $\mu g/m^3$, respectively (Figure S17). The mean submicron PN concentrations were approximately 1.2×10^5 cm^{-3} and 4.6×10^4 cm^{-3}, respectively. The corresponding calculated PM_{10} concentrations were approximately 52 $\mu g/m^3$ and 42 $\mu g/m^3$, respectively, and the coarse PN concentrations were about 1 cm^{-3}. This indicates that similar activities might have different impacts on particle concentrations depending on the indoor conditions and the way in which the activity was conducted. For example, variability in dwelling ventilation may play a role, as well as the burning intensity of the natural gas stove.

Intensive cooking activities were reported in dwelling GFA2 (Figure S17, central heating) and A2 (Figure S15, AC heating). Indoor aerosol concentrations during these intensive cooking activities were higher than those observed during light cooking activities (in the absence of combustion heating processes). For example, the mean calculated $PM_{2.5}$ concentrations were between 62 $\mu g/m^3$ and 88 $\mu g/m^3$. The mean submicron PN concentrations were between 7.4×10^4 cm^{-3} and 2.1×10^5 cm^{-3}. The corresponding mean calculated PM_{10} concentrations were between 112 $\mu g/m^3$ and 201 $\mu g/m^3$ and the mean coarse PN concentrations were between 3 cm^{-3} and 14 cm^{-3}.

Concurrent Cooking Activities and Combustion Heating Processes

Periodically, the cooking activities occurred concurrently with a combustion heating process (natural gas or kerosene heaters). All of these cooking activities, aside from two, did not report the type of cooking; therefore, it was not possible to classify them as light or intensive cooking. One of the activities was very intensive cooking (grilling burger and sausages) and the other one was a birthday party (candles burning with more than 15 people in the living room). During cooking activities accompanied by a natural gas heater, the mean calculated $PM_{2.5}$ concentrations were between 9 $\mu g/m^3$ and 70 $\mu g/m^3$ (submicron PN concentrations between 6.8×10^4 cm^{-3} and 2.7×10^5 cm^{-3}). The corresponding mean calculated PM_{10} concentrations were between 16 $\mu g/m^3$ and 81 $\mu g/m^3$.

Grilling had a significant impact on indoor aerosol concentrations: the mean calculated $PM_{2.5}$ concentration was approximately 378 $\mu g/m^3$ (submicron PN concentration of 3.8×10^5 cm^{-3}) and the mean calculated PM_{10} concentration was approximately 2100 $\mu g/m^3$ (mean coarse PN concentration of 130 cm^{-3}). The birthday party event had a clear impact on both submicron and micron aerosol concentrations: the mean calculated $PM_{2.5}$ concentration was approximately 65 $\mu g/m^3$ (submicron PN concentration of 1.7×10^5 cm^{-3}) and mean calculated PM_{10} concentration was 374 $\mu g/m^3$. Using a kerosene heater instead of a natural gas heater further elevated the concentrations of indoor aerosols. During these activities, the mean calculated $PM_{2.5}$ concentrations were between 43 $\mu g/m^3$ and 130 $\mu g/m^3$

(submicron PN concentration between 1.7×10^5 cm^{-3} and 3.2×10^5 cm^{-3}). The corresponding mean calculated PM$_{10}$ concentrations were between 90 µg/m^3 and 460 µg/m^3.

Indoor Smoking of Shisha and Tobacco

Smoking indoors is prohibited in Jordan. However, this is often violated in many indoor environments in the country. In this study, shisha smoking and/or tobacco smoking was reported in three dwellings (GFA3, H1, and D1). It was not possible to separate the smoking events from the combustion processes used for heating or cooking. Therefore, the concentrations reported here were due to a combination of smoking and heating/cooking activities.

Tobacco smoking increased indoor aerosol concentrations as follows: the mean calculated PM$_{2.5}$ concentrations were between 40 µg/m^3 and 100 µg/m^3 (submicron PN concentrations between 9×10^4 cm^{-3} and 1.5×10^5 cm^{-3}). The corresponding mean calculated PM$_{10}$ concentrations were between 160 µg/m^3 and 190 µg/m^3 (mean coarse PN concentrations between 6 cm^{-3} and 8 cm^{-3}). Shisha smoking had a more pronounced impact on indoor aerosol concentrations compared to tobacco smoking. The mean calculated PM$_{2.5}$ concentrations were between 60 µg/m^3 and 140 µg/m^3 (submicron PN concentrations between 1.2×10^5 cm^{-3} and 4×10^5 cm^{-3}). The corresponding mean calculated PM$_{10}$ concentrations were between 90 µg/m^3 and 290 µg/m^3 (mean coarse PN concentrations between 2 cm^{-3} and 15 cm^{-3}).

For shisha smoking, the tobacco is mixed with honey (or sweeteners), oil products (such as glycerin), and flavoring products. Charcoal is used as the source of heat to burn the shisha tobacco mixture. Usually, the charcoal is heated up indoors on the stove prior to the shisha smoking event. Shisha and cigarette smoking produces a vast range of pollutants in the form of primary and secondary particulate and gaseous pollution [49–58]. It was also reported that cigarette and shisha smoke may contain compounds of microbiological origin, in addition to hundreds of compounds of known carcinogenicity and inhalation toxicity [49].

3.4. Concentrations of Selected Gaseous Pollutants in Jordanian Dwellings

The indoor activities documented in the eight dwellings were associated with emissions of gaseous pollutants for which exceptionally high concentrations were observed (Figures S1–S8). For example, the shisha smoking and preceding preparation (i.e., charcoal combustion) were associated with CO concentrations that reached as high as 10 ppm in D1 and GFA3. The CO concentrations were further elevated in H1, with concentrations approaching 100 ppm. Emissions of SO$_2$ were also recorded in D1 during charcoal combustion that accompanied shisha smoking. During shisha smoking, the CO concentrations exceeded the exposure level of 6 ppm due to smoking a single cigarette, as reported by Breland et al. [56], and 2.7 ppm as reported by Eissenberg and Shihadeh [52]. Previous studies have reported CO concentrations in the range of 24–32 ppm during shisha smoking events [51–53].

The eight dwellings exhibited variable concentrations of TVOCs, NO$_2$, and HCHO. For instance, TVOC concentrations were in the range of 100–1000 ppm in A2 and H2, whereas they were in the range of 1000–10,000 ppm in all ground floor apartments (GFA1, GFA2, and GFA3). NO$_2$ concentrations were in the range of 0.01–1 ppm in the duplex apartment (D1), ground floor apartments (GFA1, GFA2, and GFA3), and houses (H1 and H2). HCHO concentrations were in the range of 0.01–1 ppm in A2 and GFA1 and reached as high as 5 ppm in H2. O$_3$ was not detected in any of the dwellings. It should be noted that the gaseous pollutant concentrations presented here are estimates and are likely uncertain due to technical limitations of the low-cost sensing module employed.

3.5. Indoor Versus Outdoor Particle Concentrations

It is important to note that the indoor aerosol measurement periods at each dwelling were short during the winter campaign. Outdoor aerosol measurements were made on a few occasions at each dwelling; however, they were not of sufficient length to make meaningful conclusions about the aerosol indoor-to-outdoor relationship. However, comprehensive measurements of ambient aerosols have been made in the urban background in Amman [40,41,59–62], for which comparisons with the indoor measurements presented in this study can be made.

In the urban background atmosphere of Amman [62], outdoor PN concentrations were typically higher during the winter compared to the summer; the ratio can be 2–3 based on the daily means. Based on the hourly mean, the outdoor PN concentration had a clear diurnal and weekly pattern, with high concentrations during the workdays, especially during traffic rush hours. For example, the PN concentration diurnal pattern was characterized by two peaks: morning and afternoon. The afternoon peak (wintertime highest concentration range of 3×10^4–3.5×10^4 cm^{-3}) was rather similar on all weekdays; however, the first peak was higher on workdays compared to weekends (wintertime highest concentration range of 4.5×10^4–6.5×10^4 cm^{-3}). The lowest outdoor concentrations were typically observed between 3:00 to 6:00 in the morning, when they are as low as 1.8×10^4 cm^{-3} during the wintertime.

When compared to the results reported in this study (Tables 4–7), the mean indoor PN concentrations were generally higher than those outdoors during the daytime, when indoor activities were taking place. For example, PN concentrations inside all dwellings were less than 1.5×10^4 cm^{-3} between midnight and early morning; i.e., in the absence of indoor activities. However, the overall mean PN concentrations during the winter campaign inside the studied dwellings were in the range of 1.6×10^4–1.3×10^5 cm^{-3}. Looking at the mean concentrations during the indoor activities, the PN concentrations were as high as 4.7×10^4 cm^{-3} during non-combustion cooking activities. During cooking activities conducted on a natural gas stove, the PN concentrations were in the range of 4.6×10^4–3.8×10^5 cm^{-3}. The combination of cooking activities and combustion processes (as the main source of heating) increased the PN concentrations to be in the range of 6.8×10^4–2.7×10^5 cm^{-3}. Grilling sausages and burger indoors was associated with a substantial increase in mean PN concentrations, with levels reaching as high as 3.8×10^5 cm^{-3} (PM$_{2.5}$ = 378 µg/m^3 and PM$_{10}$ = 2094 µg/m^3). Both tobacco and shisha smoking were also associated with significant increases in PN concentrations, with levels reaching 9.1×10^4–4.0×10^5 cm^{-3}.

It is very well documented in the literature that the temporal variation in indoor aerosol concentrations closely follows those outdoors in the absence of indoor activities [20,30,32,63–74]. As such, the aerosol indoor-to-outdoor relationship depends on the size-resolved particle penetration factor for the building envelope, the ventilation and infiltration rates, and the size-resolved deposition rate onto available indoor surfaces [20,30,64]. As can be seen here, and also reported in previous studies, indoor aerosol emission sources, which are closely connected to human activities indoors, produce aerosol concentrations that are usually several times higher than those found outdoors [17,75–77]. Indoor aerosol sources can thus have a significant adverse impact on human health given that people spend the majority of their time indoors [10,11,32].

4. Conclusions

Indoor air quality has been given very little attention in the Middle East. Residential indoor environments in Jordan have unique characteristics with respect to size, ventilation modes, occupancy, activities, cooking styles, and heating processes. These factors vary between the winter and summer. In this study, we reported the results of one of the first comprehensive indoor aerosol measurement campaigns conducted in Jordanian indoor environments. Our methodology was based on the use of portable aerosol instruments covering different particle diameter ranges, from which we could investigate particle number and mass size distributions during different indoor activities. We focused on standard particle size fractions (submicron versus micron, fine versus coarse). The study provides valuable information regarding exposure levels to a wide range of pollutant sources that are commonly found in Jordanian dwellings.

In the absence of indoor activities, indoor PN concentrations varied among the dwellings and were in the range of 6×10^3–1.5×10^4 cm^{-3} (corresponding PM$_{2.5}$ of 5–12 µg/m^3). The coarse PN concentrations were in the range of 0.4–5.2 cm^{-3} (corresponding PM$_{coarse}$ of 0.9–39.9 µg/m^3). Indoor activities significantly impacted indoor air quality by increasing exposure to particle concentrations that exceeded what could be observed outdoors. Non-combustion cooking activities (microwave, brewing coffee, and toasting bread) had the smallest impact on indoor aerosol concentrations. During such activities, the PN concentrations were in the range of 1.1×10^4–4.7×10^4 cm^{-3}, PM$_{2.5}$ concentrations were in the range of 7–25 μg/m^3, micron PN concentrations were in the range of 1–9 cm^{-3}, and PM$_{10}$ concentrations were in the range of 44–181 μg/m^3. Cooking on a natural gas stove had a more pronounced impact on indoor aerosol concentrations compared to non-combustion cooking, with measured PN concentrations in the range of 4.6×10^4–2.1×10^5 cm^{-3}, PM$_{2.5}$ concentrations in the range of 16–88 μg/m^3, micron PN concentrations in the range of 1–14 cm^{-3}, and PM$_{10}$ concentrations in the range of 42–201 μg/m^3.

The combination of cooking activities (varying in type and intensity) with heating via combustion of natural gas or kerosene had a significant impact on indoor air quality. PN concentrations were in the range of 6.8×10^4–2.7×10^5 cm^{-3}, PM$_{2.5}$ concentrations were in the range of 9–130 μg/m^3, micron PN concentrations were in the range of 1–27 cm^{-3}, and PM$_{10}$ concentrations were in the range of 16–458 μg/m^3. Grilling sausages and burgers indoors was identified as an extreme event, with mean PN concentration reaching 3.8×10^5 cm^{-3}, PM$_{2.5}$ concentrations reaching 378 μg/m^3, micron PN concentrations reaching 131 cm^{-3}, and PM$_{10}$ concentrations reaching 2094 μg/m^3.

Both tobacco and shisha smoking adversely impacted indoor air quality in Jordanian dwellings, with the latter being more severe. During tobacco smoking, the PN concentrations were in the range of 9.1×10^4–1.5×10^5 cm^{-3}, PM$_{2.5}$ concentrations were in the range of 40–98 μg/m^3, micron PN concentrations were in the range of 6–8 cm^{-3}, and PM$_{10}$ concentrations were in the range of 158–189 μg/m^3. During shisha smoking, the PN concentrations were in the range of 1.2×10^5–4.0×10^5 cm^{-3}, PM$_{2.5}$ concentrations were in the range of 61–173 μg/m^3, micron PN concentrations were in the range of 2–36 cm^{-3}, and PM$_{10}$ concentrations were in the range of 92–424 μg/m^3.

The above-mentioned concentration ranges were reported during the winter campaign, when the houses were tightly closed for heating purposes. Indoor aerosol concentrations during the summer campaign were generally lower. The overall mean PN concentrations during the summer campaign were less than 2×10^4 cm^{-3} and PM$_{2.5}$ concentrations were less than 50 μg/m^3. Some of the reported indoor activities were accompanied with high concentrations of gaseous pollutants. TVOC concentrations exceeded 100 ppm. NO$_2$ concentrations were in the range of 0.01–1 ppm. HCHO concentrations were in the range of 0.01–5 ppm. During shisha smoking and preceding preparation (e.g., charcoal combustion), the mean CO concentrations reached as high as 100 ppm.

There are a number of limitations of the present study: (1) the measurement periods were short at each dwelling during the winter campaign, (2) the sample population was small (eight dwellings), and (3) outdoor measurements were only conducted on a few occasions for short periods. These limitations can be addressed in future indoor–outdoor measurement campaigns in Jordan. However, indoor aerosol concentrations were compared to long-term outdoor PN measurements conducted in past studies in Jordan.

The results of this study can offer several practical recommendations for improving indoor air quality in Jordanian indoor environments: source control by prohibiting the smoking of tobacco and shisha indoors, improved ventilation during the use of fossil fuel combustion for heating, and cooking with a natural gas stove under a kitchen hood.

Supplementary Materials: Table S1: Average particle mass and number concentrations (mean ± stdev) during selected indoor activities. Figure S1: Aerosol concentrations inside apartment A1 during the winter campaign (23–25 December 2018). Figure S2: Aerosol concentrations inside ground floor apartment GFA1 during the winter campaign (25–27 December 2018). Figure S3: Aerosol concentrations inside duplex apartment D1 during the winter campaign (28–30 December 2018). Figure S4: Aerosol concentrations inside ground floor apartment GFA3 during the winter campaign (31 December 2018–2 January 2019). Figure S5: Aerosol concentrations inside house H1 during the winter campaign (2–4 January 2019). Figure S6: Aerosol concentrations inside apartment A2 during the winter campaign (4–5 January 2019). Figure S7: Aerosol concentrations inside house H2 during the winter campaign (6–9 January 2019). Figure S8: Aerosol concentrations inside ground floor apartment GFA2 during the winter campaign (9–12 January

2019). Figure S9: Mean particle number size distributions and corresponding particle mass size distributions in the absence of indoor activities during the winter campaign at each study site. Figure S10: Mean particle number size distributions and particle mass size distributions during selected activities reported inside Apartment A1 during the winter campaign (23–25 December2018). Figure S11: Mean particle number size distributions and particle mass size distributions during selected activities reported inside ground floor apartment GFA1 during the winter campaign (25–27 December 2018). Figure S12: Mean particle number size distributions and particle mass size distributions during selected activities reported inside duplex D1 during the winter campaign (28–30 December 2018). Figure S13: Mean particle number size distributions and particle mass size distributions during selected activities reported inside ground floor apartment GFA3 during the winter campaign (31 December 2018–2 January 2019). Figure S14: Mean particle number size distributions and particle mass size distributions during selected activities reported inside house H1 during the winter campaign (2–4 January 2019). Figure S15: Mean particle number size distributions and particle mass size distributions during selected activities reported inside apartment A2 during the winter campaign (4–5 January 2019). Figure S16: Mean particle number size distributions and particle mass size distributions during selected activities reported inside house H2 during the winter campaign (6–9 January 2019). Figure S17: Mean particle number size distributions and particle mass size distributions during selected activities reported inside ground floor apartment GFA2 during the winter campaign (9–12 January 2019).

Author Contributions: Conceptualization, T.H., M.M., A.A.-H., and O.A.; methodology, T.H., O.J., K.A., A.A., and O.A.; validation, T.H.; formal analysis, T.H., O.J., and A.A.; investigation, T.H.; resources, T.H. and M.M.; data curation, T.H., O.J., K.A., and A.A.; writing—original draft preparation, T.H. and A.A.-H.; writing—review and editing, T.H., B.E.B., A.J.K., J.L., M.M., and A.A.-H.; visualization, T.H.; supervision, T.H. and A.A.-H.; project administration, T.H. and A.A.-H.; funding acquisition, T.H. and M.M. All authors have read and agreed to the published version of the manuscript.

Acknowledgments: The first author would like to thank the occupants for allowing the indoor measurement campaigns to be conducted in their dwellings. Some of them also helped in follow-up aerosol measurements and reporting of indoor activities. This manuscript was written and completed during the sabbatical leave of the first author (T.H.) that was spent at the University of Helsinki and supported by the University of Helsinki during 2019. Open access funding was provided by the University of Helsinki.

References

1.   World Health Organization (WHO). Ambient Air Pollution: A Global Assessment of Exposure and Burden of Disease. 2016. Available online: http://apps.who.int/iris/bitstream/10665/250141/1/9789241511353-eng.pdf?ua=1 (accessed on 3 December 2019).

2.   World Health Organization (WHO). Household Air Pollution and Health. 2018. Available online: https://www.who.int/news-room/fact-sheets/detail/household-air-pollution-and-health (accessed on 3 December 2019).

3.   Health Effects Institute (HEI). State of Global Air. In *Special Report*; Health Effects Institute: Boston, MA, USA, 2017.

4.   World Health Organization (WHO). The Right to Healthy Indoor Air. 2000. Available online: http://www.euro.who.int/en/health-topics/environment-and-health/air-quality/publications/pre2009/the-right-to-healthy-indoor-air (accessed on 3 December 2019).

5.   Odeh, I.; Hussein, T. Activity pattern of urban adult students in an Eastern Mediterranean Society. *Int. J. Environ. Res. Public Health* **2016**, *13*, 960. [CrossRef] [PubMed]

6.   Hussein, T.; Paasonen, P.; Kulmala, M. Activity pattern of a selected group of school occupants and their family members in Helsinki-Finland. *Sci. Total Environ.* **2012**, *425*, 289–292. [CrossRef] [PubMed]

7.   Schweizer, C.; Edwards, R.; Bayer-Oglesby, L.; Gauderman, W.; Ilacqua, V.; Jantunen, M.; Lai, H.K.; Nieuwenhuijsen, M.; Künzli, N. Indoor time-microenvironment-activity patterns in seven regions of Europe. *J. Expo. Sci. Environ. Epidemiol.* **2007**, *17*, 170–181. [CrossRef]

8.   Klepeis, N.E.; Nelson, W.C.; Ott, W.R.; Robinson, J.P.; Tsang, A.M.; Switzer, P.; Behar, J.V.; Hem, S.C.; Engelmann, W.H. The national human activity pattern survey NHAPS: A resource for assessing exposure to environmental pollutants. *J. Expo. Anal. Environ. Epidemiol.* **2001**, *11*, 231–252. [CrossRef] [PubMed]

9.   McCurdy, T.; Glen, G.; Smith, L.; Lakkadi, Y. The National Exposure Research Laboratory's Consolidated Human Activity Database. *J. Expo. Anal. Environ. Epidemiol.* **2000**, *10*, 566–578. [CrossRef] [PubMed]

10.   Koivisto, A.J.; Kling, K.I.; Hänninen, O.; Jayjock, M.; Löndahl, J.; Wierzbicka, A.; Fonseca, A.S.; Uhrbrand, K.; Boor, B.E.; Jiménez, A.S.; et al. Source specific exposure and risk assessment for indoor aerosols. *Sci. Total Environ.* **2019**, *668*, 13–24. [CrossRef] [PubMed]

11.   Jones, A.P. Indoor air quality and health. *Atmos. Environ.* **1999**, *33*, 4535–4564. [CrossRef]

12. Streets, D.G.; Yan, F.; Chin, M.; Diehl, T.; Mahowald, N.; Schultz, M.; Wild, M.; Wu, Y.; Yu, C. Anthropogenic and natural contributions to regional trends in aerosol optical depth, 1980–2006. *J. Geophys. Res.* **2009**, *114*, D00D18. [CrossRef]

13. Karagulian, F.; Belis, C.A.; Francisco, C.; Dora, C.; Prüss-Ustün, A.M.; Bonjour, S.; Adair-Rohani, H.; Amann, M. Contributions to cities' ambient particulate matter (PM)—A systematic review of local source contributions at global level. *Atmos. Environ.* **2015**, *120*, 475–483. [CrossRef]

14. Abadie, M.O.; Blondeau, P. PANDORA database: A compilation of indoor air pollutant emissions. *HVAC&R Res.* **2011**, *17*, 602–613.

15. Boor, B.E.; Spilak, M.P.; Laverge, J.; Novoselac, A.; Xu, Y. Human exposure to indoor air pollutants in sleep microenvironments: A literature review. *Build. Environ.* **2017**, *125*, 528–555. [CrossRef]

16. Chen, C.; Zhao, B. Review of relationship between indoor and outdoor particles: I/O ratio, infiltration factor and penetration factor. *Atmos. Environ.* **2011**, *45*, 275–288. [CrossRef]

17. Hussein, T.; Wierzbicka, A.; Löndahl, J.; Lazaridis, M.; Hänninen, O. Indoor aerosol modeling for assessment of exposure and respiratory tract deposited dose. *Atmos. Environ.* **2015**, *106*, 402–411. [CrossRef]

18. Koivisto, J.; Hussein, T.; Niemelä, R.; Tuomi, T.; Hämeri, K. Impact of particle emissions of new laser printers on a modeled office room. *Atmos. Environ.* **2010**, *44*, 2140–2146. [CrossRef]

19. Lin, C.-H.; Lo, P.-Y.; Wu, H.-D.; Chang, C.; Wang, L.-C. Association between indoor air pollution and respiratory disease in companion dogs and cats. *J. Vet. Intern. Med.* **2018**, *32*, 1259–1267. [CrossRef] [PubMed]

20. Morawska, L.; Afshari, A.; Bae, G.N.; Buonanno, G.; Chao, C.Y.H.; Hänninen, O.; Hofmann, W.; Isaxon, C.; Jayaratne, E.R.; Pasanen, P.; et al. Indoor aerosols: From personal exposure to risk assessment. *Indoor Air* **2013**, *23*, 462–487. [CrossRef] [PubMed]

21. Morawska, L.; He, C.; Johnson, G.; Jayaratne, R.; Salthammer, T.; Wang, H.; Uhde, E.; Bostrom, T.; Modini, R.; Ayoko, G.; et al. An Investigation into the characteristics and formation mechanisms of particles originating from the operation of laser printers. *Environ. Sci. Technol.* **2009**, *43*, 1015–1022. [CrossRef]

22. Sangiorgi, G.; Ferrero, L.; Ferrini, B.S.; Lo Porto, C.; Perrone, M.G.; Zangrando, R.; Gambaro, A.; Lazzati, Z.; Bolzacchini, E. Indoor airborne particle sources and semi-volatile partitioning effect of outdoor fine PM in offices. *Atmos. Environ.* **2013**, *65*, 205–214. [CrossRef]

23. Wensing, M.; Schripp, T.; Uhde, E.; Salthammer, T. Ultra-fine particles release from hardcopy devices: Sources, real-room measurements and efficiency of filter accessories. *Sci. Total Environ.* **2008**, *407*, 418–427. [CrossRef]

24. He, C.; Morawska, L.; Hitchins, J.; Gilbert, D. Contribution from indoor sources to particle number and mass concentrations in residential houses. *Atmos. Environ.* **2004**, *38*, 3405–3415. [CrossRef]

25. He, C.; Morawska, L.; Taplin, L. Particle emission characteristics of office printers. *Environ. Sci. Technol.* **2007**, *41*, 6039–6045. [CrossRef] [PubMed]

26. Ren, Y.; Cheng, T.; Chen, J. Polycyclic aromatic hydrocarbons in dust from computers: One possible indoor source of human exposure. *Atmos. Environ.* **2006**, *40*, 6956–6965. [CrossRef]

27. Afshari, A.; Matson, U.; Ekberg, L.E. Characterization of indoor sources of fine and ultrafine particles: A study conducted in a full-scale chamber. *Indoor Air* **2005**, *15*, 141–150. [CrossRef] [PubMed]

28. Weschler, C.J. Changes in indoor pollutants since the 1950s. *Atmos. Environ.* **2009**, *43*, 153–169. [CrossRef]

29. Madanat, H.; Barnes, M.D.; Cole, E.C. Knowledge of the effects of indoor air quality on health among women in Jordan. *Health Educ. Behav.* **2008**, *35*, 105–118. [CrossRef] [PubMed]

30. Hussein, T. Particle size distributions inside a university office in Amman, Jordan. *Jordan J. Phys.* **2014**, *7*, 73–83.

31. Hussein, T.; Dada, L.; Juwhari, H.; Faouri, D. Characterization, fate, and re-suspension of aerosol particles (0.3–10 µm): The effects of occupancy and carpet use. *Aerosol Air Qual. Res.* **2015**, *15*, 2367–2377. [CrossRef]

32. Hussein, T. Indoor-to-Outdoor relationship of aerosol particles inside a naturally ventilated apartment—A comparison between single-parameter analysis and indoor aerosol model simulation. *Sci. Total Environ.* **2017**, *596–597*, 321–330. [CrossRef]

33. Wang, Y.; Xing, Z.; Zhao, S.; Zheng, M.; Mu, C.; Du, K. Are emissions of black carbon from gasoline vehicles overestimated? Real-time, in situ measurement of black carbon emission factors. *Sci. Total Environ.* **2016**, *547*, 422–428. [CrossRef]

34. Jiang, R.T.; Acevedo-Bolton, V.; Cheng, K.C.; Klepeis, N.E.; Ott, W.R.; Hildemann, L.M. Determination of

response of real-time SidePak AM510 monitor to secondhand smoke, other common indoor aerosols, and outdoor aerosol. *J. Environ. Monit.* **2011**, *13*, 1695–1702. [CrossRef]

35. Matson, U.; Ekberg, L.E.; Afshari, A. Measurement of ultrafine particles: A comparison of two handheld condensation particle counters. *Aerosol Sci. Technol.* **2004**, *38*, 487–495. [CrossRef]

36. Lin, C.; Gillespie, J.; Schuder, M.D.; Duberstein, W.; Beverland, I.J.; Heal, M.R. Evaluation and calibration of Aeroqual series 500 portable gas sensors for accurate measurement of ambient ozone and nitrogen dioxide. *Atmos. Environ.* **2015**, *100*, 111–116. [CrossRef]

37. Cai, J.; Yan, B.; Ross, J.; Zhang, D.; Kinney, P.L.; Perzanowski, M.S.; Jung, K.; Miller, R.; Chillrud, S.N. Validation of MicroAeth® as a black carbon monitor for fixed-site measurement and optimization for personal exposure characterization. *Aerosol Air Qual. Res.* **2014**, *14*, 1–9. [CrossRef] [PubMed]

38. Cheng, Y.H.; Lin, M.H. Real-time performance of the micro-aeth AE51 and the effects of aerosol loading on its measurement results at a traffic site. *Aerosol Air Qual. Res.* **2013**, *13*, 1853–1863. [CrossRef]

39. Chung, A.; Chang, D.P.Y.; Kleeman, M.J.; Perry, K.; Cahill, T.A.; Dutcher, D.; McDougal, E.M.; Stroud, K. Comparison of real-time instruments used to monitor airborne particulate matter. *J. Air Waste Manag. Assoc.* **2001**, *51*, 109–120. [CrossRef] [PubMed]

40. Hussein, T.; Boor, B.E.; Dos Santos, V.N.; Kangasluoma, J.; Petäjä, T.; Lihavainen, H. Mobile aerosol measurement in the eastern Mediterranean—A utilization of portable instruments. *Aerosol Air Qual. Res.* **2017**, *17*, 1775–1786. [CrossRef]

41. Hussein, T.; Saleh, S.S.A.; dos Santos, V.N.; Abdullah, H.; Boor, B.E. Black carbon and particulate matter concentrations in eastern Mediterranean urban conditions—An assessment based on integrated stationary and mobile observations. *Atmosphere* **2019**, *10*, 323. [CrossRef]

42. Hämeri, K.; Koponen, I.K.; Aalto, P.P.; Kulmala, M. The particle detection efficiency of the TSI3007 condensation particle counter. *Aerosol Sci.* **2002**, *33*, 1463–1469. [CrossRef]

43. Maricq, M.M. Monitoring motor vehicle PM emissions: An evaluation of three portable low-cost aerosol instruments. *Aerosol Sci. Technol.* **2013**, *47*, 564–573. [CrossRef]

44. Nyarku, M.; Mazaheri, M.; Jayaratne, R.; Dunbabin, M.; Rahman, M.M.; Uhde, E.; Morawska, L. Mobile phones as monitors of personal exposure to air pollution: Is this the future? *PLoS ONE* **2018**, *13*, e0193150. [CrossRef]

45. Wang, X.; Chancellor, G.; Evenstad, J.; Farnsworth, J.; Hase, A.; Olson, G.; Sreenath, A.; Agarwal, J. A novel optical instrument for estimating size segregated aerosol mass concentration in real time. *Aerosol Sci. Technol.* **2009**, *43*, 939–950. [CrossRef]

46. Viana, M.; Rivas, I.; Reche, C.; Fonseca, A.S.; Pérez, N.; Querol, X.; Alastuey, A.; Álvarez-Pedrerol, M.; Sunyer, J. Field comparison of portable and stationary instruments for outdoor urban air exposure assessments. *Atmos. Environ.* **2015**, *123*, 220–228. [CrossRef]

47. Hussein, T.; Saleh, S.S.A.; dos Santos, V.N.; Boor, B.E.; Koivisto, A.J.; Löndahl, J. Regional inhaled deposited dose of urban aerosols in an eastern Mediterranean city. *Atmosphere* **2019**, *10*, 530. [CrossRef]

48. Huang, L.; Pui, Z.; Sundell, J. Characterizing the indoor-outdoor relationship of fine particulate matter in non-heating season for urban residences in Beijing. *PLoS ONE* **2015**, *10*, e0138559. [CrossRef]

49. Markowicz, P.; Löndahl, J.; Wierzbicka, A.; Suleiman, R.; Shihadeh, A.; Larsson, L. A study on particles and some microbial markers in waterpipe tobacco smoke. *Sci. Total Environ.* **2014**, *499*, 107–113. [CrossRef]

50. Daher, N.; Saleh, R.; Jaroudi, E.; Sheheitli, H.; Badr, T.; Sepetdjian, E.; Al Rashidi, M.; Saliba, N.; Shihadeh, A. Comparison of carcinogen, carbon monoxide, and ultrafine particle emissions from narghile waterpipe and cigarette smoking: Sidestream smoke measurements and assessment of second-hand smoke emission factors. *Atmos. Environ.* **2010**, *44*, 8–14. [CrossRef]

51. Maziak, W.; Rastam, S.; Ibrahim, I.; Ward, K.D.; Shihadeh, A.; Eissenberg, T. CO exposure, puff topography, and subjective effects in waterpipe tobacco smokers. *Nicotine Tobacco Res.* **2009**, *11*, 806–811. [CrossRef]

52. Eissenberg, T.; Shihadeh, A. Waterpipe tobacco and cigarette smoking direct comparison of toxicant exposure. *Am. J. Prev. Med.* **2009**, *37*, 518–523. [CrossRef]

53. El-Nachef, W.N.; Hammond, S.K. Exhaled carbon monoxide with waterpipe use in US students. *JAMA* **2008**, *299*, 36–38. [CrossRef]

54. Al Rashidi, M.; Shihadeh, A.; Saliba, N.A. Volatile aldehydes in the mainstream smoke of the narghile waterpipe. *Food Chem. Toxicol.* **2008**, *46*, 3546–3549. [CrossRef]

55. Sepetdjian, E.; Alan Shihadeh, A.; Saliba, N.A. Measurement of 16 polycyclic aromatic hydrocarbons in narghile waterpipe tobacco smoke. *Food Chem. Toxicol.* **2008**, *46*, 1582–1590. [CrossRef] [PubMed]

56. Breland, A.B.; Kleykamp, B.A.; Eissenberg, T. Clinical laboratory evaluation of potential reduced exposure products for smokers. *Nicotine Tobacco Res.* **2006**, *8*, 727–738. [CrossRef] [PubMed]

57. Shihadeh, A.; Saleh, R. Polycyclic aromatic hydrocarbons, carbon monoxide, "tar", and nicotine in the mainstream smoke aerosol of the narghile water pipe. *Food Chem. Toxicol.* **2005**, *43*, 655–661. [CrossRef] [PubMed]

58. Shihadeh, A. Investigation of mainstream smoke aerosol of the argileh water pipe. *Food Chem. Toxicol.* **2003**, *41*, 143–152. [CrossRef]

59. Hussein, T.; Halayka, M.; Abu Al-Ruz, R.; Abdullah, H.; Mølgaard, B.; Petäjä, T. Fine particle number concentrations in Amman and Zarqa during spring 2014. *Jordan J. Phys.* **2016**, *9*, 31–46.

60. Hussein, T.; Rasha, A.; Tuukka, P.; Heikki, J.; Arafah, D.; Kaarle, H.; Markku, K. Local air pollution versus short-range transported dust episodes: A comparative study for submicron particle number concentration. *Aerosol Air Qual. Res.* **2011**, *11*, 109–119. [CrossRef]

61. Saleh, S.S.A.; Shilbayeh, Z.; Alkattan, H.; Al-Refie, M.R.; Jaghbeir, O.; Hussein, T. Temporal variations of submicron particle number concentrations at an urban background site in Amman—Jordan. *Jordan J. Earth Environ. Sci.* **2019**, *10*, 37–44.

62. Hussein, T.; Dada, L.; Hakala, S.; Petäjä, T.; Kulmala, M. Urban aerosols particle size characterization in eastern Mediterranean conditions. *Atmosphere* **2019**, *10*, 710. [CrossRef]

63. Hussein, T.; Kulmala, M. Indoor aerosol modeling: Basic principles and practical applications. *Water Air Soil Pollut. Focus* **2008**, *8*, 23–34. [CrossRef]

64. Nazaroff, W.W. Indoor particle dynamics. *Indoor Air* **2004**, *14*, 175–183. [CrossRef]

65. Wierzbicka, A.; Bohgard, M.; Pagels, J.H.; Dahl, A.; Löndahl, J.; Hussein, T.; Swietlicki, E.; Gudmundsson, A. Quantification of differences between occupancy and total monitoring periods for better assessment of exposure to particles in indoor environments. *Atmos. Environ.* **2015**, *106*, 419–428. [CrossRef]

66. Buonanno, G.; Fuoco, F.C.; Marini, S.; Stabile, L. Particle re-suspension in school gyms during physical activities. *Aerosol Air Qual. Res.* **2012**, *12*, 803–813. [CrossRef]

67. Hussein, T.; Hämeri, K.; Kulmala, M. Long-term indoor-outdoor aerosol measurement in Helsinki, Finland. *Boreal Environ. Res.* **2002**, *7*, 141–150.

68. Shaughnessy, R.; Vu, H. Particle loadings and resuspension related to floor coverings in chamber and in occupied school environments. *Atmos. Environ.* **2012**, *55*, 515–524. [CrossRef]

69. Hussein, T.; Hämeri, K.; Aalto, P.; Asmi, A.; Kakko, L.; Kulmala, M. Particle size characterization and the indoor-to-outdoor relationship of atmospheric aerosols in Helsinki. *Scand. J. Work Health Environ.* **2004**, *30* (Suppl. 2), 54–62.

70. Ferro, A.R.; Kopperund, R.J.; Hildemann, L.M. Source strengths for indoor human activities that resuspend particulate matter. *Environ. Sci. Technol.* **2004**, *38*, 1759–1764. [CrossRef]

71. Hussein, T.; Korhonen, H.; Herrmann, E.; Hämeri, K.; Lehtinen, K.; Kulmala, M. Emission rates due to indoor activities: Indoor aerosol model development, evaluation, and applications. *Aerosol Sci. Technol.* **2005**, *39*, 1111–1127. [CrossRef]

72. Kubota, Y.; Higuchi, H. Aerodynamic Particle re-suspension due to human foot and model foot motions. *Aerosol Sci. Technol.* **2013**, *47*, 208–217. [CrossRef]

73. Hirsikko, A.; Kulmala, M.; Yli-Juuti, T.; Nieminen, T.; Hussein, T.; Vartiainen, E.; Laakso, L. Indoor and outdoor air ion and particle number size distributions in the urban background atmosphere of Helsinki, Finland. *Boreal Environ. Res.* **2007**, *12*, 295–310.

74. Lazaridis, M.; Eleftheriadis, K.; Ždímal, V.; Schwarz, J.; Wagner, Z.; Ondráček, J.; Drossinos, Y.; Glytsos, T.; Vratolis, S.; Torseth, K.; et al. Number concentrations and modal structure of indoor/outdoor fine particles in four European Cities. *Aerosol Air Qual. Res.* **2017**, *17*, 131–146. [CrossRef]

75. Hussein, T.; Hämeri, K.; Heikkinen, M.S.A.; Kulmala, M. Indoor and outdoor particle size characterization at a family house in Espoo—Finland. *Atmos. Environ.* **2005**, *39*, 3697–3709. [CrossRef]

76. Hussein, T.; Glytsos, T.; Ondráček, J.; Ždímal, V.; Hämeri, K.; Lazaridis, M.; Smolik, J.; Kulmala, M. Particle size characterization and emission rates during indoor activities in a house. *Atmos. Environ.* **2006**, *40*, 4285–4307. [CrossRef]

77. Maragkidou, A.; Jaghbeir, O.; Hämeri, K.; Hussein, T. Aerosol particles (0.3–10 µm) inside an educational workshop-emission rate and inhaled deposited dose. *Build. Environ.* **2018**, *140*, 80–89. [CrossRef]

9

A Promising Technological Approach to Improve Indoor Air Quality

Thomas Maggos [1,*], **Vassilios Binas** [2], **Vasileios Siaperas** [3], **Antypas Terzopoulos** [3], **Panagiotis Panagopoulos** [1] and **George Kiriakidis** [2]

[1]   Environmental Research Laboratory, NCSR "Demokritos", 15310 Ag. Paraskevi, Athens, Greece; ppanag.uoi@gmail.com

[2]   Institute of Electronic Structure and Laser, Foundation for Research and Technology, 70013 Heraclion, Crete, Greece; binasbill@iesl.forth.gr (V.B.); kiriakid@iesl.forth.gr (G.K.)

[3]   691 Industrial Base Factory, 19011 Avlonas, Greece; bsiaper@gmail.com (V.S.); antyterzo@gmail.com (A.T.)

*   Correspondence: tmaggos@ipta.demokritos.gr.

Abstract: Indoor Air quality (IAQ) in private or public environments is progressively recognized as a critical issue for human health. For that purpose the poor IAQ needs to be mitigated and immediate drastic measures must be taken. In environmental science and especially in advanced oxidation processes and technologies (AOPs-AOTs), photocatalysis has gained considerable interest among scientists as a tool for IAQ improvement. In the current study an innovative paint material was developed which exhibits intense photocatalytic activity under direct and diffused visible light for the degradation of air pollutants, suitable for indoor use. A laboratory and a real scale study were performed using the above innovative photo-paint. The lab test was performed in a special design photo-reactor while the real scale in a military's medical building. Nitrogen Oxide (NO) and Toluene concentration was monitored between "reference" rooms (without photo paint) and "green" rooms (with photo-paint) in order to estimate the photocatalytic efficiency of the photo-paint to degrade the above pollutants. Results of the study showed a decrease up to 60% and 16% for NO and toluene respectively under lab scale tests while an improvement of air quality up to 19% and 5% under real world conditions was achieved.

Keywords: IAQ improvement; photo-paint; NO; Toluene degradation

1. Introduction

Indoor air quality (IAQ) is an important determinant of human health, comfort and productivity. For that purpose, high quality indoor air is desirable. Indoor air pollution can be addressed through the two approaches of prevention and removal. The latter includes the use of air cleaning technologies especially in buildings where ventilation rates are being reduced in order to save energy. Indicative air cleaning technologies which have been developed during the recent years are: filtration and adsorption, electrostatic air purification, air filtration and gas adsorption filtration [1–3], ozonation [4–7], non-thermal plasma [8,9] and photocatalytic oxidation (PCO) [10–13]. PCO is a general air cleaning technology, which is able to degrade Volatile Organic Compounds (VOCs), such as aromatics, alkanes, odor compounds etc. Air cleaning photocatalytic technology is based on the principle that radiation of suitable wave-lengths can be absorbed by semiconductors, which leads to the creation of reactive oxygen species (ROS) that can degrade air pollutants. TiO_2 is the most commonly-used semiconductor in PCO research. However, over the last years, scientists have combined TiO_2 with other materials such as activated carbon and zeolite hybrid catalysts in order to enhance the PCO degradation of air pollutants [14–16]. Furthermore, numerous TiO_2 photocatalysts with different morphological designs have been developed: nanoparticles, nanotubes, hollow fibers and mesoporous. To this end, the need to evaluate the photocatalytic performance of the above materials in a common methodological approach has been raised and extensive research efforts have been devoted to it [17,18].

However, in most of these studies, only a single compound was tested, using a photocatalytic reactor and a methodological approach, which have been developed by the same lab that produced the material. It is well known that indoor air contains numerous contaminants; thus, tests of only one compound may be misleading. Furthermore, many studies proved the generation of by-products during the photocatalytic processes, such as formaldehyde, ozone, benzaldehyde, acetaldehyde etc. It is obvious that in some cases, by-products could be more harmful than the target pollutant [19–22]. For that purpose, and although photocatalytic technology is promising, the synthesis route of a photocatlytic material should be designed carefully in order to avoid contaminants which could lead to the formation and emission in the gas phase of intermediate products as they can be more hazardous than the target pollutant. The latter is clearly demonstrated during the evaluation of two photocatalytic air-purification in a mock-up air cabin. Although two symptoms, dizziness and claustrophobia, were reported to decrease when either one of the photocatalytic air-purification devices was operated, intermediates (acetaldehyde and formaldehyde) were detected as a result of ethanol photodegradation [23,24].

The present research addresses the indoor air purification study using a photocatalytic paint, which was tested under both laboratory and real world indoor conditions (application in building walls). A modified TiO_2 was chosen as photocatalyst in order to be activated by visible light which is the dominant spectrum in indoor environments. Traditional TiO_2 is activated only under UV light. More specifically a Mn-doped TiO_2 photocatalyst (powder) was used in the synthetic route of a photocatalytic paint production. The powder named TCM-1 [25] has already been successfully tested for the oxidation of air pollutants under indoor-like illumination conditions and when mixed either with calcareous or cementitious base matrices demonstrated a unique ability to efficiently degrade volatile organic compounds (VOCs) such as BTX (Benzene, Toluene, Xylene), formaldehyde and nitrogen oxides (NOx) [26–28]. Additionally, it has been proven effective on eliminating bacteria such as *E. coli* and *Klebsiella pneumoniae* and phages such as MS2 [29–32]. Furthermore, no dangerous by-products were produced during the degradation process. In a previous work, in order to demonstrate the photocatalytic effectiveness of TCM-1, the powder has been incorporated in/on different construction matrices such as glass, plywood, wood, ceramic and concrete substances, which were, subsequently, tested indicating very promising results [33].

In the current study TCM-1 incorporated into paint production process in order to produce paint with the ability to photocatalytically improve IAQ by degraded indoor air pollutants. NO and toluene were chosen as the target pollutants since they are typical indoor air pollutants that can be emitted from various indoor sources, such as cooking, tobacco smoke, furniture, building materials and fireplaces, as well as from outdoor sources, e.g., traffic, domestic heating. These indoor pollutants can have significant health impacts and for that purpose their elimination from the indoor environment should be appropriately addressed. In order to simulate the real indoor air conditions, visible light was used as the source for the PCO reaction in the current study. The latter is different from what we usually find in the literature. More specifically, UV light is the most commonly used parameter in the experimental procedures, which usually applied to test the efficiency of a photo-material.

The experimental study of the current work was carried out using the European Committee for Standardization (CEN) Technical Specification (TS) 16980:2016 as a basic reference to perform the lab scale experiments, while the real scale experiments were based on the comparison between the concentration level of a pollutant in a reference and the "green" room, respectively. The latter was performed at the military medical center of the Cadets Training Camp in Heraklion Crete.

2. Materials and Methods

2.1. Materials

Chemicals for the preparation of the photocatalytic powder such as Titanium (IV) oxysulfate hydrate (TiOSO4·xH2O), manganese (II) acetate tetrahydrate Mn (CH3COO)2) and ammonium hydroxide (25% NH4OH) purchased from Aldrich were applied.

Chemicals for the preparation of the photo-paint were purchased from VINAVIL EGYPT (New Cairo, Egypt), DOW chemicals (Midland, Michigan, MI, United States), BYK (Wesel, Germany), DuPont (Midland, Michigan, MI, United States) and Dionyssos Marbles (Penteli, Greece) and are readily available raw materials mainly used at the coating industry.

2.2. Methods

2.2.1. Preparation of Photo-Material (powder)

In the present study, an optimum powder (0.1% Mn-doped) named TCM-1 was used for the preparation of photocatalytic paint. TCM-1 was synthesized by a co-precipitation method with 0.1% of manganese. TCM-1 was precipitated at pH ~ 7 from aqueous solution of titanium (IV) oxysulfate hydrate and manganese by the addition of ammonia. After aging the suspension overnight, the precipitate was filtered and dried under air at 373 K. The residue was crushed to a fine powder and calcined in a furnace at 973 K for 3 h. More details in the synthesis procedure and the characterization are given in previous work along with the preparation details [25].

2.2.2. Preparation of Photo-Paint

TCM-1 powder (0.1% Mn-doped) was added to a specially formulated architectural coating. The paint was consisted 10% w/w of the TCM-1 powder partly replacing the Titanium Oxide (normally used as a white pigment). The raw materials for the preparation of the photo-paint are commonly used in the coating industry. A polyvinyl acetate copolymer binder was used and the Pigment Volume Content (PVC) was adjusted, but kept below its critical value in order for the paint to have a high quality, matte finish.

The production process of the photo-paint was consisted of three discreet phases. At phase 1, the mill-base (the minimum amount of liquid that can wet the solid particles of the paint, added at phase 2) was produced. Water was used as solvent with additives, such as sodium hexametaphosphate (dispersion agent), propylene glycol (antifreezing agent), silica modified surfactant (defoaming agent), cellulose (rheology modifier), isothiazolinone based mixture (biocide agent) and a mixture of Alkanol-amines (pH adjusting agent). At phase 2, we applied high shear forces to the mill-base by using a disperser equipped with a blade dissolver disc and gradually added the solid phase of the paint [Calcium carbonate powder (20 μm), Titanium Oxide (0.2–0.5 μm), Talc Powder (6 μm) and TCM-1 powder]. The goal of this phase is to maximize dispersion by eliminating the presence of any agglomerate in the final paint. At phase 3, we added the Polyvinyl Acetate copolymer binder emulsion and a coalescent solvent mixture, necessary for the final product to have a smooth surface (filmer). The dispersion was achieved by using low shearing forces and the pigment dispersion was stabilized in order to prevent the formation of uncontrolled flocculates by using non-ionic urethane rheology modifiers.

2.3. Characterization

Powder X-ray diffraction patterns were obtained by a Rigaku D/MAX-2000H rotating anode diffractometer (CuKα radiation) equipped with a secondary pyrolytic graphite monochromator operating at 40 kV and 80 mA over the 2θ collection range of 20–80° (scan rate was 0.05° s^{-1}). The grain size (nm) of TCM-1 was calculated from the line broadening of the X-ray diffraction peak according to the following Scherrer formula:

$$D = k\lambda/\beta cos\theta \tag{1}$$

where k is the Scherrer contact (~ 0.9), λ is the wavelength of the X-ray radiation (1.54 Å for CuKα), β is the full width at half maximum (FWHM) of the diffraction peak measured at 2θ, and θ is the Bragg angle. The morphology and elemental analysis were performed using scanning electron microscopy (SEM) and energy dispersive X-ray spectroscopy (EDX), on a JSM-6390LV microscope (Jeol, Tokyo, Japan).

The photo-paint was evaluated through typical quality control measurements. Viscosity was measured using a Brookfield KU-2 viscometer (Brookfield Engineering Laboratories, INC. Middleboro, MA, USA) at a temperature of 25 °C with the help of a laboratory water bath and the measurements were conducted according to ASTM D 562-10 (2018) standard test method.. Density was measured using a 100 ml pyknometer (density cup) by Elcometer (Manchester, United Kingdom) and an analytical scale, and the measurements were conducted according to ISO 2811 -1:2016 standard. Fineness of Grind was evaluated using a Hegman gauge and the evaluation was conducted according to ASTM D 1210-05 (2014) standard test method. Lastly, pH was measured using a Hanna HI 83141V pH meter (Hanna Instruments, Greece) and the measurements were conducted according to ISO 787-7:2009 standard.

2.4. Photocatalytic Evaluation_Lab-Scale

The experimental methodology that was applied and the required scientific equipment were based on CEN Technical Specification (TS) 16980:2016. The photocatalytic effect of the optimized material studied in a continuous flux photocatalytic reactor (Figure 1), which consists of a) a gas transfer and mixing unit in order to adjust the concentration and humidity levels; b) the photocatalytic reactor main body made of special plastic so as radiation intensity and wavelength of the radiation is not affected; (c) the sample irradiation system (OMICRON FS LED Rodgau-Dudenhofen, Germany) consisting of an LED device connected via software to a computer in order to achieve the optimum efficiency and accuracy in measuring the radiation; d) a NOx and VOCs analyzer installed on line with the reactor for continuous monitoring of the pollutant concentration.

Furthermore, in order to ensure optimum mixing of atmospheric pollutants in the reactor, a fan is installed inside the chamber, while its intensity is adjusted externally to ensure the stability of the experimental conditions throughout the experiments. The NO concentration is set to (0.50 ± 0.05) ppmv, while the relative humidity to 40 ± 5%. The illumination provided an average irradiance to the test specimen surface within the range of wavelengths that are mostly adsorbed by the photocatalyst, equal to (10.0 ± 5%) W/m^2.

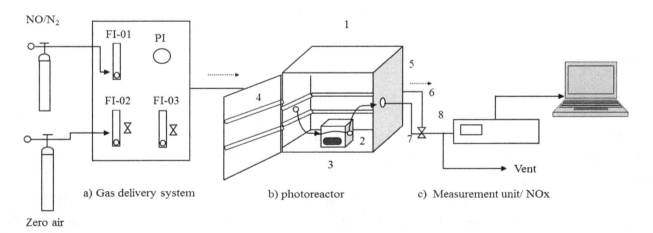

Figure 1. Photocatalytic reactor.

2.5. Photocatalytic Evaluation_Real-Scale

A building of the Hellenic Army in Crete was used to test the de-polluting efficiency of the photo-paint under real scale application. In general, the photocatalytic efficiency of a material while applied in a real scale environment and more specifically in indoor building environment could be estimated through two approaches:

a) A first approach could be the installation of air quality and environmental monitoring systems (passive inorganic (NO) and organic (BTX) samplers, temperature and humidity recorders) in the buildings prior to the application of the photo-paint. To that end, a reliable record on the concentration levels of air pollutants in the case study building (without the photo paint) for at least 12 months should be obtained. Accordingly, the photo-paint should be applied in the buildings and environmental parameters should be monitored through passive sampling techniques for another 12 months. The potential changes in the indoor air quality due to the photocatalytic action of the photo-paint will be recorded and quantified to illustrate the capability of the photo-paint to improve IAQ. A restriction of this approach could be the variations in outdoor air quality and meteorological conditions during the different sampling periods. In order to eliminate the effect of the above restrictions, the experiments has to take place the same season (e.g., winter) and the outdoor concentration should be considered on the final results.

b) A second approach to evaluate the efficiency of the photo-paints is to estimate the IAQ differences between "reference" rooms (without photo paint) and "green" rooms (with photo-paint), which are located on the same level (in a raw) and where the same activities take place. The current approach overcomes the restrictions of the previous one and for that purpose was used in the current study. More specifically a room of 120 m^2 in the ground floor was paint with the innovative photocatalytic paint ("green room") and compared with a same size and usage room ("reference room") located very close to the "green" one. The outcome of this approach was compared with the outcome of the lab tests in order to estimate the differences of the photocatalytic performance of a material when studied in a control experimental reactor and under real world environmental conditions. Passive samplers for NO and Toluene were applied. More specifically, four passive samplers/pollutants were installed in each of the rooms for 30 days and then analyzed in the lab using the well-established Saltzmann spectrophotometric method for NO, while the Toluene samplers were and desorbed by carbon disulphide and analyszed by gas chromatography (GC/FID).

3. Results

3.1. Physical and Chemical Properties

Figure 2 shows the X-Ray Difraction (XRD) pattern of TCM-1(dopant concentration 0.1 wt%), calcinated at 700 °C for 3 h 1. The peaks at 2θ values of 25.3°, 37.6°, and 48.2° correspond to the (101), (004) and (200) planes, respectively, and they are all anatase signature peaks. The grain size for the TCM-1was 38.69 nm and it was determined from the full width at half maximum (FWHM) of the (101) anatase peak according to the Scherrer's formula. In our previous work, it was proven that the doping and the role of Mn in TiO_2 in the case of 0.1 wt% manganese shows only 2+ oxidation state in comparison with high concentrations [34].

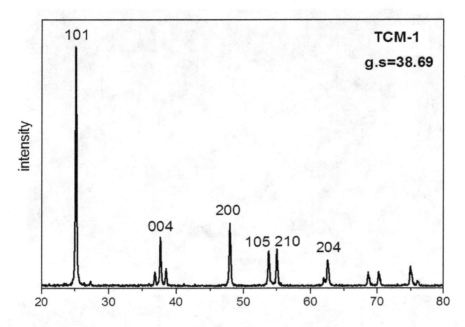

Figure 2. Powder XRD patterns of TCM-1.

Figure 3 shows the UV-Vis absorption as a function of wavelength for TCM-1 and exhibited an absorption edge in the visible light range (400–800 nm). The band gap energy was 2.87 eV for Mn doped catalyst.

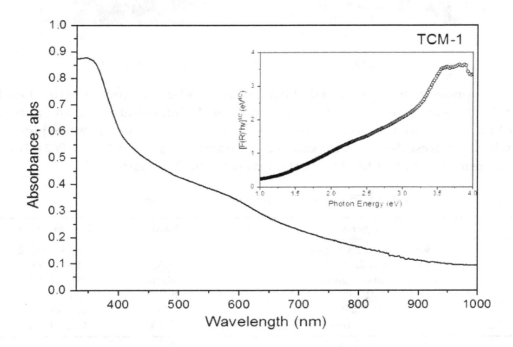

Figure 3. UV-vis absorption of TCM-1and energy gaps calculated from Kubenka–Munk plots.

Figure 4a shows the morphology of the photocatalyst were investigated with SEM, without specific morphology, while the spherical shape particles of all the samples demonstrated some degree of agglomeration and the diameter ranged from 0.1 to 40 μm. Figure 4b shows the characteristic peaks of Mn and Ti atoms.

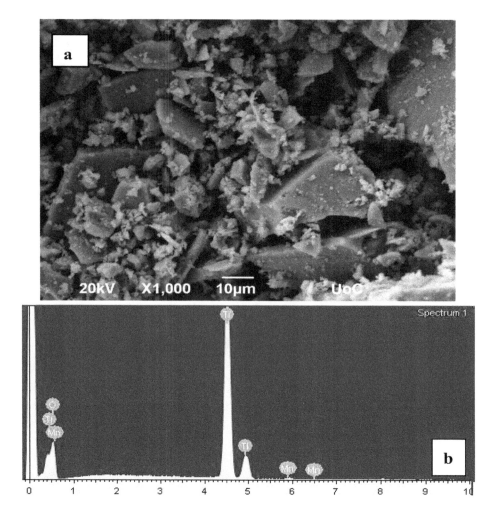

Figure 4. (**a,b**) SEM image of TCM-1 and EDX analysis.

As far as the photo-paint is concerned, the main physicochemical properties are listed in Table 1. The measured values are typical for an indoor emulsion architectural paint, and although TCM-1 was used at 10% w/w, TCM-1 behaved as expected from a Titanium Oxide white pigment. A paint formulated with the process described in Section 2.2.2 and with the use of Ti Pure 902+ Titanium Oxide white pigment instead of TCM-1 has identical physicochemical properties.

Table 1. Physicochemical Properties of photo-paint.

Property	Value	Test Method
Viscosity at 25°C (KU)	100–110	ASTM D 562
Density at 25°C (kg/l)	1.55	ISO 2811
Fineness and Dispersion	<40 μm	ASTM D1210
pH at 25°C	8.5–9	ISO 787-9
PVC (%)	66	Calculated
Usage rate for a 50 μm dry film thickness (m²/kg)	5.95	Calculated

3.2. Lab and Real-Scale Photocatalytic Performance

In order to estimate the background, experiments in the absence of the photocatlatyic paint were performed. The photocatalytic experiments determine the total pollutant degradation involving both UV photolysis and photocatalysis on the photo-paint. The net photocatalytic effect is calculated by the subtraction of the background contribution from the photocatalytic experiments. More specifically, blank tests were carried out by polluting the reactor with NO and Toluene in the absence of the photocatalyst without and with irradiation, respectively; then, the same experiments were carried out in the presence of the photocatalysts.

Figure 5a,b present the elimination of NO and toluene under the irradiation of the photo-paint from Vis-light. It is obvious that just after the irradiation of the sample (Time 30), a sharp decrease in NO and a smaller but significant decrease in toluene concentration is observed, which demonstrate the immediate response of the photocatalytic system and provide the photo-efficiency of the paint.

Adsorption of NO onto the chamber's wall area and photolysis are the main sinks of NO during blank tests. Calculations have shown that both these mechanisms did not contributed to the total NO and toluene removal during the photocatalytic experiments.

The photocatalytic activity was evaluated by the calculation photocatalytic yield (% n, Equation (2)) and photodegradation rate (r, Equation (3)). The corresponding equations were used for toluene. The results in Table 2 showed the possibility of developing a very promising and highly active to air pollutants photocalatylic paint (Table 2).

$$\% \, \eta_{NO}^{total} = \frac{C_{NO}^{IN} - C_{NO}^{OUT,light}}{C_{NO}^{IN}} \times 100 \tag{2}$$

where:

C_{NO}^{IN}: the concentration of NO at reactor inlet

$C_{NO}^{OUT,light}$: the concentration of NO at reactor outlet under stable conditions with irradiation (lamp on)

The rate of photocatalytic yield of the material is calculated by the formula below (Equation (3)) and expressed in $\mu g m^{-2} \, s^{-1}$:

$$r_{NO}^{photo} = \frac{613F}{S} \left(\frac{\eta_{NO}^{total}}{\left(1 - \eta_{NO}^{total}\right)} - \frac{\eta_{NO}^{dark}}{\left(1 - \eta_{NO}^{dark}\right)} \right) \tag{3}$$

where

F: the gas flow ($m^3 \, h^{-1}$)

S: The area of the test surface (m^2)

The photodegradation rate provides a more accurate measure of the photocatalytic activity of the material in comparison with the % photocatalytic decomposition. The latter is attributed to the fact that r is taking into consideration the initial concentration of the pollutant, the sample's area and the irradiation time. It is expressed as μg of converted NO/toluene per m^2 of material per second of irradiation. The current parameter was calculated only for the lab tests, where the pollutant flow rate in the chamber was known.

The losses in the system are minimal, and as a consequence the fraction $\frac{\eta_{NO}^{dark}}{(1 - \eta_{NO}^{dark})}$ is zero.

Table 2. Photocatalytic parameters for lab and real scale tests.

Parameter	Lab Scale		Real Scale	
	NO	Toluene	NO	Toluene
% η	59.08	16.7	18.8	5.26
rphoto (μg/m²s)	3.89	1.05	-	-

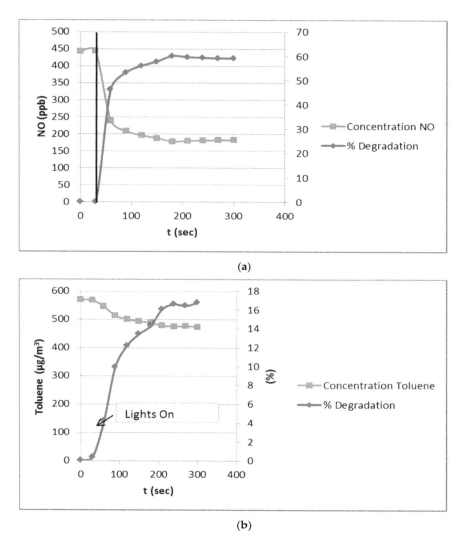

Figure 5. (a) Photocatalytic yields (% η) concentration profile versus time for the photo-oxidation of NO in the presence of the catalyst. **(b)** Photocatalytic yields (% η) concentration profile versus time for the photo-oxidation of Toluene in the presence of the catalyst.

The variations on NO and Toluene concentrations during the experimental procedure of photocatalysis are presented in Figure 5a and Figure 5b, respectively. At first, the pollutant was introduced into the photo-reactor and the system remained in equilibrium for 30 min. Then, irradiation was followed and a sharp reduction in NO and significant but less intense than NO elimination in toluene concentration is observed, which stabilized after 3 h and remained stable for approximately 2 h where the irradiation stopped. The results from the laboratory tests showed that for both pollutants, the photocatalytic paint gave very promising results. It is worth to note that in the case of toluene, the degradation efficiency is significantly lower than NO. However, a reduction of VOCs even at that level could have significant effects in the improvement of IAQ. Furthermore, the lower degradation efficiency in VOCs compare with NO is in line with the results of various studies [35–38]. The latter could be attributed to the low adsorption capacity of toluene molecules at the catalyst surface. The extremely hydroxylated surface of the paint material due to physically or chemically bound water, could constrain its interaction with the active radicals through surface diffusion [39,40].

3.3. Real-Scale Photocatalytic Performance

As far as the real scale application is concerned, the deppolution efficiency was calculated by the absolute difference in the concentration levels of NO and toluene, which was measured in the two rooms of the military building: the "green" and the "reference" rooms. Results are shown in Table 2. It is observed that real scale tests showed significant lower values of NO and toluene degradation than in the lab-scale application. The latter was also observed in other studies [41–45]. The control environment of the laboratory versus the more complex and polyparametric environment of a real application could be the main reason that led to significant differences in the values that characterize the photocatalytic efficiency of the paint. However, they provide an indication of the photocatalytic efficiency of the paints to degrade pollutants in situ and a basis for photo-paint applications in order to improve IAQ.

4. Discussion

The effect of photocatalytic paint in the improvement of IAQ has been studied in both laboratory and real scale tests. It is interesting to note that NO could be effectively converted by the photocatalytic paint, while toluene showed lower photocatalytic removal in the same photocatalytic process (lab and real scale). A possible explanation of the different photocatalytic performance of NO and toluene under the same application could be explained as follows: the dominant oxidants in a photocatalytic reaction are hydroxyl radicals ($OH_\bullet$) and hydroperoxyl radicals ($HO_{2\bullet}$), which are generated from the redox reactions of positive holes (h^+) and electrons (e^-) with O_2, H_2O or OH^-. The kinetic coefficients of gas phase reactions for HO_x radicals and targeted gas (NO and toluene in this case) have only minor differences [46,47]. However, NO is better adsorbed by the alkaline constituents of the paint due to its acidic property. This significantly increases their reaction potential with HO_x radicals. On the other hand, the adsorption capacity of toluene is lower than NO due to the extremely hydroxylated surface of the paint, which constrain its interaction with the active radicals. Toluene is absorbed more easily in a less hydrophilic TiO_2 surface than a more hydroxylated one [41]. It is obvious that different gaseous pollutants present significant variations in their photocatalytic activity due to the difference in their diffusion in the paint matrix. Efficient adsorption of the pollutant molecules at the catalyst surface could promote the photocatalytic reaction.

Comparing lab versus real scale experiments results, it is observed that laboratory tests showed significant higher values of both pollutants (NO and toluene) degradation than in real scale application. The translation from the laboratory results to "real" site efficiency is difficult because of the great number of parameters involved such as traffic and environmental parameters (temperature, light intensity, relative humidity wind speed). For that purpose, the more complex and polyparametric environment of a real application lead to significant differences in the values, which characterize the photocatalytic efficiency of the paint. In any case, precaution has to be taken with the interpretation of data obtained from the real scale experiments since these results are limited over time. However, they provide an indicative picture of the efficiency of the photocatalytic paints to eliminate air pollutants under real world conditions and a basis on which to improve their photocatalytic capacity for future applications. The need for large scale applications is more imperative nowadays, as many photo-materials have indicated very promising results during lab scale tests, but their capacity under real world conditions has not been proven. The latter is critical and proven by the outcome of the current work, which showed significant differences between the photo-efficiency of the same material under two different scales. Additionally, the demonstration of the effectiveness of photocatalytic materials on site should also include negative effects, such as the formation and emission of by-products during the photocatalytic reactions, as well as the durability of the photo-paint mechanical properties. The latter is of high importance, as in most of the cases, TiO_2 oxidizes not only the air pollutants on the photo-material surface, but also their organic and inorganic components. For that purpose, special attention is given in the formulation of such materials in order to avoid it. As it is not in the scope of the current manuscript to study the durability of the photo-paint mechanical properties, measurements to characterize the mechanical durability of the paint were not performed. However, it is worth mentioning that almost two years after the photo-paint application, cracks or other surface damages have not been observed. Nevertheless, beyond the visual observation, the latter should be certified by analytical measurements, which should be the topic of future study. Furthermore, better results could be obtained by using mathematical models to simulate the photocatalytic processes, validating the model using the outcomes of the lab measurements and then implement various parameters in order to assess the real life effects [48].

5. Conclusions

This article addresses the effect of a photocatalytic paint on the elimination of air pollutants, and more specifically NO and toluene, for application in indoor environments. The following conclusions can be drawn:

- The physicochemical properties (including the mechanical parameters) of the photo-paint does not seem to be affected from the introduction of the photocatalytic powder (TCM-1) in the synthesis route.
- The photocatalytic efficiency of the paint on NO removal was significantly higher than toluene. The potential of a pollutant removal depends on the intrinsic properties of gas and the chemical nature of the paint in which the TiO_2 particles are embedded. However, the removal rate of toluene was very promising for the improvement of IAQ while using the studied photo-paint.
- Lab tests showed better photocatalytic properties of the paint than the results from the real scale application due to the great number of parameters involved in the case of real scale application.
- There is a need for large scale applications to demonstrate the effectiveness of photocatalytic materials on site, including any negative effects of the application, such as the emission of by-products (e.g., carbonyl compounds, O_3 etc.).

Author Contributions: Conceptualization, T.M., V.B. and G.K.; Data curation, T.M. and P.P.; Investigation, G.K.; Methodology, T.M., V.B., V.S. and A.T.; Project administration, V.B.; Writing—original draft, T.M.; Writing—review & editing, T.M., V.B., V.S., A.T., P.P. and G.K.

References

1. Ren, H.; Koshy, P.; Chen, W.-F.; Qi, S. Charles Christopher Sorrell Photocatalytic materials and technologies for air purification. *J. Hazard. Mater.* **2017**, *325*, 340–366. [CrossRef] [PubMed]
2. Zhang, Y.; Moa, J.; Li, Y.; Sundell, J.; Wargocki, P.; Zhang, J.; Little John, C.; Corsi, R.; Deng, Q.; Leung, M.H. Can commonly-used fan-driven air cleaning technologies improve indoor air quality? A literature review. *Atmos. Environ.* **2011**, *45*, 4329–4343. [CrossRef]
3. Bolashikov, Z.D.; Melikov, A.K. Methods for air cleaning and protection of building occupants from airborne pathogens. *Build. Environ.* **2009**, *44*, 1378–1385. [CrossRef]
4. Mamaghani, A.H.; Haghighat, F.; Lee, C.-S. Photocatalytic oxidation technology for indoor environment air purification: The state-of-the-art. *Appl. Catal. B Environ.* **2017**, *203*, 247–269. [CrossRef]
5. Fan, Z.; Lioy, P.; Weschler, C.; Fiedler, N.; Kipen, H.; Zhang, J. Ozone-initiatedreactions with mixtures of volatile organic compounds under simulatedindoor conditions. *Environ. Sci. Technol.* **2003**, *37*, 1811–1821. [CrossRef] [PubMed]
6. Boeniger, M.F. Use of ozone generating devices to improve indoor airquality. *Am. Ind. Hyg. Assoc. J.* **1995**, *56*, 590–598. [CrossRef] [PubMed]
7. Zhong, L.; Haghighat, F. Ozonation air purification technology in HVACapplications. *ASHRAE Trans.* **2014**, *120 (Pt 1)*, 8.
8. Bahri, M.; Haghighat, F.; Rohani, S.; Kazemian, H. Impact of design parameterson the performance of non-thermal plasma air purification system. *Chem. Eng. J.* **2016**, *302*, 204–212. [CrossRef]
9. Bahri, M.; Haghighat, F. Plasma-based indoor air cleaning technologies: Thestate of the art-review. *CLEAN-Soil Air Water* **2014**, *42*, 1667–1680. [CrossRef]
10. Zhong, L.; Haghighat, F.; Lee, C.-S. Ultraviolet photocatalytic oxidation for indoor environment applications: Experimental validation of the model. *Build. Environ.* **2013**, *62*, 155–166. [CrossRef]
11. Farhanian, D.; Haghighat, F. Photocatalytic oxidation air cleaner: Identification and quantification of by-products. *Build. Environ.* **2014**, *72*, 34–43. [CrossRef]

12. Mo, J.; Zhang, Y.; Xu, Q. Effect of water vapor on the by-products anddecomposition rate of ppb-level toluene by photocatalytic oxidation. *Appl. Catal. B Environ.* **2013**, *132–133*, 212–218. [CrossRef]

13. Debono, O.; Thévenet, F.; Gravejat, P.; Héquet, V.; Raillard, C.; Le, L.; Locoge, N. Gas phase photocatalytic oxidation of decane at ppb levels: Removal kinetics, reaction intermediates and carbon mass balance. *J. Photochem. Photobiol. A Chem.* **2013**, *258*, 17–29. [CrossRef]

14. Ao, C.H.; Lee, S.C.; Mak, C.L.; Chan, L.Y. Photodegradation of volatile organic compounds (VOCs) and NO for indoor air purification using TiO2: Promotion versus inhibition effect of NO. *Appl. Catal. B Environ.* **2003**, *42*, 119–129. [CrossRef]

15. Ao, C.H.; Lee, S.C.; Zou, S.C.; Mak, C.L. Inhibition effect of SO$_2$ on NOx and VOCs during the photodegradation of synchronous indoor air pollutants at parts perbillion (ppb) level by TiO2. *Appl. Catal. B Environ.* **2004**, *49*, 187–193. [CrossRef]

16. Selishchev, D.S.; Kolinko, P.A.; Kozlov, D.V. Adsorbent as an essential participant in photocatalytic processes of water and air purification: Computer simulation study. *Appl. Catal. A Gen.* **2010**, *377*, 140–149. [CrossRef]

17. Verbruggen, S.W. TiO$_2$ photocatalysis for the degradation of pollutants ingas phase: From morphological design to plasmonic enhancement. *J. Photochem. Photobiol. C Photochem.* **2015**, *24*, 64–82. [CrossRef]

18. Ismail, A.A.; Bahnemann, D.W. Mesoporous titania photocatalysts: Preparation, characterization and reaction mechanisms. *J. Mater. Chem.* **2011**, *21*, 11686–11707. [CrossRef]

19. Hodgson, A.T.; Destaillats, H.; Sullivan, D.P.; Fisk, W.J. Performance of ultraviolet photocatalytic oxidation for indoor air cleaning applications. *Indoor Air* **2007**, *17*, 305–316. [CrossRef] [PubMed]

20. Muggli, D.S.; McCue, J.T.; Falconer, J.L. Mechanism of the photocatalytic oxidation of ethanol on TiO2. *J. Catal.* **1998**, *173*, 470–483. [CrossRef]

21. Mo, J.H.; Zhang, Y.P.; Xu, Q.J.; Zhu, Y.F.; Lamson, J.J.; Zhao, R.Y. Determination and risk assessment of by-products resulting from photocatalytic oxidation of toluene. *Appl. Catal. B Environ.* **2009**, *89*, 570–576. [CrossRef]

22. Kovalevskiy, N.S.; Lyulyukina, M.N.; Selishcheva, D.S.; Kozlova, D.V. Analysis of air photocatalytic purification using a total hazard index: Effect of the composite TiO2/zeolite photocatalyst. *J. Hazard. Mater.* **2018**, *358*, 302–309. [CrossRef] [PubMed]

23. Denny, F.; Permma, E.; Scott, J.; Wang, J.; Pui, D.Y.H.; Amal, R. Integrated Photocatalytic Filtration Array for Indoor Air Quality Control. *Environ. Sci. Technol.* **2010**, *44*, 5558–5563. [CrossRef] [PubMed]

24. Yu, H.; Zhang, K.; Rossi, C. Experimental Study of the Photocatalytic Degradation of Formaldehyde in Indoor Air using a Nano-particulate Titanium Dioxide Photocatalyst Indoor. *Built Environ.* **2007**, *16*, 529–537. [CrossRef]

25. Binas, V.D.; Sambani, K.; Maggos, T.; Katsanaki, A.; Kiriakidis, G. Synthesis and photocatalytic activity of Mn-doped TiO$_2$ nanostructured powders under UV and visible light. *Appl. Catal. B Environ.* **2012**, *113–114*, 79–86. [CrossRef]

26. Cacho, C.; Geiss, O.; Barrero-Moreno, J.; Binas, V.D.; Kiriakidis, G.; Botalico, L.; Kotzias, D. Studies on photo-induced NO removal by Mn-doped TiO$_2$ under indoor-like illumination conditions. *J. Photochem. Photobiol. A Chem.* **2011**, *222*, 304–306. [CrossRef]

27. Karafas, E.S.; Romanias, M.N.; Stefanopoulos, V.; Binas, V.; Zachopoulos, A.; Kiriakidis GPapagiannakopoulos, P. Effect of metal doped and co-doped TiO$_2$ photocatalysts oriented to degrade indoor/outdoor pollutants for air quality improvement. A kinetic and product study using acetaldehyde as probe molecule. *J. Photochem. Photobiol. A-Chem.* **2019**, *371*, 255–263. [CrossRef]

28. Binas, V.; Stefanopoulos, V.; Kiriakidis, G.; Papagiannakopoulos, P. Photocatalytic oxidation of gaseous benzene, toluene and xylene under UV and visible irradiation over Mn-doped TiO2 nanoparticles. *J. Mater.* **2019**, *5*, 56–65. [CrossRef]

29. Venieri, D.; Tournas, F.; Gounaki, I.; Binas, V.; Zachopoulos, A.; Kiriakidis, G.; Mantzavinos, D. Inactivation of Staphylococcus aureus in water by means of solar photocatalysis using metal doped TiO$_2$ semiconductors. *J. Chem. Technol. Biotechnol.* **2017**, *92*, 43–51. [CrossRef]

30. Venieri, D.; Gounaki, I.; Bikouvaraki, M.; Binas, V.; Zachopoulos, A.; Kiriakidis, G.; Mantzavinos, D. Solar photocatalysis as disinfection technique: Inactivation of Klebsiella pneumoniae in sewage and investigation of changes in antibiotic resistance profile. *J. Environ. Manag.* **2017**, *195*, 140–147. [CrossRef] [PubMed]

31. Venieri, D.; Gounaki, I.; Binas, V.; Zachopoulos, A.; Kiriakidis, G.; Mantzavinos, D. Inactivation of MS2 coliphage in sewage by solar photocatalysis using metal-doped TiO$_2$. *Appl. Catal. B Environ.* **2015**, *178*, 54–64. [CrossRef]

32. Venieri, D.; Fraggedaki, A.; Kostadima, M.; Chatzisymeon, E.; Binas, V.; Zachopoulos, A.; Kiriakidis, G.;

Mantzavinos, D. Solar light and metal-doped TiO$_2$ to eliminate water-transmitted bacterial pathogens: Photocatalyst characterization and disinfection performance. *Appl. Catal. B Environ.* **2014**, *154–155*, 93–101. [CrossRef]

33. Binas, V.; Papadaki, D.; Maggos, T.; Katsanaki, A.; Kiriakidis, G. Study of innovative photocatalytic cement based coatings: The effect of supporting materials. *Constr. Build. Mater.* **2018**, *168*, 923–930. [CrossRef]

34. Binas, V.; Venieri, D.; Kotzias, D.; Kiriakidis, G. Modified TiO$_2$ based photocatalysts for improved air and health quality. *J. Mater.* **2017**, *3*, 3–16.

35. Ramirez, A.M.; Demeestere, K.; De Belie, N.; Mäntylä, T.; Levänen, E. Titanium dioxide coated cementitious materials for air purifying purposes: Preparation, characterization and toluene removal potential. *Build. Environ.* **2010**, *45*, 832–838. [CrossRef]

36. Tsoukleris, D.S.; Maggos, T.; Vassilakos, C.; Falaras, P. Photocatalytic degradation of volatile organics on TiO$_2$ embedded glass spherules. *Catal. Today* **2007**, *129*, 96–101. [CrossRef]

37. Kannangara, Y.Y.; Wijesena, R.; Rajapakse, R.G.M.; Nalin de Silva, K.M. Heterogeneous photocatalytic degradation of toluene in static environment employing thin films of nitrogen-doped nano-titanium dioxide. *Int. Nano Lett.* **2018**, *8*, 31–39. [CrossRef]

38. Pei, C.C.; Leung, W.W.-F. Photocatalytic oxidation of nitrogen monoxide and o-xylene by TiO$_2$/ZnO/Bi$_2$O$_3$ nanofibers: Optimization, kinetic modeling and mechanisms. *Appl. Catal. B Environ.* **2015**, *174–175*, 515–525. [CrossRef]

39. Chen, J.; Kou, S.-c.; Poon, C.-s. Photocatalytic cement-based materials: Comparison of nitrogen oxides and toluene removal potentials and evaluation of self-cleaning performance. *Build. Environ.* **2011**, *46*, 1827–1833. [CrossRef]

40. Ardizzone, S.; Bianchi, C.L.; Cappelletti, G.; Naldoni, A.; Pirola, C. Photocatalytic degradation of toluene in the gas phase: Relationship between surface species and catalyst features. *Environ. Sci. Technol.* **2008**, *42*, 6671–6676. [CrossRef] [PubMed]

41. Boonen, E.; Beeldens, A. Photocatalytic roads: From lab tests to real scale applications. *Eur. Transp. Res. Rev.* **2013**, *5*, 79–89. [CrossRef]

42. Beeldens, A. Air purification by pavement blocks: Final results of the research at the BRRC. In Proceedings of the Transport Research Arena Europe–TRA, Ljubljana, Slovenia, 21–24 April 2008.

43. Gignoux, L.; Christory, J.P.; Petit, J.F. Concrete roadways and air quality–Assessment of trials in Vanves in the heart of the Paris region. In Proceedings of the 12th International Symposium on Concrete Roads, Sevilla, Spain, 13–15 October 2010.

44. Maggos, T.; Plassais, A.; Bartzis, J.G.; Vasilakos, C.; Moussiopoulos, N.; Bonafous, L. Photocatalytic degradation of NOx in a pilot street canyon configuration using TiO$_2$-mortar panels'. *Environ. Monit. Assess.* **2007**, *136*, 35–44. [CrossRef] [PubMed]

45. Maggos, T.; Bartzis, J.G.; Liakou, M. Gobin Photocatalytic degradation of NOx gases using TiO$_2$-containing paint: A real scale study. *J. Hazard. Mater.* **2007**, *146*, 668–673. [CrossRef] [PubMed]

46. Atkinson, R.; Baulch, D.L.; Cox, R.A.; Crowley, J.N.; Hampson, R.F.; Hynes, R.G.; Jenkin, M.E.; Rossi, M.J.; Troe, J. Evaluated kinetic and photochemical data for atmospheric chemistry: Volume I-gas phase reactions of O$_x$, HO$_x$, NO$_x$ and SO$_x$ species. *Atmos. Chem. Phys.* **2004**, *4*, 1461–1738. [CrossRef]

47. Finlayson-Pitts, B.J.; Pitts, J.N. *Chemistry of the Upper and Lower Atmosphere*; Academic Press: San Diego, CA, USA, 2000.

48. Hunger, M.; Hüsken, G.; Brouwers, H.J.H. Photocatalytic degradation of air pollutants–from modeling to large scale application. *Cem. Concr. Res.* **2010**, *40*, 313–320. [CrossRef]

10

An Investigation of the Effects of Changes in the Indoor Ambient Temperature on Arousal Level, Thermal Comfort and Physiological Indices

Jongseong Gwak [1],*, Motoki Shino [2], Kazutaka Ueda [3] and Minoru Kamata [2]

[1]   Institute of Industrial Science, The University of Tokyo, Tokyo 153-8505, Japan
[2]   Department of Human and Engineered Environment Studies, Graduate School of Frontier Sciences,
      The University of Tokyo, Chiba 277-8563, Japan; motoki@k.u-tokyo.ac.jp (M.S.);
      mkamata@k.u-tokyo.ac.jp (M.K.)
[3]   Department of Mechanical Engineering, The University of Tokyo, Tokyo 113-8656, Japan;
      ueda@design-i.t.u-tokyo.ac.jp
*   Correspondence: js-gwak@iis.u-tokyo.ac.jp.
†   This paper is an extended version of paper published in the 2015 IEEE International Conference on Systems,
    Man, and Cybernetics, held in Hong Kong, 9–12 October 2015.

Abstract: Thermal factors not only affect the thermal comfort sensation of occupants, but also affect their arousal level, productivity, and health. Therefore, it is necessary to control thermal factors appropriately. In this study, we aim to design a thermal environment that improves both the arousal level and thermal comfort of the occupants. To this end, we investigated the relationships between the physiological indices, subjective evaluation values, and task performance under several conditions of changes in the indoor ambient temperature. In particular, we asked subjects to perform a mathematical task and subjective evaluation related to their thermal comfort sensation and drowsiness levels. Simultaneously, we measured their physiological parameters, such as skin temperature, respiration rate, electroencephalography, and electrocardiography, continuously. We investigated the relationship between the comfort sensation and drowsiness level of occupants, and the physiological indices. From the results, it was confirmed that changes in the indoor ambient temperature can improve both the thermal comfort and the arousal levels of occupants. Moreover, we proposed the evaluation indices of the thermal comfort and the drowsiness level of occupants using physiological indices.

Keywords: thermal comfort; arousal level; physiological indices; electroencephalography; electrocardiography

---

1. Introduction

The control of indoor environmental quality (IEQ, which consists of visual elements, olfactory elements, and thermal factors, etc.) is important for improvements of the comfort and productivity of occupants. Among the factors of IEQ, thermal factors, such as ambient temperature, radiant temperature, humidity, and air velocity, are especially related with not only the thermal sensation and thermal comfort of occupants, but also productivity and health. With the high technology of heating, ventilation, and air conditioning systems (HVAC systems), the relationship between thermal factors and thermal sensation and thermal comfort has received attention from several researchers worldwide. In previous research, indices of thermal comfort, such as the predicted mean vote (PMV), predicted percent dissatisfied (PPD), and standard new effective temperature (SET*), have been proposed based on the relationship between thermal factors and the subjective evaluation of thermal comfort sensation [1,2]. Many studies have analyzed the indoor thermal quality based on the evaluation of thermal factors using these indices [3–7].

In addition, improvements of the productivity of occupants who are working, studying, and driving in indoor spaces, such as offices, classrooms, and vehicles, are needed. It is necessary to maintain a high arousal state of the occupants to improve their productivity. In previous studies, it was shown that productivity improved when the arousal level of occupants was high, and this was related to the indoor ambient temperature that occupants felt as cool or cold [8,9]. From the results of previous studies, the conditions of the ambient temperature to improve task performance were different from the conditions of the ASHRAE (American Society of Heating, Refrigerating and Air-Conditioning Engineers)'s thermal comfort zone [10]. Therefore, the conditions of the constant ambient temperature were not appropriate to improve both the task performance and thermal comfort levels. In previous studies, it was shown that the thermal comfort of occupants was immediately changed in accordance with changes in the indoor ambient temperature [11]. The arousal level was increased by outer stimulation and was maintained at a high state for several minutes [12,13]. Based on these results of previous studies, we hypothesized that thermal stimulation due to cooling can improve the arousal level of occupants. Furthermore, thermal comfort can be improved while maintaining high arousal levels due to the removal of thermal stimulation. When considering the possibility that changes in thermal factors can improve both arousal and thermal comfort levels, the physiological effects associated with such changes in thermal factors, and how they affect the arousal level and thermal comfort of occupants are not clear. Therefore, the design requirements for changes in ambient temperature to improve both the arousal level and thermal comfort are also not clear. Therefore, a continuous and quantitative evaluation of the thermal comfort and arousal level of occupants using indices, which can be measured both continuously and quantitatively, such as physiological signals, is needed to clarify the design requirements of changes in the indoor ambient temperature to improve both the arousal level and thermal comfort.

In this study, we aimed to investigate the characteristics of changes in the arousal level and feelings of thermal comfort of occupants, and the relationship between them when thermal factors are changed. There are many factors that affect the thermal comfort and arousal level of occupants, such as the ambient temperature, indoor air velocity, mean air radiant temperature, and metabolic activity [5–7]. Especially, we focused on the changes in the indoor ambient temperature as a fundamental investigation in this study. In addition, to propose evaluation indices that can evaluate the thermal comfort and arousal level of occupants continuously and quantitatively, we investigated the relationship between the arousal level, feelings of thermal comfort, and physiological indices, which can be measured continuously and quantitatively.

2. Strategy

2.1. Investigation of the Effects of Changes in the Indoor Ambient Temperature on Arousal Levels, Thermal Comfort, and Task Performance

To verify the hypothesis that thermal stimulation, due to cooling, can improve the arousal level of occupants, and that thermal comfort can be improved while maintaining high arousal levels due to the removal of thermal stimulation, several thermal conditions were set. Subjects were asked to conduct a mathematical task and periodically evaluate their sensation values (using a subjective sensation vote) in relation to their arousal level and thermal comfort. We attempt to clarify the effect of changes in the indoor ambient temperature on the arousal level, thermal comfort, and task performance of occupants by conducting an analysis of the results of the subjective sensation vote and task performance.

2.2. Investigation of the Relation between the Subjective Evaluation Value and Physiological Parameters, and a Recommendation for Evaluation Indices

Indices that can evaluate the arousal and thermal comfort levels of occupants continuously and quantitatively are necessary to clarify the design requirements of a temperature control that can improve the arousal and thermal comfort levels of occupants. However, as it was not possible to perform a subjective sensation voting continuously, we considered clarifying the design requirements for thermal environments using changes over time. This relied on the possibility to evaluate the arousal level and feelings of thermal comfort using physiological parameters that could be measured quantitatively and continuously. Previous studies have shown that the physiological indices measured utilizing electroencephalograms (EEG), electrocardiograms (ECG), and respiration rates were effective for the evaluation of arousal levels [14,15], and the skin temperature, EEGs, and ECGs of occupants were effective for the evaluation of the thermal comfort [16,17]. Therefore, in this study, we assumed a flow of physiological responses when the occupant was stimulated by thermal factors, as shown in Figure 1, based on the above-mentioned previous studies to clarify the characteristics of the physiological parameters under the condition of changes in both the arousal level and thermal comfort. A thermal stimulation is transmitted from the sensory organ, such as warm spots and cold spots of skin, to the central nervous system, which affects thermal and comfort sensation. This change in comfort sensation then affects the autonomic nervous system, and the response is then conducted through the locomotive organ. Based on this process, we selected the skin temperature as an indicator of heat transfer between the environment and the human body, EEGs as an indicator of the reaction of the central nervous system, and ECGs as indicators of the reaction of the autonomic nervous system. Next, we attempted to search for indices that could separately evaluate the arousal level and thermal comfort by performing a multiple regression analysis utilizing the physiological parameters and subjective evaluation values of the arousal level and feelings of thermal comfort. In Figure 1, the flow of physiological responses with the thermal stimulus is shown.

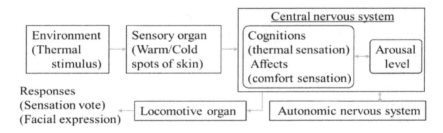

Figure 1. Flow of physiological responses with the thermal stimulus.

3. Methodology

3.1. Subjects

Ten subjects with vital statistics of a height of 173.5 ± 4.4 cm, weight of 61 ± 4.8 kg, age of 22.1 ± 1.2 years old, and who were right-handed participated in the experiment. The experimental contents and procedures, which were approved by the ethics committee of the University of Tokyo, were explained to the subjects before conducting the experiment. The subjects were then asked to avoid intense physical activity, alcohol, and caffeine for 24 h prior to the experimental session.

3.2. Experimental Task

The subjects were asked to conduct mental arithmetic tasks known as "MATH", which is based on the algorithm proposed by Tuner et al. [18]. A 1–3-digit addition or subtraction question is displayed for 2 s on the monitor, and then "equals to" is displayed for 1.5 s. Lastly, the answer is displayed for 1 s, and the next question is displayed after 0.5 s. Subjects had to determine if the answer was correct or incorrect when the answer was displayed and click the left mouse button if the answer was correct or the right mouse button if the answer was incorrect. The levels of questions in the original version of MATH consisted of 1–5 levels. In this study, the beginning level was 3 (which is 2-digit addition or subtraction), and the level of the next question was raised if the responses of the subjects were correct, and was reduced if answers were incorrect. We deduced that changing the levels affected the physiological indices. Thus, we did not change the level of the question, and fixed it to level 3. The MATH task included 50 questions for 250 seconds per set.

3.3. Experimental Conditions and Experimental Procedure

To evaluate the effects of the duration and degree of thermal stimulation, three environmental conditions (A–C) were set as follows.

- Condition A: The indoor ambient temperature was maintained at 27 °C.
- Condition B: The indoor ambient temperature was decreased from 27 °C to 20 °C, and then increased from 20 °C to 27 °C.
- Condition C: The indoor ambient temperature was decreased from 27 °C to 20 °C, and then maintained at 20 °C.

We set condition A as a thermally comfort condition [10], and condition C as a high arousal condition [8,9]. Condition B was set to verify the hypothesis that thermal stimulation due to cooling can improve the arousal levels of occupants, after which thermal comfort can be improved while maintaining high arousal levels due to the removal of thermal stimulation. There are many parameters which affect the thermal comfort and arousal levels of occupants, such as the indoor velocity, mean air radiant temperature, metabolic activity, and amount of clothing. In this study, we focused on only the indoor ambient temperature as a fundamental investigation, thus those parameters except the indoor ambient temperature were controlled in the experiment. The subjects wore short sleeves and short pants and were asked to remain in the pre-room. The room temperature was set at 27 ± 0.5 °C for approximately 1 h so that subjects could adjust to the thermal environment; they were asked to practice the "MATH" at least twice during this time. Sensors were then attached to the bodies of the subjects to measure their physiological indices. One set of tasks consisted of completing the subjective sensation vote and the "MATH" task. The subjects were asked to perform seven sets of tasks at time intervals of 10 min and to rest between each task for approximately 3 min. Physiological indices were measured from the start time (0 min) till the end time (70 min). The experimental procedure and environmental conditions are shown in Figure 2.

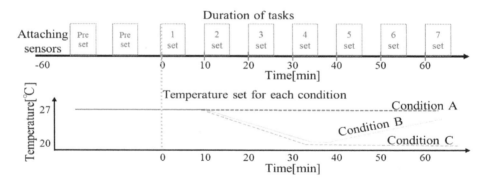

Figure 2. Experimental conditions and environmental procedure.

3.4. Measurement

3.4.1. Subjective Evaluation of the Drowsiness Level and the Thermal Comfort Sensation

Before completing the "MATH" task, the subjects were asked to complete a questionnaire related to their thermal sensation, thermal comfort sensation, and arousal level. The scale of thermal sensation was based on ASHRAE/ISO (International Organization for Standardization) [10], and was denoted using integral numbers from -3 to 3 (where -3, -2, -1, 0, 1, 2, and 3 are the meanings of cold, cool, slightly cool, neutral, slightly warm, warm, and hot, respectively.). The scale of comfort sensation was based on ISO10551 [19], and was denoted using integral numbers from -3 to 0 (where -3,

−2, −1, and 0 are the meanings of very uncomfortable, uncomfortable, slightly uncomfortable, and comfortable, respectively.). The scale of the arousal level was based on the drowsiness level of Zilberg's indicators [20], and was denoted using integral numbers from 0 to 4 (where 0, 1, 2, 3, and 4 are the meanings of alert, slightly drowsy, moderately drowsy, significantly drowsy, and extremely drowsy, respectively.).

3.4.2. Physiological Indices

(a) EEG

EEGs were recorded using an EEG-measuring instrument (EEG-1200, Nihonkohden Co., Japan) at a sampling rate of 500 Hz. EEG electrodes were attached on 16 channels (based on the internationally accepted 10–20 system, Fp1, Fp2, F7, F3, F4, F8, T7, C3, C4, T8, P7, P3, P4, P8, O1, O2). Next, raw data were processed using the Fourier transform method, and the spectral power of each frequency band, such as the content of theta wave (4–8 Hz), low-alpha wave (8–10 Hz), high-alpha wave (10–13 Hz), low-beta wave (13–20 Hz), high-beta wave (20–30 Hz), and SMR (12–15 Hz) bands, was calculated in addition to the values of the beta per alpha and alpha per high-beta for each channel.

(b) ECG

ECGs were recorded using an ECG-measuring instrument (WEB-7000 and ECG picker, Nihonkohden Co., Japan) at a sampling rate of 1000 Hz. Three electrodes were attached to the chests of subjects using the precordial leads method. The R-R interval (RRI) was calculated from the ECG waveform using MATLAB (Mathwork Co.) programs. The values of the mean of the RRI and the coefficient of variance of RRI [CVRR (100*SD/Mean of RRI)] were calculated from the RRI data. In addition, the spectral power of each frequency band, such as the very low frequency (VLF, 0.001–0.04 Hz), low frequency (LF, 0.04–0.15 Hz), and high frequency (HF, 0.15–0.45 Hz) bands, was calculated from the time series of the RRI using the fast Fourier transform method.

(c) Respiration

A thermal picker (WEB-7000, Nihonkohden Co., Japan) was used to measure the temperature of the breath of subjects. The peak values in the time series graph of the temperature were detected, and the mean and standard deviation of the respiratory cycle time were calculated.

(d) Skin Temperature

Thermocouples were attached at 7 places on the bodies of subjects based on the Hardy–Du Bois method [16] to measure the skin temperature (LT8, GRAM Co., Japan). Finally, the mean skin temperature was calculated using following Equation (1):

$$MST = 0.07\, T_1 + 0.14\, T_2 + 0.05\, T_3 + 0.35\, T_4 + 0.19\, T_5 + 0.13\, T_6 + 0.07\, T_7 \tag{1}$$

where MST is the mean skin temperature based on the Hardy-DuBois method [16], and T_1, T_2, T_3, T_4, T_5, T_6, and T_7 are the temperatures of the forehead, forearms, hands, abdomen, thighs, legs, and feet, respectively.

3.4.3. Facial Expression and Task Performance

The activities of subjects were recorded using the video camera, and the drowsiness levels were recorded by an observer at intervals of 10 s based on Zilberg's criteria [20]. After the experiment was completed, the mean of the data recorded when subjects were performing the MATH task was calculated. The task performance of MATH included the number of correct answers and the mean of the reaction time taken to solve 50 questions. Figure 3 shows a participant completing the experiment. Figure 3 shows photos of the field of experimental scene.

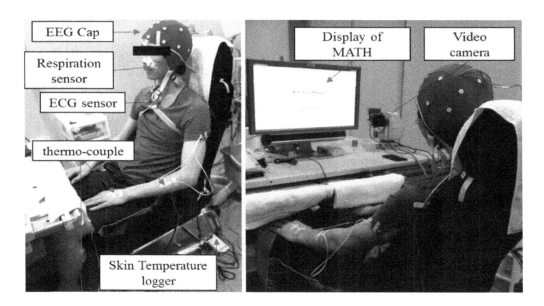

Figure 3. Field photos of the experimental scene.

4. Results and Discussion

4.1. Characteristics of the Arousal Level and Thermal Comfort Corresponding to Changes in the Indoor Ambient Temperature

4.1.1. Subjective Evaluation Value

The results of the subjective evaluation value corresponding to the drowsiness levels and the thermal comfort sensation vote of subjects under conditions A, B, and C are shown in Figures 4 and 5. We used multiple comparison based on the Bonferroni method to investigate significant differences of the subjective evaluation value due to changes in the indoor ambient temperature. There was a small change in the comfort sensation vote, but the drowsiness level increased under condition A, where the indoor temperature was maintained at 27 °C. Under condition B, the comfort sensation vote decreased to an uncomfortable state when the indoor ambient temperature dropped, and increased to a comfortable state when the indoor ambient temperature increased. Under condition C, the comfort sensation vote decreased to an uncomfortable state when the indoor temperature decreased and was then maintained in this uncomfortable state. Under conditions B and C, the drowsiness levels decreased and were maintained at an alert state, corresponding to the drop in the indoor temperature. After the completion of set 4, the comfort sensation vote increased, corresponding to the increase in the indoor temperature under condition B, but was maintained at an uncomfortable state under condition C. The drowsiness level was maintained at a low state under both conditions. According to this result, the arousal level was maintained at a high state even when the temperature increased, and subjects felt comfortable even when the temperature dropped, leading to an increased arousal level. This suggests that there is not always a dependence relationship between the arousal levels and feelings of thermal comfort. Figures 4 and 5 show that when condition B is applied in sets 6 and 7, it becomes possible to improve both the arousal level and thermal comfort of occupants by changing the indoor ambient temperature. These results show that the hypothesis that thermal stimulation due to cooling can improve the arousal level of occupants, after which thermal comfort can also be improved while maintaining a high arousal levels due to the removal of thermal stimulation was verified for condition B. Figure 4 shows the result of the thermal comfort sensation vote. Figure 5 shows the result of the drowsiness level sensation vote.

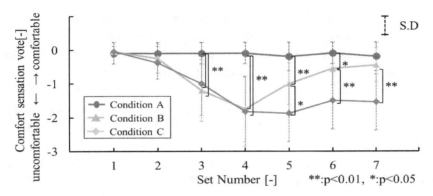

Figure 4. Results of the value of the thermal comfort sensation vote.

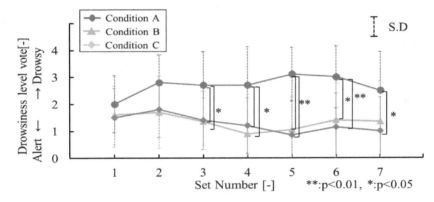

Figure 5. Results of the value of the drowsiness level sensation vote.

4.1.2. Rating Value of Zilberg's Drowsiness Level by Observer

The rating values of Zilberg's drowsiness level for each condition are shown in Figure 6, and it is evident from the figure that the results are similar to those of the subjective sensation vote. Multiple comparison based on the Bonferroni method was conducted to investigate significant differences of the drowsiness level due to the changes in the indoor ambient temperature. After the completion of set 3, the drowsiness levels under conditions B and C decreased significantly compared to that under condition A. From the result of condition B, it was confirmed that the drowsiness level was maintained at a low state after the completion of set 3.

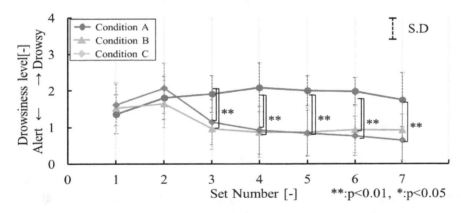

Figure 6. Drowsiness level reached when calculating MATH (analysis of facial expressions based on Zilberg's method [20]).

4.1.3. Task Performance of MATH

The results of the MATH score under each condition are shown in Figure 7, and the results of the response time taken to calculate MATH questions under each condition are shown in Figure 8. Multiple comparison based on the Bonferroni method was conducted to investigate significant differences in the task performance due to changes in the indoor ambient temperature. It can be seen from Figures 7 and 8 that the MATH score was high after set 4, and the response time taken to calculate MATH questions increases in the order of conditions C, B, and then A. Differences were observed between the conditions in the results of the subjective sensation vote and the rating value of Zilberg's drowsiness level. However, significant differences were observed only between the conditions in the result of the response time taken to calculate MATH for set 4, and the result of the MATH score for set 5. It was assumed that the MATH work was easy enough for the subjects to perform, even at a low state of arousal. Figure 6 shows the drowsiness level reached when calculating MATH (analysis of facial expressions based on Zilberg's method [20]). Figures 7 and 8 show the result of the MATH score and the response time from the display of the MATH question to the participant's click.

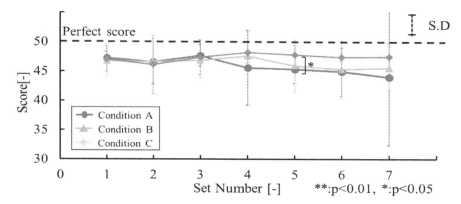

Figure 7. Result of MATH score.

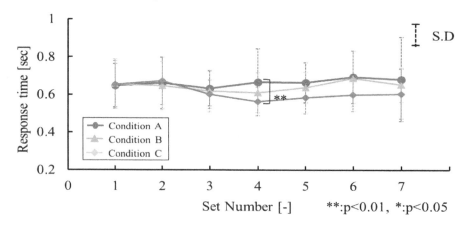

Figure 8. Result of response time.

4.1.4. Consideration of Thermal Stimulation for High Arousal Levels

In the previous studies of Yoshida et al. [12] and Mohri et al. [13], to maintain a high arousal level, the effects of several stimulation methods, such as the use of a fragrance, alarm, and shoulder oscillation, were confirmed. As a result, the sleeping rebound occurred 10 min after the stimulation in those studies. As in our results, high arousal states were maintained for over 10 min after the removal of thermal stimulation. Though further investigation of these comparisons of the time to rebound is necessary, our results showed that thermal stimulation is effective in improving the arousal levels of occupants. Especially, our method is useful in spaces where air conditioning systems are installed, because the installation of additional devices in the space to make stimulations, such as fragrances, alarms, and shoulder oscillation, is unnecessary.

4.2. Relationship between the Subjective Evaluation Value and Physiological Indices

Three hundred and twenty-nine physiological indices were calculated by using the data of the EEGs, ECGs, respiration rate, and mean skin temperature in the experiment (320 indices from EEGs, 7 indices from ECGs, 2 indices from respiration rate, and 1 index from skin temperature). By considering the result of the subjective evaluation value as the standard for the perception of state, we used correlative analysis to investigate the relationship between the subjective evaluation value and the physiological indices. According to the flow of the physiological response shown in Figure 1, changes in both the arousal level and thermal comfort affect the physiological response. To find indices (out of the 329 physiological parameters) that corresponded to the arousal level, we used data from sections when the arousal level changed, but the value of the thermal comfort sensation was maintained constantly. On the other hand, to find indices that corresponded to thermal comfort, we used data from sections when the value of the thermal comfort sensation changed and the arousal level was maintained constantly.

4.2.1. Relationship between the Drowsiness Level and Physiological Indices

As a result of the correlation analysis, the correlation coefficients between values of the drowsiness level vote and physiological indices were calculated using data from 57 sets in which the value of the drowsiness level vote varied, but the value of the comfort sensation vote remained at 0 (comfortable). There was a significant correlation between the drowsiness level and 102 physiological indices (data $N = 57$, $r > 0.339$, $p < 0.01$). The indices in relation to the ECGs were not included in these parameters; however, the indices in relation to the EEGs were included. Thus, this result showed that the relationship between the reaction of the central nervous system and changes in arousal levels is more significant than that between the reaction of the autonomic nervous system and changes in arousal levels.

4.2.2. Relationship between Thermal Comfort and Physiological Indices

As a result of the correlation analysis, the correlation coefficients between the values of the comfort sensation vote and physiological indices were calculated using the data from 98 sets in which the value of the comfort sensation vote varied, but the value of the drowsiness level vote was between 0 (alert) and 1 (slightly drowsy). There was a significant correlation between the value of the comfort sensation vote and seven physiological indices in relation to skin temperature, EEGs, and ECGs (data $N = 98$, $r > 0.259$, $p < 0.01$). The change in the mean skin temperature is related to thermal stimulation from the outer environment, which is transmitted to the central nervous system and perceived as thermal comfort, thus affecting the parameters of the EEGs. It was observed that changes in the thermal comfort affected the autonomic nervous system, which in turn affected the indices in relation to the ECGs. It appears that this process affected the result of the correlation analysis in this experiment.

4.2.3. Calculation of the Evaluation Index Using Multiple Regression Analysis

To propose an index that evaluates the changes in each arousal level and thermal comfort states, multiple regression analysis was performed using the physiological indices that have a significant correlation with the value of subjective evaluation. It was found that 102 parameters had a significant correlation with the drowsiness level vote, and seven parameters with the comfort sensation vote. These were thus considered as explanatory variables. Because an explanatory variable should have a high correlation with the subjective sensation vote, to obtain a high multiple regression coefficient, we selected an explanatory variable for use in the multiple regression analysis using the following process.

- Indices that had a significant correlation with the value of subjective evaluation were sorted according to their correlation coefficient from high to low ($x1$, $x2$, ... $x102$).
- $x1$ was selected as the explanatory variable, since it had the highest correlation coefficient with the value of subjective evaluation.

- $x2$ was selected as the explanatory variable if there was no significant correlation between $x2$ and $x1$.
- xn was selected as the explanatory variable if there was no significant correlation between xn and all the parameters selected previously as explanatory variables.

(a) Evaluation Index indicating the Arousal Level

The result of the multiple regression analysis 1 (Yd) is shown in Table 1, and the regression equation obtained from the result is expressed in Equation (2). The variables of the regression equation include the indices of the EEGs. As a result, the coefficient of determination, R^2, in relation to Yd was 0.750.

$$Yd = -3.31 + 0.134\, Xd_1 + 2.520\, Xd_2 + 0.088\, Xd_3 - 0.268\, Xd_4 + 0.271\, Xd_5 \qquad (2)$$

where Xd_1, Xd_2, Xd_3, Xd_4, and Xd_5 are the high alpha content of T7, beta per alpha content of F7, beta content of F3, low beta content of F7, and alpha content of Fp1, respectively.

Table 1. Result of multiple regression 1 (Yd).

Model Summary				
R	0.8657		Std. Error	0.7358
R^2	0.7495		data N	57
Adjusted R^2	0.7249			

Index	Coefficient	Std.Error	t	p-value
Intercept	−3.3123	0.4959	−4.6631	2.28×10^{-5}
T7_High-alpha	0.1338	0.0201	6.6478	1.96×10^{-8}
F7_Beta/Alpha	2.5196	0.3073	8.1996	7.09×10^{-11}
F3_Beta	0.0878	0.0245	3.5817	0.0008
F7_Low beta	−0.2684	0.0481	−5.5791	9.29×10^{-7}
Fp1_Alpha	0.2708	0.0315	8.5969	1.72×10^{-11}

We defined Yd as the index of the arousal level. To confirm the possibility that Yd can evaluate the arousal level even when the thermal comfort changed, we calculated Yd from the data obtained from five subjects for whom the arousal level and thermal comfort changed considerably, and then performed a correlation analysis between Yd and the evaluation value of the drowsiness level. As a result of the correlation analysis, the correlation coefficient (R) was calculated to be 0.726, and there was a significant correlation between Yd and the evaluation value of the drowsiness level. Therefore, this result suggests that Yd can be used to evaluate the drowsiness level, even when there are changes in the thermal comfort.

(b) Evaluation Index Indicating Thermal Comfort

The result of the multiple regression analysis 2 (Yc) is shown in Table 2, and the regression equation obtained from the result is expressed in Equation (3). The variable for the regression equation included the indices of the skin temperature and EEGs. As a result, the coefficient of determination, R^2, in relation to Yc was 0.528.

$$Yc = -23.372 + 0.697\, Xc_1 + 0.172\, Xc_2 + 0.142\, Xc_3 \qquad (3)$$

where Xc_1, Xc_2, and Xc_3 are the MST (Mean Skin Temperature), alpha per high-beta content of C4, and alpha per high beta content of P4, respectively.

Table 2. Result of multiple regression 2 (Yc).

Model Summary				
R	0.7267		Std. Error	0.6859
R^2	0.5276		data N	98
Adjusted R^2	0.5125			

Index	Coefficient	Std.Error	t	p-value
Intercept	−23.3721	2.5461	−9.1795	1.02×10^{-14}
MST	0.6967	0.0765	9.1119	1.42×10^{-14}
C4_Alpha/HB	0.1718	0.0586	2.9295	0.0043
P4_Alpha/HB	−0.1418	0.0296	−4.7885	6.25×10^{-6}

We defined Yc as the index of thermal comfort. To confirm the possibility that Yc can evaluate thermal comfort even when there were changes in the drowsiness level, we calculated Yc from the data of nine subjects, excluding the data of participant D whose thermal comfort vote and drowsiness level vote changed slightly. We thus confirmed the relationship between Yc and the value of thermal comfort. As a result, the correlation coefficient (R) was calculated to be 0.728, and there was a significant correlation between Yc and the value of thermal comfort ($p < 0.01$). Therefore, this result suggests that Yc can be used to evaluate thermal comfort even when there are changes in the drowsiness level.

Table 1 shows the result of the multiple regression 1 (Yd), and Table 2 shows the result of the multiple regression 2 (Yc). Figure 9 shows the relationship between the drowsiness level vote and Yd(a), and the relationship between the comfort sensation and Yc (b)

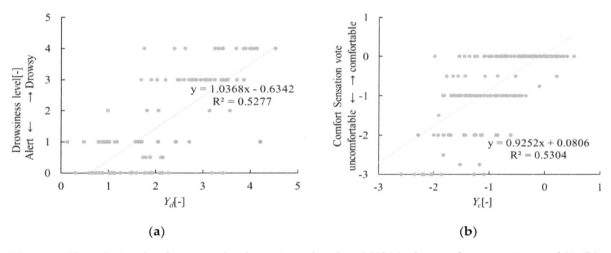

(a)   (b)

Figure 9. The relationship between the drowsiness level and Yd (**a**), the comfort sensation and Yc (**b**).

(c) Review of Continuous Evaluation using Yd and Yc

As mentioned in Section 2.2., indices that can continuously evaluate the drowsiness level and thermal comfort of occupants are needed to clarify the design requirements related to thermal factors that can improve these states of the occupants. Therefore, we conducted a time-series analysis with Yd and Yc to confirm the possibility that Yd and Yc can continuously evaluate arousal levels and thermal comfort. We calculated the time series of Yd using data from subject C, who seemed to be drowsy frequently throughout the experiment, and compared these data with those of the drowsiness level, which was recorded at intervals of 10 s by analyzing the facial expressions of the subject based on Zilberg's method [20]. The time series analyses of the normalized Yd and the drowsiness level of subject C are shown in Figure 10a,b. The figures clearly show the trend of changes in both indicators. It can be seen that Yd and values of the drowsiness level show a similar trend. For example, the red circles in Figure 10a,b indicate that values of the drowsiness level and the normalized Yd are higher than 3 (significantly drowsy) and 0, respectively, at the same time. Furthermore, it can be seen from Figure 11 that there is a similar trend between Yc and the value of the comfort sensation vote. In Figure 11a,b, the value of the comfort sensation drops to -3 (very uncomfortable) at 30 min, and at the same time, the value of Yc drops and becomes a negative number. Thereafter, the value of the comfort sensation vote remains at an uncomfortable state (-3: Very uncomfortable, -2: Uncomfortable), and similarly, the value of Yc remains at a negative value. These results suggest that Yd and Yc can be used to evaluate the drowsiness level and thermal comfort of occupants continuously. Figures 10 and 11 show, respectively, changes in the normalized Yd (a) and drowsiness level (b), (subject C) and changes in Yc (a) and the comfort sensation vote (b) (subject B).

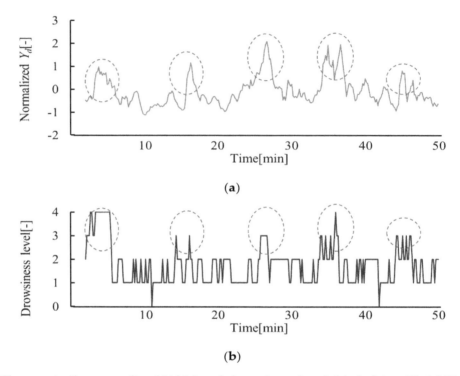

Figure 10. Changes in the normalized Yd (**a**) and drowsiness level (**b**), (subject C): (**a**) Time series of normalized Yd, (**b**) Time series of the value of the drowsiness level.

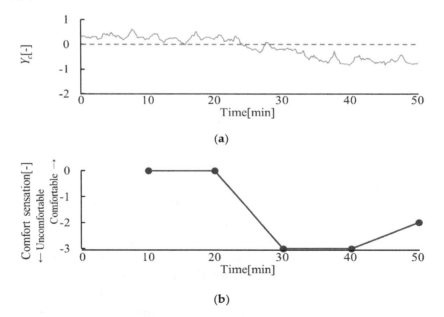

Figure 11. Changes in Yc (**a**) and the value of the comfort sensation vote (**b**): Time series of the value of the comfort sensation vote, (subject B).

(d) Review of the Utility and Validity of the Evaluation Index

To review the utility and validity of the evaluation index, as an example, we set the threshold value in relation to Yd and the drowsiness level. When the threshold value of Yd is sets as 2.3, as shown in Figure 12, Yd can be used as the classifier, which can classify the drowsiness level between level 0, 1, 2, or more with an 86.7% accuracy. This result suggests that there is a significant difference in the physiological reaction between the drowsiness level of 1 (slightly drowsy) and 2 (moderately drowsy), and is similar trend to the results of a previous study about the relation between the arousal levels and physiological indices of a driver [21], though the task of the subjects is different from that of the previous study (arithmetic task vs driving). This result also suggests that Yd is a valid indicator of the drowsiness level of occupants to classify between the drowsiness level of 1 (slightly drowsy) and 2 (moderately drowsy). Figure 12 shows the setting of the threshold value in relation to Yd.

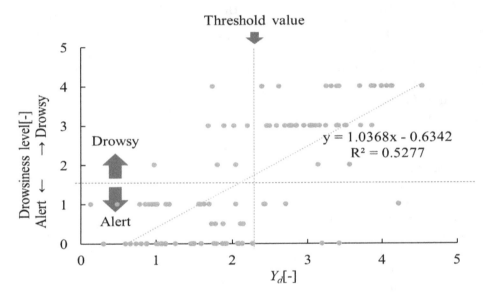

Figure 12. Setting of the threshold value in relation to Yd.

5. Conclusions

In this study, we aimed to design a thermal environment that can improve the indicators of both the arousal level and feelings of thermal comfort. We hypothesized that thermal stimulation due to cooling can improve the arousal level of occupants, after which thermal comfort can be improved while maintaining high arousal levels due to the removal of thermal stimulation. To verify the hypothesis, we measured physiological indices, values of subjective evaluation, and performances of arithmetic tasks throughout several thermal conditions in which the indoor ambient temperature was changed. In addition, we investigated the relationships between them to identify the indices that can be used to evaluate the arousal levels and thermal comfort of occupants. As a result, the following findings were noted:

- When the indoor ambient temperature decreased and then increased, both the arousal level and thermal comfort of occupants remained at high levels. This result suggests that the hypothesis of this study was verified and changes in the indoor ambient temperature can be used to improve both thermal comfort and the arousal level of occupants.
- We proposed the evaluation indices of thermal comfort and the drowsiness level of occupants. It was observed that the drowsiness level and thermal comfort of occupants can be evaluated quantitatively and continuously using Yd and Yc, which were obtained from the equation consisting of physiological indices in relation to EEGs, ECGs, and skin temperature.

In future work, we will investigate the relationship between the change patterns of the indoor ambient temperature, thermal comfort, and arousal levels of occupants. After that, we aim to design a novel thermal environment, considering all comfort parameters based on these findings, with the aim of improving both the arousal levels and feelings of thermal comfort of occupants, and then carry out an evaluation of the validity of the designed thermal environment.

Author Contributions: Conceptualization, J.G., M.S. and M.K.; methodology, J.G., M.S., K.U. and M.K.; software, J.G.; validation, J.G.; formal analysis, J.G.; investigation, J.G.; resources, M.S., K.U. and M.K.; data curation, J.G.; writing—original draft preparation, J.G.; writing—review and editing, J.G., M.S., K.U. and M.K.; visualization, J.G.; supervision, M.S. and M.K.; project administration, M.S. and M.K.; funding acquisition, M.S. and M.K.

Formula Symbols

Index	Meaning	Unit
MST	Mean skin temperature	°C
T_1	Temperature of the forehead	°C
T_2	Temperature of the forearms	°C
T_3	Temperature of the hands	°C
T_4	Temperature of the abdomen	°C
T_5	Temperature of the thighs	°C
T_6	Temperature of the legs	°C
T_7	Temperature of the feet	°C
Yc	Index of thermal comfort	
Yd	Index of drowsiness level	
Xd_1	High alpha content of T7	
Xd_2	Beta content per alpha content of F7	
Xd_3	Beta content of F3	
Xd_4	Low beta content of F7	
Xd_5	Alpha content of Fp1	
Xc_1	Mean skin temperature	°C
Xc_2	Alpha content per high beta content of C4	
Xc_3	Alpha content per high beta content of P4	

References

1. ISO7730:2005. Ergonomics of the thermal environment-Analytical determination and interpretation of thermal comfort using calculation of the PMV and PPD indices and local thermal comfort criteria
2. De Dear, R.; Brager, G.S. Developing an adaptive model of thermal comfort and preference. *ASHRAE Trans.* **1998**, *104*, 145–167.
3. Yang, Y.; Li, B.; Liu, H.; Tan, M.; Yao, R. A study of adaptive thermal comfort in a well-controlled climate chamber. *Appl. Therm. Eng.* **2015**, *76*, 283–291. [CrossRef]
4. Cannistraro, M.; Cannistraro, G.; Restivo, R. Some Observations on the Influence on Exchanges Radiative. *Int. J. Heat Techonol.* **2015**, *33*, 115–122. [CrossRef]
5. Cannistraro, M.; Lorenzini, E. The Applications of the New Techonologies E-Sensing in Hospitals. *Int. J. Heat Techonol.* **2016**, *34*, 551–557. [CrossRef]
6. Cannistraro, M.; Cannistraro, G.; Restivo, R. Smart Controll of Air Climatization System in Function on the Values of the Mean Local Radiant Temperature. *Smart Sci.* **2015**, *3*, 157–163. [CrossRef]
7. Liu, J.; Yao, R.; McCloy, R. A method to weight three categories of adaptive thermal comfort. *Energy Build.* **2012**, *47*, 312–320. [CrossRef]
8. McIntyre, D.A. *Indoor Climate*; Applied science publishers: London, UK, 1980.
9. Gohara, T.; Iwashita, G. Disscussion of relationship between indoor environmental comfort and performance in simple task. *J. Environ. Eng.* **2003**, *572*, 75–80. [CrossRef]

10.	ASHRAE-55. *Thermal Environmental Conditions for Human Occupancy*; American Society of Heating, Refrigerating and Air-conditioning Engineers: Atlanta, GA, USA, 2004.

11.	Kondo, E.; Kurazumi, Y.; Horikoshi, T. Human physiological and psychological reactions to thermal transients with air temperature step changes: A Case of Climacteric Aged Females in Summer. *J. Hum. Living Environ. Jpn.* **2014**, *21*, 75–84.

12.	Yoshida, M.; Kato, C.; Kakamu, Y.; Kawasumi, M.; Yamasaki, H.; Yamamoto, S.; Nakano, T.; Yamada, M. Study on Stimulation Effects for Driver Based on Fragrance Presentation. In Proceedings of the MVA2011 IAPR Conference on Machine Vision Applications, Nara, Japan, 13–15 June 2011; pp. 332–335.

13.	Mohri, Y.; Kawaguchi, M.; Kojima, S.; Yamada, M.; Nakano, T.; Mohri, K. Arousal retention effect of magnetic stimulation to car drivers preventing drowsy driving without sleep rebound. *Trans. Inst. Electr. Eng. Jpn.* **2016**, *136*, 383–389. [CrossRef]

14.	Mardi, Z.; Naghmeh, S.; Ashtiani, M.; Mikaili, M. EEG-based drowsiness detection for safe driving using chaotic features and statistical tests. *J. Med. Signals Sens.* **2011**, *1*, 130–137. [PubMed]

15.	Furman, G.D.; Baharav, A.; Cahan, C.; Akselrod, S. Early detection of falling asleep at the wheel: A heart rate variability approach. *Comput. Cardiol.* **2008**, *35*, 1109–1112.

16.	Hardy, J.D.; Dubois, E.F. The technic of measuring radiation and convection. *Nutrition* **1938**, *5*, 461–475. [CrossRef]

17.	Yao, Y.; Lian, Z.; Liu, W.; Jiang, C.; Liu, Y.; Lu, H. Heart rate variation and electroencephalograph-the potential physiological factors for thermal comfort study. *Indoor Air* **2009**, *19*, 93–101. [CrossRef] [PubMed]

18.	Turner, J.R.; Hewitt, J.K.; Morgan, R.K.; Sims, J.; Carroll, D.; Kelly, K.A. Graded mental arithmetic as an active psychological challenge. *Int. J. Psychophysiol.* **1986**, *3*, 307–309. [CrossRef]

19.	SO10551. First edition 1995-05-15, Ergonomics of the thermal environment–Assessment of the influence of the thermal environment using subjective judgement scales.

20.	Zilberg, E.; Burton, D.; Xu, Z.M.; Karrar, M. Methodology and initial analysis for development of non-invasive and hybrid driver drowsiness detection systems. In Proceedings of the 2nd IEEE International Conference Wireless Broadband and Ultra Wideband Communications, AusWireless, Sydney, Australia, 27–30 August 2007; p. 16.

21.	Gwak, J.S.; Shino, M.; Hirao, A. Eary detection of driver drowsiness using machine learning based on physiological signals, behavioral measures and driving performance. In Proceedings of the 21th IEEE International Conference on Intelligent Transportation Systems, Maui, Hawaii, USA, 4–7 November 2018; pp. 1794–1800.

11

Indoor Air Quality: A Focus on the European Legislation and State-of-the-Art Research

Gaetano Settimo [1], Maurizio Manigrasso [2] and Pasquale Avino [3],*

[1] Italian Institute of Health, viale Regina Elena 299, I-00185 Rome, Italy; gaetano.settimo@iss.it
[2] Department of Technological Innovations, INAIL, via Roberto Ferruzzi 38, I-00143 Rome, Italy; m.manigrasso@inail.it
[3] Department of Agriculture, Environmental and Food Sciences, University of Molise, via F. De Sanctis, I-86100 Campobasso, Italy
* Correspondence: avino@unimol.it.

Abstract: The World Health Organization (WHO) has always stressed the importance of indoor air quality (IAQ) and the potential danger of pollutants emitted from indoor sources; thus, it has become one of the main determinants for health. In recent years, reference documents and guidelines have been produced on many pollutants in order to: i) decrease their impact on human health (as well as the number of pollutants present in indoor environments), and ii) regulate the relevant levels of chemicals that can be emitted from the various materials. The aim of this paper is to discuss and compare the different legislations present in the European Union (EU). Furthermore, a focus of this paper will be dedicated at Italian legislation, where there is currently no specific reference to IAQ. Although initiatives in the pre-regulatory sector have multiplied, a comprehensive and integrated policy on the issue is lacking. Pending framework law for indoor air quality, which takes into account WHO indications, the National Study Group (GdS) on Indoor Air Pollution by the Italian Institute of Health (IIS) is committed to providing shared technical-scientific documents in order to allow actions harmonized at a national level. An outlook of the main Italian papers published during these last five years will be reported and discussed.

Keywords: indoor air quality; legislation; Europe; focus; residential; pollutants; TLV; health; workers; school

1. Introduction

Indoor air quality (IAQ) has been a well-known problem since the late 1970s. Its significant impact on human health has been addressed several times by the World Health Organization (WHO) in various documents and meetings, and has been carried out at various levels [1–3]. Further, economic studies and researches have highlighted the great importance that IAQ now has in all environments, e.g., houses, schools, banks, post offices, offices, hospitals, and public transport, just to name a few [4]. IAQ also has strong repercussions in the competitiveness of an organization, considering the increase in difficulty in carrying out its job in the best way, its performance, and the social and economic competitiveness between countries, due to the influence on the attention, degree, and number of days lost [5].

Scientific literature contains large documentation in terms of articles, conference papers, reviews, books, editorials, letters, and public articles on chemical contaminants in indoor environments. A search on the Scopus literature database, using the keyword "indoor air quality", led to a total of 7287 publications between 2000 and 2020 in the European Union (EU) (search executed on 19 January 2020), including Norway, Switzerland, and Turkey. According to this search, Italy and the United Kingdom (UK) are major contributors to this total amount of European publications, 12.3% and

10.4% of the total, respectively, followed by France and Germany with 9.5% and 9.1%. Figure 1 shows the relative percentage contribution of each Member State of the European Union (EU) including Norway, Switzerland, Turkey, as well as United Kingdom, which is expected to leave the EU on 31 January 2021. The United Kingdom, Italy, France, and Germany contribute more than 41% of the total amount of publications in the European IAQ field.

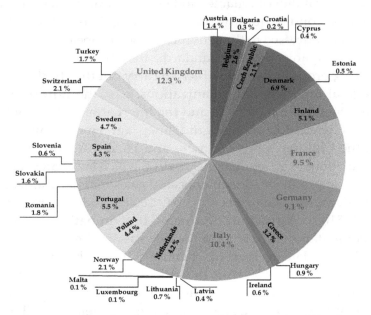

Figure 1. Percentage of country contributions to the total amount of publications on indoor air quality (IAQ) in Europe from 2000 to 2020. (Source: Scopus, search: 19 January 2020); includes Norway, Switzerland, Turkey, and the UK, which is to be expected to leave UE on 31 January 2021.

This continuous and growing attention on IAQ has evidenced, over time, the need for a profound cultural change, according to WHO indications, in order to develop organic health prevention and promotion actions, and cope with the complexity of such an issue.

Noteworthy, at the European Community level, the resolution of 13 March 2019, defends clean air for everyone and highlights that people spend almost 90% of their time in indoors [6]. In these environments, the air can be significantly more polluted compared to outside [7] and, therefore, considered mandatory to issue indoor air quality certificates for both new and old buildings. From this perspective, it urges member states to adopt and implement measures to combat air pollution at the source.

There are specific cases, such as schools, healthcare, or office environments, where the permanence of workers (e.g., medical, administrative, teaching, and non-teaching staff) is supposed to last for a relatively long period, and where "users", as well, are present (e.g., patients, students, vulnerable and/or fragile subpopulation, some of which with physical and psychological disabilities, etc.). In these situations, it is essential to consider the very close relationships between the various work activities and the quality of the building structure, finish, furnishings, and the degree of crowding of such environments. This includes the presence of technological systems or interventions for energy purposes only, without forgetting the ventilation needs of the environment for aspects related to health, performance, and staff and student performance [8].

The combination of these actions is fundamental for developing and implementing plans for the protection and promotion of health safety for citizens and workers [9]. This represents the priority and the common objectives of both national and European prevention plans (National Prevention Plans (NPPs) and programs from the United Nations (UN) Sustainability Development Agenda).

Nonetheless, several European countries have had to overcome the absence of specific legislation, or legislative acts already developed, due to generic definitions of the characteristics of air quality.

For instance, in closed workplaces, such as closed offices where employees have individual working areas that are distinctly divided—either by walls, cubicles, or panels—it is necessary to ensure that workers have healthy air in sufficient quantity, which is also obtained with ventilation systems [10]. It is necessary for updated laws and regulations to be adopted to improve the indoor air quality.

Another fundamental requirement for correct understanding of the air quality pollution phenomena indoors is the availability of reliable (and systematically collected) information, according to well-established protocols, on the quality, quantity, and origin of the pollutants. In this regard, particular attention should be addressed to the activities of the European Committee for Standardization (CEN) and the International Organization for Standardization (ISO), which provide a series of specific indications on the operating procedures with which to carry out the checks.

In recent years, several international organizations, e.g., the European Collaborative Action (ECA), the World Health Organization (WHO), and the International Agency for Research on Cancer (IARC) have produced reference documents, guidelines, agreements, and protocols. For example, the Parma declaration, the Children's Environment and Health Action Plan for Europe CEHAPE), European Union (EU) regulations (e.g., regulation 305/2011, which lays down harmonized conditions for the marketing of construction products); documents, and rules for characterization and determination on many pollutants (e.g., European Standards (EN) ISO 16000—Indoor air quality, European technical specification (CEN/TS) 16516: construction products—determination of emission into indoor air). The purpose of this documentation is to tend to the decrease in the number of pollutants present in indoor environments and to regulate the levels of chemicals that can be emitted from different materials, in order to contain the negative impacts on IAQ. In particular, the activities carried out by the European Committee for Standardization (CEN) and the International Organization for Standardization (ISO) represent important references, because harmonized methods of detection allow for better comparison between the different indoor air quality data produced at the European level. Such methods should be implemented by laboratories that carry out environmental surveys.

Within this context, the aim of this paper is to summarize the entire legislation on IAQ present in the EU (the UK included), in February 2020, along with reference values, guide values, and unitary risks for many kinds of indoor air pollutants present. Particularly, the foundations of the different legislations will be compared for evidencing the main characteristics of each one, and the levels of the main pollutants will be presented and discussed. The focus is to highlight the strengths and weaknesses to deal with this important topic. According to the authors' knowledge, this is the first critical revision of the European legislation. Further, a section will be dedicated to the state-of-the-art research in Italy, from a legislative and scientific point of view. Although there have been many scientific papers and studies performed on IAQ, and a methodic and analytical review of the papers published in the last five years concerning the indoor field will be documented, it will be highlighted that, in Italy, the main problem is the lack of reference standards for residential indoor air quality.

2. The Main European Legislation on Indoor Air Quality

The WHO has developed guidelines for IAQ, relating to a certain number of pollutants, present indoor, for which scientific knowledge relating to human health effects were considered robust enough. The substances considered are benzene (C_6H_6, CAS number 71-43-2), nitrogen dioxide (NO_2, 10102-44-0), polycyclic aromatic hydrocarbons (especially benzo[a]pyrene BaP, $C_{20}H_{12}$, 50-32-8) (PAHs), naphthalene ($C_{10}H_8$, 91-20-3), carbon monoxide (CO, 630-08-0), radon, trichlorethylene (C_2HCl_3, 79-01-6), and tetrachloroethylene (C_2Cl_4, 127-18-4). For carcinogenic pollutants (such as benzene, BaP, trichloroethylene), a unitary risk (UR) is defined for the general population associated with their presence in the air. Alongside these guidelines, mention should be made of those relating to the risks associated with the presence of humidity and biological agents. Furthermore, for the purpose of risk assessment, it is of particular importance to consider not only the guide value or reference parameter, but also other fundamental elements, such as the vulnerability of the population and the exposure conditions.

There is no specific reference directive on IAQ in European legislation, although pre-legislative initiatives have multiplied over the years. For example, indoor air quality and its impact on human activities within the European Collaborative Action (ECA), e.g., Urban Air, Indoor Environment, and Human Exposure, as well as funded studies, EN standards, etc.); however, to date, there is still no integrated policy on indoor air quality in all of those indoor places.

Some EU Member States, such as France, Portugal, Finland, Austria, Belgium, Germany, the Netherlands, and Lithuania, have started, through a series of actions, to adopt specific guide values, reference values, and action values for IAQ—in some cases enforced in the legislative acts of these countries. These actions can be summarized as follows:

- definition and imposition of reference concentration values on selected pollutants, in line with those developed by the WHO for some time;
- national plans on IAQ;
- drafting of legislative acts for indoor environments;
- setting up and planning mandatory indoor air monitoring activities;
- training and information programs dedicated to technical offices, managers, and staff on IAQ issues;
- protocols and guides for self-diagnostic activities based on scientific knowledge and practical experience on indoor air quality.

One particular aspect is to give indications for the IAQ evaluation at workplaces other than industrial ones. Currently, in order to evaluate the IAQ in environments where work is carried out (e.g., in offices, schools, hospitals, banks, post offices, etc.), the occupational exposure limit values (OELs) present in the regulations, or the threshold limit values (TLVs) of the American Conference of Governmental Industrial Hygienists (ACGIH), or the Scientific Committee For Occupational Exposure Limits (SCOEL-RAC) are used—albeit reduced by 1/10 or 1/100. This approach is overcome, as indicated, by specific documents elaborated by different national and European working groups on the indoor topic [11–14]. Such recommendations, as given by the WHO in the early 1980s in the document "Indoor air pollutants exposure and health effects" [15], reported that it was incorrect to use the industrial occupational exposure limit values for non-industrial indoor environments, and that for such environments, it was necessary to develop specific references. It should be remembered that these values represent the parameters to which references must be made for an assessment of the inhalation risk of workers and the population. They are not the only ones, because specific exposure and vulnerability conditions are fundamental elements to be considered for a correct risk assessment. In the document "Opinion on risk assessment on indoor air quality" [16], the Scientific Committee on Health and Environmental Risks (SCHER) of the European Commission recommends that risk assessment should always be focused on the most vulnerable groups, represented by children, pregnant women, elderly people (over 65), people suffering from asthma, and other respiratory and cardiovascular diseases, following a case-by-case approach. In fact, for groups of particularly sensitive and vulnerable individuals, who are potentially being exposed to the risk factors under consideration, the problem of the simultaneous presence of multiple risk factors may require the need to carry out specific in-depth assessments, which must be based on adequate knowledge of the context. In fact, it should be remembered that the reference values for confined spaces are more severe than the corresponding values in industrial environments (TLVs) whose hygienic-sanitary references are based on a working life of 8 h a day, 5 days a week, for a maximum period of 40 years, and are aimed at protecting workers against occupational diseases.

In this context, the efforts carried out by bodies such as ISO and CEN, which have long been involved in the development of the specific standard "EN ISO 16000: Indoor air" [17], which describes the procedures for performing sampling activities and analyzes the main pollutants indoors, should not be forgotten. The adoption of these rules constitutes a significant improvement compared to what has been achieved so far in the study and control activities. The standardization of the methods also increases the possibility of a correct comparison between the different indoor air quality data produced at the European level [11]. The advantage is, also, in terms of the possibility of the correct comparison between the various IAQ data produced at the European level, underlining the need for timely application of the rules. This is particular so for the sampling phase (e.g., choice of the sampling point and height, distance from walls, preliminary activities, etc.), which represents the beginning of the control procedure and, therefore, conditions the final result. Table 1 shows the 40 parts of the ISO 16000 standard [17].

Table 1. List of International Organization for Standardization (ISO) 16000 series for IAQ. EN = European Standard.

	EN ISO 16000 « Indoor Air »
Part 1	General aspects of sampling strategy
Part 2	Sampling strategy for formaldehyde
Part 3	Determination of formaldehyde and other carbonyl compounds—active sampling method
Part 4	Determination of formaldehyde—diffusive sampling method
Part 5	Sampling strategy for volatile organic compounds (VOCs)
Part 6	Indoor air Determination of volatile organic compounds in indoor and test chamber air by active sampling on Tenax TA sorbent, thermal desorption, and gas chromatography using MS or MS-flame ionization detector (FID)
Part 7	Sampling strategy for determination of airborne asbestos fiber concentrations
Part 8	Determination of local mean ages of air in buildings for characterizing ventilation conditions
Part 9	Determination of the emission of volatile organic compounds from building products and furnishing—emission test chamber method
Part 10	Determination of the emission of volatile organic compounds from building products and furnishing—emission test cell method
Part 11	Determination of the emission of volatile organic compounds from building products and furnishing—sampling, storage of samples, and preparation of test specimens
Part 12	Sampling strategy for polychlorinated biphenyls (PCBs), polychlorinated dibenzo-p-dioxins (PCDDs), polychlorinated dibenzofurans (PCDFs), and polycyclic aromatic hydrocarbons (PAHs)
Part 13	Determination of total (gas and particle-phase) polychlorinated dioxin-like biphenyls and polychlorinated dibenzo-p-dioxins/dibenzofurans—collection on sorbent-backed filters with high resolution gas chromatographic/mass spectrometric analysis
Part 14	Determination of total (gas and particle-phase) polychlorinated dioxin-like biphenyls (PCBs) and polychlorinated dibenzo-p-dioxins/dibenzofurans (PCDDs/PCDFs)—extraction, clean up, and analysis by high-resolutions gas chromatographic and mass spectrometric analysis)
Part 15	Sampling strategy for nitrogen dioxide (NO_2)
Part 16	Detection and enumeration of molds—sampling of molds by filtration
Part 17	Detection and enumeration of molds—culture-based method
Part 18	Detection and enumeration of molds—sampling by impaction
Part 19	Sampling strategy for molds
Part 20	Detection and enumeration of molds—determination of total spore count
Part 21	Detection and enumeration of molds—sampling from materials
Part 22	Detection and enumeration of molds—molecular methods
Part 23	Performance test for evaluating the reduction of formaldehyde concentrations by sorptive building materials
Part 24	Performance test for evaluating the reduction of volatile organic compound (except formaldehyde) concentrations by sorptive building material
Part 25	Determination of the emission of semi-volatile organic compounds by building products—micro-chamber method
Part 26	Sampling strategy for carbon dioxide (CO_2)
Part 27	Determination of settled fibrous dust on surfaces by SEM (scanning electron microscopy) (direct method)
Part 28	Determination of odor emissions from building products using test chambers

Table 1. *Cont.*

	EN ISO 16000 « Indoor Air »
Part 29	Test methods for VOC detectors
Part 30	Sensory testing of indoor air
Part 31	Measurement of flame retardants and plasticizers based on organophosphorus compounds—phosphoric acid ester
Part 32	Investigation of buildings for pollutants and other injurious factors—inspections
Part 33	Determination of phthalates with gas chromatography/mass spectrometry (GC/MS)
Part 34	General strategies for the measurement of airborne particle
Part 35	Measurement of polybrominated diphenylether, hexabromocyclododecane, and hexabromobenzene
Part 36	Test method for the reduction rate of airborne bacteria by air purifiers using a test chamber ISO 16000
Part 37	Strategies for the measurement of $PM_{2.5}$
Part 38	Determination of amines in indoor and test chamber air—active sampling on samplers containing phosphoric acid impregnated filters
Part 39	Determination of amines—analysis of amines by (ultra-)high-performance liquid chromatography coupled to high resolution or tandem mass spectrometry
Part 40	Indoor air quality management system

In some EU countries such as France, Belgium, Portugal, etc., there are specific legislations for each pollutant and the relative reference ISO standards to be used. The great confusion of these years has been precisely the absence of sampling and analysis standards dedicated to IAQ. Standards for industrial environments were often used, i.e., National Institute for Occupational Safety and Health (NIOSH), Occupational Safety and Health Administration (OSHA), etc.), which have mg m^{-3} sensitivities (and have nothing to do with indoor μg m^{-3} concentrations). Against this background, the adoption of the ISO 16000 standard represented a significant improvement as to the study and control activities.

Now, following behavior consolidated in several countries, it is, therefore, appropriate to develop indoor specific harmonized reference values in order to better manage particularly problematic situations in such environments. In the absence of specific national references to be used for a comparison, those reported by ad hoc working groups, or in the legislation of other European countries, are currently used.

Several EU countries, in recent years, have set up working groups with a specific mandate to develop guide values for air quality in confined spaces. Table 2 shows a series of guide values, present in the official documents, for selected pollutants, including those considered in the WHO guidelines.

For instance, Germany, by the German Working Group on Indoor Guideline Values of the Federal Environmental Agency and the States' Health Authorities (AG IRK/AOLG) [18], used a methodology starting from Lowest Observed Adverse Effect Level (LOAEL), or lower level of exposure to a toxic pollutant, for which negative health effects have been observed, introducing safety factors, such as inter- and intra-species. On the other hand, the UK adopted a different approach. In particular, the commission on the effects of air pollution on human health, i.e., the Committee on the Medical Effects of Air Pollutants (COMEAP) (updated in 2020) [19] and the Royal College of Pediatrics and Child Health (RCPCH) [20], developed guide values on the basis of WHO studies. France did the same, thanks to the collaboration between the French Scientific and Technical Center for Construction (CSTB) and the French Agency for Environmental and Occupational Health Safety (AFSSET) [21–23]. The working group developed a long series of studies to arrive at the elaboration of guide values for eight pollutants, such as hydrogen cyanide, carbon monoxide, benzene, formaldehyde, trichlorethylene, tetrachlorethylene, naphthalene, PM_{10} and $PM_{2.5}$. This activity was part of the National Health and Environment Plan PNSE 2004–2008 [24], followed by the second National Plan for Health and Environment (PNSE 2) that was published for the period 2009–2013. Alongside the AFSSET indications, the authors would like to mention those identified by the High Council of Public Health-Haut Conseil de la Santé

Publique (HCSP), which elaborated a series of documents on the values of action, and long-term for the evaluation of IAQ [25].

France implemented a plan of targeted interventions with the enactment of law no. 2010-788 of 12 July 2010, which is continuously updated (the last one in 2016), and establishes the obligation of periodic monitoring of the air quality in confined spaces, as well as the responsibility of the owners or occupants, gradually in force:

-       from 1 January 2018 for confined spaces, such as leisure centers, swimming pools, health facilities, social services, and nurseries with children under 6 years of age;
-       from 1 January 2018 for elementary education institutions;
-       before 1 January 2020 for juvenile detention facilities and first- and second-degree education or vocational training institutions;
-       from 1 January 2023 for all other indoor spaces.

For structures open to the public after these dates, the first periodic monitoring must be carried out no later than 31 December of the year, following the opening of the structure. Failure to comply with the terms of implementation of this obligation is punished with a fine. The control of the indoor environment through the monitoring of pollutants must be repeated every seven years, except in the case in which at least one of the pollutants measured during the monitoring shows levels higher than that foreseen in the aforementioned decrees. In this case, monitoring of the confined environment must be carried out within two years.

In addition, the Netherlands, using the studies performed by the National Institute for Public Health and the Environment (RIVM) [26], achieved guiding values starting from the Maximum Permissible Risk (MPR), which represents the level of exposure to a toxic substance for which there are no negative health effects.

Among the Nordic countries, in Finland, for example, the working group (coordinated by the Ministry of Social Affairs and Health (MSAH)), developed guide values for five pollutants: ammonia, carbon monoxide, carbon dioxide, hydrogen sulfide, and PM_{10}. They were proposed in the decrees of the Ministry of the Environment Housing and Building Department D2 National Building Code of Finland—Indoor Climate and Ventilation of Buildings Regulations and Guidelines [27], which entered into force 1 October 2003. For the other pollutants, it is possible to derive guide values using 1/10 of the limits for industrial work environments (Occupational Exposure Limit, OEL). If more pollutants are present, the formula is to be applied: $\Sigma\ (C_i/(HTP)_i) > 0.1$, where C_i is the measured concentration of a single pollutant and (HTP) is the occupational exposure limit of the pollutant in question. The guide values for confined spaces apply to buildings that are occupied for at least six months and where the ventilation system is kept constantly on. Alongside these references are those developed by the Finnish Society of Indoor Air Quality and Climate Classification [27]. It is an initiative desired and financed by the Ministry of the Environment, in collaboration with the experts of the manufacturers and stakeholders of the materials sector, which led to the identification of the target values defined as S1 (individual indoor environment), S2 (good indoor environment), and S3 (satisfactory indoor environment) categories [27].

Belgium, on the other hand, in the Flanders region, established by decree that entered into force on 1 October 2004, reference values for 15 pollutants: acetaldehyde, formaldehyde, total aldehydes, benzene, asbestos, carbon dioxide, nitrogen dioxide, toluene, ozone, carbon monoxide, volatile organic compounds, trichlorethylene, tetrachloroethylene, PM_{10} and $PM_{2.5}$. For five of these pollutants, a category of concentration levels was also identified, defined as intervention values or concentrations of the pollutants corresponding to a level of maximum permissible risk that cannot be exceeded. Another interesting aspect present in the decree is that, in the event that an intervention on the field is requested by experts from the health inspectorate, and that the analytical results of this investigation highlights critical conditions linked to the negligence of the owner or occupant, the inspectorate charges the intervention costs to the applicant [28]. In 2019, further legislative acts were issued for office workplaces intended to welcome the public (decrees 31 January 2019, 2019/201064, and 21 May 2019, 2019/201857).

In the late 1990s, in Austria, the Ministry of the Environment in collaboration with the Academy of Sciences established an interdisciplinary working group for the drafting of guiding values for indoor environments, using a methodology starting from No-Observed-Adverse-Effect-Level (NOAEL) [29]. Using this approach, guide values of six substances were developed: formaldehyde, styrene, toluene, carbon dioxide, volatile organic compounds (VOCs), and trichlorethylene.

Portugal, in April 2006, by decree no. 79 of the Ministry of Public Works, Transport, and Communications [30], and in 2013 by decree no. 60 [31], set maximum reference concentrations for six pollutants: PM_{10}, carbon dioxide, carbon monoxide, ozone, formaldehyde, total VOCs. The decree, in force since June 2006, also establishes the mandatory monitoring of the type and size of the building, and provides corrective actions within 30 days, if after the monitoring, the concentrations of pollutants present levels higher than reported in article 29 paragraph 8 of the decree. Further, the owner or tenant must also provide, within the following 30 days, the results obtained from the new measurements made. In case one of the above conditions is not met, the owner or tenant is subject to the penalties provided for in the decree, such as, for example, the immediate closure of the apartment or the payment of a fine.

In all countries, the proposed guide values are correlated by the relative sampling and analysis methods developed or implemented by the various national training bodies for correct evaluation (e.g., sampling and analysis strategies). These training bodies include the German Institute for Standardization (Deutsches Institut für Normung, DIN), Association Française de Normalization (AFNOR), Bureau de Normalization (NBN), Finnish Standards Association (SFS), Austrian Standards Institute (ASI), Nederlands Normalisatie Instituut (NEN), and the British Standards Institution (BSI). It should be noted that the guide or reference value must always be related to the sampling and analysis method to be adopted for its verification.

For all these countries, except Belgium, Finland, Lithuanian, Portugal, and France (for benzene, formaldehyde, carbon dioxide, and tetrachloroethylene), the recommended guide values have no legal value, even though, in practice they have reached considerable importance. These values, if properly used, can allow for better assessment of the IAQ.

Finally, IAQ is also important for protecting vulnerable materials, including cultural heritage in museums. Inside museums, libraries, and cultural environments—or storage of materials of historical and artistic interest—the quality of indoor air, together with the microclimate (temperature and relative humidity, which must mainly take into account the nature of the materials and goods), and the lighting (another important parameter that can enhance the phenomena of degradation of materials and goods), is fundamental for the management, conservation, and enhancement of goods and finds, and for the choice of measures to contain energy consumption and improve the quality of museum environments, for the health of workers and visitors. There are several reference sources for museum environments, such as the United Nations Educational, Scientific, and Cultural Organization (UNESCO), International Council of Museums (ICOM), International Center for the study of the preservation and restoration of cultural property (ICCROM), National Information Standards Organization (NISO), Getty Conservation Institute, Environmental Conditions for Exhibiting Library and Archival Materials, WHO, and the Ministry of Cultural Heritage and Activities (MIBACT, Italy), just to cite a few. Among the different documents, the authors would like to highlight the following:

o   WHO guidelines for some chemical and biological pollutants and the risks associated with the presence of humidity;

○   EN 15758:2010 Conservation of Cultural Heritage—procedures and instruments for measuring the temperature of the air and that of the surface of objects;

○   EN 15759:2011 Conservation of cultural heritage—indoor climate—part 2: management of ventilation for the protection of buildings belonging to the cultural heritage and collections;

○   EN 15759-2:2018 Conservation of cultural heritage—indoor climate—part 2: ventilation management for the protection of cultural heritage buildings and collections;

○   EN 15898:2019 Conservation of cultural heritage—general terms and definitions;

○   EN 16141:2012 Conservation of cultural heritage—guidelines for the management of environmental conditions in the storage areas of museum collections and plant engineering: definition and characteristics of collection centers for the preservation and management of cultural heritage;

○   EN 16242:2012 Conservation of cultural heritage—procedures and instruments for measuring the humidity of the air and the exchange of steam between the air and the assets cultural heritage;

○   EN 16682:2017 Conservation of cultural heritage—methods of measurement of moisture content, or water content, in materials constituting immovable cultural heritage;

○   EN 16853:2017 Conservation of cultural heritage—conservation process—decision making, planning, and implementation;

○   EN 16883:2017 Conservation of cultural heritage—guidelines for improving the energy performance of historic buildings;

○   EN 16893:2018 Conservation of Cultural Heritage—specifications for location, construction, and modification of buildings or rooms intended for the storage or use of heritage collections.

The adoption of these rules constitutes a significant improvement compared to what has been achieved so far in the study and control activities; the standardization of the methods also increases the possibility of a correct comparison between the different data produced at the European level.

Table 2. Indoor air contaminants: reference values used in some European countries, guide values, and unitary risk of the World Health Organization (WHO).

Contaminant (Unit of Measurement)	WHO Guidelines (Outdoor [a])	WHO Guidelines (Indoor [a])	France	Germany	Netherlands	United Kingdom	Belgium (Flanders)	Finland [c]	Austria	Portugal	Norway	Lithuania	Poland (Residential)	Poland (Public Offices)
Reference	[32,33]	[34]	[21–25]	[18]	[26]	[20]	[28]	[27]	[29]	[30]	[35]	[36]	[37]	[37]
Benzene [b] ($\mu g\ m^{-3}$)	0.17 (UR/lt) 10^{-6} / 1.7 (UR/lt) 10^{-5}	0.17 (UR/lt) 10^{-6} / 1.7 (UR/lt) 10^{-5}	30 (24 h) / 10 (1 y) / RA: 10 / LP: 2 / 0.2 (UR/lt) 10^{-6} / 2 (UR/lt) 10^{-5}	-	20	5 (1 y)	GV ≤ 2 / IV 10	-	-	5 (8 h)	-	-	10 (24 h)	20 (8 h)
Formaldehyde ($\mu g\ m^{-3}$)	100 (30 m)	100 (30 m)	50 (2 h) / 10 (1 y) / 30 (10 from 2023) / RA: 100 / LP: 10	120	120 (30 m) / 10 (1 y) / 1.2 (LP)	100 (30 m)	GV10 (30 m) / IV100 (30 m)	50	100 (30 m) / 60 (24 h)	100 (8 h)	100 (30 m)	100	50 (24 h)	100 (8 h)
CO ($mg\ m^{-3}$)	100 (15 m) / 60 (30 m) / 30 (1 h) / 10 (8 h)	100 (15 m) / 35 (1 h) / 10 (8 h) / 7 (24 h)	100 (15 m) / 60 (30 m) / 30 (1 h) / 10 (8 h)	1.5 (8 h) RWI / 6 (30 m) RWI / 60 (30 m) RWII / 15 (8 h) RWII	100 (15 m) / 60 (30 m) / 30 (1 h) / 10 (8 h)	100 (15 m) / 60 (30 m) / 30 (1 h) / 10 (8 h)	GV 5.7 (24 h) / IV 30 (1 h)	8	-	10 (8 h)	25 (1 h) / 10 (8 h)	10	25 (1 h)	10 (8 h)
NO$_2$ ($\mu g\ m^{-3}$)	200 (1 h) / 40 (1 y)	200 (1 h) / 40 (1 y)	200 (1 h) / 40 (1 y)	350 (30 m) RWII / 60 (7 d) RWII	200 (1 h) / 40 (1 y)	300 (1 h) / 40 (1 y)	GV 135 (1 h) / IV 200 (1 h)	-	-	-	200 (1 h) / 100 (24 h)	100	-	-
Naphthalene ($\mu g\ m^{-3}$)	-	10 (1 y)	10 (1 y)	20 (7 d) RWI / 200 (7 d) RWII	25	-	-	-	-	-	-	-	100 (24 h)	150 (8 h)
Styrene ($\mu g\ m^{-3}$)	260 (7 d) 70 (30 m)	-	-	30 (7 d) RWI / 300 (7 d) RWII	900	-	-	1	40 (7 d) / 10 (1 h)	-	-	-	20 (24 h)	30 (8 h)
PAHs (as BaP) [b] ($ng\ m^{-3}$)	0.012 ng m^{-3} (UR/lt) 10^{-6} / 0.12 ng m^{-3} (UR/lt) 10^{-5}	0.012 ng m^{-3} (UR/lt) 10^{-6} / 0.12 ng m^{-3} (UR/lt) 10^{-5}	-	-	1.2	0.25 (1 y)	-	-	-	-	-	-	-	-
Tetrachloroethylene ($\mu g\ m^{-3}$)	250 (1 y) / 8000 (30 m)	250 (1 y)	1380 (1–14 d) / 250 (1 y) / RV: 250 / LP: 250	1 (7 d)	250	-	100	-	250 (7 d)	250 (8 h)	-	-	-	-
Trichloroethylene [b] ($\mu g\ m^{-3}$)	2.3 µg m^{-3} (UR/lt) 10^{-6} / 23 µg m^{-3} (UR/lt) 10^{-5}	2.3 µg m^{-3} (UR/lt) 10^{-6} / 23 µg m^{-3} (UR/lt) 10^{-5}	800 (14 d^{-1} y) / RA: 10 / RV: 2 / LP: 2.0 (UR/lt) 10^{-6} / 20 (UR/lt) 10^{-5}	1 (7 d)	-	-	200	-	-	25 (8 h)	-	-	150 (24 h)	200 (8 h)
Dichloromethane ($\mu g\ m^{-3}$)	3000 (24 h) / 450 (7 d)	-	-	200 (24 h) RWI / 2000 (24 h) RWII	200 (1 y)	-	-	-	-	-	-	-	-	-

Table 2. *Cont.*

Contaminant (Unit of Measurement)	WHO Guidelines (Outdoor)[a]	WHO Guidelines (Indoor)[a]	France	Germany	Netherlands	United Kingdom	Belgium (Flanders)	Finland [c]	Austria	Portugal	Norway	Lithuania	Poland (Residential)	Poland (Public Offices)
Toluene ($\mu g\,m^{-3}$)	260 (7 d) 1000 (30 m)	-	-	300 (1–14 d) RWI 3000 (1–14 d) RWII	200 (1 y)	-	260	-	75 (1 h)	250 (8 h)	-	-	200 (24 h)	250 (8 h)
Total VOCs ($\mu g\,m^{-3}$)	-	-	-	-	200 (1 y)	-	200	-	-	600 (8 h)	400	600	400	-
PM$_{10}$ ($\mu g\,m^{-3}$)	50 (24 h) 20 (1 y)	-	50 (24 h) 20 (1 y) RA: 75 LP: 15	-	50 (24 h) 20 (1 y)	-	40 (24 h)	50	-	50 (8 h)	90 (8 h)	100	90 (8 h)	-
PM$_{2.5}$ ($\mu g\,m^{-3}$)	25 (24 h) 10 (1 y)	-	25 (24 h) 10 (1 y) RA: 50 LP: 10	25 (24 h)	25 (24 h) 10 (1 y)	-	15 (1 y)	-	-	25 (8 h)	40 (8 h)	-	40 (8 h)	-

[a] the indoor air quality guide values indicate the concentration levels in the air of the pollutants, associated with the exposure times, to which adverse health effects are not expected, in regards to non-carcinogenic pollutants; [b] the upper-bound excess lifetime cancer risk estimated to result from continuous exposure to an agent at a concentration of 1 $\mu g\,m^{-3}$ in air (or 1 $\mu g\,L^{-1}$ in water); [c] the guide values for indoor environments apply to buildings that are occupied for at least six months and where the ventilation system is kept constantly on. Abbreviations: UR unit risk; lt lifetime; RA rapid action; LP long period; RW I (all-day use) and RW II (danger threshold) German guide values (Richtwert); GV guideline value; IV intervention value; RV reference value; PAHs Polycyclic Aromatic Hydrocarbons; BaP Benzo[a]pyrene; VOCs Volatile Organic Compounds; y year; d day; h hour; m minute.

3. The Italian Situation

Among the Member States of the EU, Italy plays an important role, as evidenced by the 10.4% of publications in the IAQ field during the last two decades (Figure 1). The Italian situation is particularly interesting because, unlike the other countries, in the Italian legislation, there is no specific reference relating to residential IAQ, even if pre-regulatory initiatives have multiplied.

In relation to IAQ, in almost all European countries, a legislative delay has been compulsorily and quickly filled. This delay has to be covered, with the issue of specific acts containing suitable references for chemical and biological pollutants, in line with those developed by the WHO, with the most recent and user-friendly specific protocols and procedures provided by the ISO 16000 indoor air standard in its various parts. For these reasons, in 2010, the National Study Group (GdS) on Indoor Air Pollution was established at the Italian Institute of Health (IIS), in which the various ministerial components are represented (Ministry of Health, Ministry of the Environment and Protection of the Territory and the Sea, Ministry of Labor and Social Policies), regions, local authorities and research institutes (IIS, National Research Council (CNR), Italian National Agency for New Technologies, Energy, and the Sustainable Economic Development (ENEA), Italian Institute for Environmental Protection and Research (ISPRA), National System for Environmental Protection (SNPA), and the National Institute for Insurance against Accidents at Work (INAIL). The GdS-ISS is working to provide shared technical–scientific documents in order to allow harmonized actions at national level in order to improve the correct assessment of indoor air pollution. The documents of the GdS-ISS, published as Rapporti ISTISAN, or dissemination documents, include recommendations to prevent indoor air pollution, to improve behavior, cultural awareness, training, to reduce exposure and effects on health, and to increase economic competitiveness.

The GdS-ISS has developed eight reference documents for the monitoring strategies of the main indoor chemical and biological pollutants, the role of the different sources, the energy efficiency activities, and the different indoor combustion [38–48]. Table 3 shows the list of ISTISAN reports already published by the GdS-ISS. Some of the technical indications can already be used for the definition of a national plan on indoor air quality and constitute an important reference for the country.

Table 3. Rapporti ISTISAN just published by the National Study Group (GdS) on Indoor Pollution.

	Title	Ref.
Rapp. ISTISAN	Monitoring strategies for volatile organic compounds (VOCs) in indoor environments	[38]
Rapp. ISTISAN	Monitoring strategies of biological air pollution in indoor environment	[39]
Rapp. ISTISAN	Proceedings of the Workshop "Indoor air pollution: current situation in Italy". Rome, 25 June 2012	[40]
Rapp. ISTISAN	Proceedings of the Workshop "Indoor air quality: current national and European situation. The expertise of the National Working Group on indoor air". Rome, 28 May 2014	[41]
Rapp. ISTISAN	Monitoring strategies to assess the concentration of airborne asbestos and man-made vitreous fibers in the indoor environment	[42]
Rapp. ISTISAN	Microclimate parameters and indoor air pollution	[43]
Rapp. ISTISAN	Presence of CO_2 and H_2S in indoor environments: current knowledge and scientific field literature	[44]
Rapp. ISTISAN	Monitoring strategies to PM_{10} and $PM_{2.5}$ in indoor environments: characterization of inorganic and organic micro-pollutants	[45]
Rapp. ISTISAN	Natural radioactivity in building materials in the European Union: a database of activity concentrations, radon emanations and radon exhalation rates [1]	[46]
Rapp. ISTISAN	Indoor air quality in healthcare environments: strategies for monitoring chemical and biological pollutants	[47]
		[48]

[1] This publication is not authored by the GdS, but it contains issues related to the IAQ.

The results of this activity have been included in the Directive of the President of the Council of Ministers 1 June 2017 published in the Official Journal on 17 July 2017, among the mandatory training activities that the employer must provide to workers. Such activities of the GdS-ISS have also been taken up in the Air Pollution Strategy of the WHO Country Profile for Italy. In this way, an informative booklet entitled "Air in our home: how to improve it" was prepared, which illustrates the origin of indoor air pollution, the role of sources (household cleaning products, construction products, furniture, fabrics, incense sticks, scented candles, stoves, etc.), and the contribution of individual behaviors, providing specific recommendations to reduce indoor pollution levels.

Work is currently in progress for the preparation of two new documents on indoor air quality in office environments and on contaminated sites. On the other hand, ISTISAN reports are being published that address the problems of indoor air quality in school and health facilities, with the identification of specific environmental detection methodologies and possible sanitary implications.

In 2018, Pierpaoli and Ruello published a paper on the bibliometric study on the IAQ [49]: the authors asked the question "What are the actual trends in Indoor Air Quality (IAQ), and in which direction is academic interest moving?" Starting from that, the authors analyzed the worldwide literature from 1990 to 2018, using the Web of Science as a database. They identified past trends and current advances in IAQ, as well as the issues that were expected to be pertinent in the future. In this section, we would like to show state-of-the-art research in the IAQ sector in Italy from previous years, considering what is shown in Figure 1: 243 scientific papers in specialized journals have been published in Italy since 2015. The topics cover different subjects, i.e. environmental science, engineering, medicine, social sciences, energy, physics and astronomy, biochemistry, genetics and molecular biology, materials science, chemistry, chemical engineering, earth and planetary sciences, agricultural and biological sciences, immunology and microbiology, pharmacology, toxicology and pharmaceutics, computer science, mathematics, business, management and accounting, economics, econometrics and finance, arts and humanities, decision sciences, multidisciplinary, and nursing. This means there is a large interest by the scientific community in this field.

Most papers are addressed to investigate the IAQ in schools. Such topics are important because, based on the subpopulation interested, such as suggested by Manigrasso et al. [50], which estimated the particle regional respiratory doses for both combustion and non-combustion aerosol sources currently encountered in microenvironments, with special regards to the age of subjects. Recent papers on school environments are related to monitoring PM, NO_x, VOCs, and CO_2, with regard to the ventilation efficiency and the energy consumption [51–56]. As to the radon exposure, according to two papers, the schools are vulnerable targets due to the long daily childhood presence, and the radon risk could be reduced by low-cost interventions (e.g., implementation of natural air ventilation and school maintenance) [57,58]. Over the last five years, several papers were published on residential IAQ: the authors would like to highlight the main papers of interest. Different research groups dealt the problems related to wood or biomass burning, evidencing the emissions and the related risk assessment [59–61]. Particular attention has been addressed to hospitals and healing places for defining protocol for inpatient rooms, to understand the state-of-the-art research and for suggesting design and management strategies for improving process quality [62,63]. Indoors, there are different combustion and non-combustion sources. Manigrasso et al. revised all of the possible sources and investigated the ultrafine particle emissions and relative doses deposited in the human respiratory tract [64–66]. The importance of the micro-climatic parameters was discussed by Zanni et al., which monitored the IAQ in the airport of Bologna (Italy) as a prototypal example of a large regional airport [67]. Siani et al. applied the cluster analysis on a long time series of temperature and relative humidity measurements for identifying the thermo-hygrometric features in a museum [68]. Cincinelli et al. characterized the IAQ in libraries and archives in Florence (Italy), evidencing that benzene, toluene, ethylbenzene, and xylenes (BTEXs) are the most abundant VOCs, along with cyclic volatile methylsiloxanes, aldehydes, terpenes, and organic acids. In particular, the authors detected presence of acetic acid, which is a chemical that can oxidize books and other exposed objects, and furfural, which is a known marker of paper degradation [69]. Tirler and Settimo discussed the increasing use of incense, magic candles, and other flameless products that may represent a health risk for humans. Pollutants, such as benzene and PM_{10} are mainly affected when these products are used indoors (for instance, the benzene concentration ranged from background levels to over 200 μg m^{-3} after the incense sticks had been tested) [70].

As can be seen, one of the main focal points of the authors is the relationship between IAQ and energy consumption, which is very important. However, it should be considered that the plans and/or interventions of restructuring or renovation cannot be only oriented to the theme of insulation, containment, and energy efficiency, which can alter or worsen air quality, microclimatic conditions, and natural ventilation. They should follow approaches allowing an overall improvement in air quality, with criteria to promote and guarantee health, primarily, to offer all of the maximum benefits of the most current quality educational and training models, and to obtain savings in management costs. Similarly, the same approach should be followed in cases of complete plant adaptations or restructuring (water, electricity, heat, fire, etc.).

4. Conclusions

The IAQ determinants on human health and the potential presence of harmful contaminants released from indoor sources have always been stressed by WHO in its technical documents and position papers. In Europe, specific directive legislative framework on the quality of indoor air is not yet available. Despite an increasing number of pre-legislative initiatives, guidelines, and documents, a harmonized and global approach is still missing. Pending a European directive on indoor air quality, as already done with outdoor air (e.g., 2008/50), which takes into account the WHO indications, this paper aims to provide an overview of the main technical–scientific references in order to allow harmonized actions and to cope with the main issues in such environments. In fact, often in the absence of specific national references to be used for comparison, surveillance actions in indoor environments are limited. The paper, gathering the main references to be used (reported by ad hoc working groups, or in legislation of other European countries, or, by analogy, with other standards, such as those relating to ambient air), means to assist operators engaged in prevention actions to implement interventions in different indoor environments. It should be remembered that these values represent the parameters to which reference must be made for an assessment of the inhalation risk of workers and the population. They are not the only ones, because specific exposure and vulnerability conditions are fundamental elements to consider for a correct risk assessment.

There is an urgent need for a change that is innovative, with a systemic, multidisciplinary approach based on skills. Nowadays, in the various member states, apart from the strong national differences, this situation entails a hygiene–health and environmental protection gap among the various countries (e.g., absence of standards and controls). To fill this gap, harmonization initiatives must be carried out, simultaneously establishing the elements (e.g., strategies, sampling, and analytical methods) and the parameters that must be considered for the control of pollutants indoors. There is no doubt that the heterogeneity of this current regulation system has led to a lack of comparability among the EU member states, both in terms of technical procedures and of health evaluation. Nonetheless, in some EU countries, regulations have been drawn up or recommendations have been developed on IAQ that can allow proper exposure assessment of the general population and the related health risks. Recently, the EU has also taken on a series of new commitments on the energy efficiency and the construction quality. In this regard, in the "Report from the Commission to the European Parliament and the Council: financial support for energy efficiency in buildings" [71] it is emphasized that improving the energy efficiency of buildings also entails important collateral benefits, including greater health.

In Italian territory, there is no reference legislation, but several commissions and working groups are at work. Among these, there is the National Study Group (activated by the ISS), which is working to provide concrete technical contribution for operators in the public and private sectors engaged in the indoor theme, in order to allow a homogeneous action at a national level. The results may lead to appropriate public health strategies aimed at reducing exposure in indoor environments.

Author Contributions: Conceptualization, G.S. and P.A.; methodology, G.S.; software, M.M.; validation, M.M.; data curation, M.M.; writing—original draft preparation, G.G. and P.A.; writing—review and editing, G.S. and P.A.; visualization, P.A.; supervision, G.G. All authors have read and agreed to the published version of the manuscript.

References

1.	Suess, M.J. The Indoor Air Quality programme of the WHO regional office for Europe. *Indoor Air* **1992**, *2*, 180–193. [CrossRef]

2.	Mølhave, L.; Krzyzanowski, M. The right to healthy indoor air: Status by 2002. *Indoor Air* **2003**, *13*, 50–53. [CrossRef] [PubMed]

3.	Braubach, M.; Krzyzanowski, M. Development and status of WHO indoor air quality guidelines. In Proceedings of the 9th International Healthy Buildings Conference and Exhibition HB2009, Syracuse, NY, USA, 13–17 September 2009. Abstract Code 94942.

4.	Tham, S.; Thompson, R.; Landeg, O.; Murray, K.A.; Waite, T. Indoor temperature and health: A global systematic review. *Public Health* **2020**, *179*, 9–17. [CrossRef] [PubMed]

5.	Smith, A.; Pitt, M. Sustainable workplaces and building user comfort and satisfaction. *J. Corp. Real Estate* **2011**, *13*, 144–156. [CrossRef]

6.	Mitova, M.I.; Cluse, C.; Goujon-Ginglinger, C.G.; Kleinhans, S.; Rotach, M.; Tharin, M. Human chemical signature: Investigation on the influence of human presence and selected activities on concentrations of airborne constituents. *Environ. Pollut.* **2020**, *257*, 113518. [CrossRef] [PubMed]

7.	Sekar, A.; Varghese, G.K.; Ravi Varma, M.K. Analysis of benzene air quality standards, monitoring methods and concentrations in indoor and outdoor environment. *Heliyon* **2019**, *5*, e02918. [CrossRef] [PubMed]

8.	Simanic, B.; Nordquist, B.; Bagge, H.; Johansson, D. Indoor air temperatures, CO_2 concentrations and ventilation rates: Long-term measurements in newly built low-energy schools in Sweden. *J. Build. Eng.* **2019**, *25*, 100827. [CrossRef]

9.	Kane, S.; Mahal, A. Cost-effective treatment, prevention and management of chronic respiratory conditions: A continuing challenge. *Respirology* **2018**, *23*, 799–800. [CrossRef]

10.	Tang, H.; Ding, Y.; Singer, B. Interactions and comprehensive effect of indoor environmental quality factors on occupant satisfaction. *Build. Environ.* **2020**, *167*, 106462. [CrossRef]

11.	Fabianova, E.; Fletcher, T.; Koppova, K.; Hruba, F.; Houthuijs, D.; Antonova, T.; Volf, J.; Rudnai, P.; Zejda, J.; Niciu, E. On indoor air in Central Europe. 2001. Available online: https://researchonline.lshtm.ac.uk/id/eprint/16961 (accessed on 29 January 2020).

12.	Olesen, B.W. Revision of EN 15251: Indoor Environmental Criteria. *REHVA J.* August 2012. Available online: https://www.rehva.eu/rehva-journal/chapter/revision-of-en-15251-indoor-environmental-criteria (accessed on 29 January 2020).

13.	Settimo, G.; D'Alessandro, D. European community guidelines and standards in indoor air quality: What proposals for Italy. *Epidemiol. Prev.* **2014**, *38*, 36–41.

14.	Kunkel, S.; Kontonasiou, E.; Arcipowska, A.; Mariottini, F.; Atanasiu, B. *Indoor Air Quality, Thermal, Comfort and Daylight*; Buildings Performance Institute Europe (BPIE): Brussels, Belgium, 2015; ISBN 9789491143106. Available online: http://bpie.eu/uploads/lib/document/attachment/121/BPIE_IndoorAirQuality2015.pdf (accessed on 29 January 2020).

15.	World Health Organization (WHO). *Indoor Air Pollutants Exposure and Health Effects Report on a WHO Meeting Nördlingen, 8–11 June 1982. (EURO reports and studies; 78)*; WHO: Copenhagen, Denmark, 1982.

16.	Scientific Committee on Health and Environmental Risks (SCHER). Preliminary Report on Risk Assessment on Indoor Air Quality. 31 January 2007. Available online: https://ec.europa.eu/health/archive/ph_risk/committees/04_scher/docs/scher_o_048.pdf (accessed on 29 January 2020).

17.	EN ISO 16000:2006. *Indoor air*; European Committee for Standardization: Brussels, Belgium, 2006.

18.	Innenraumlufthygiene-Kommission and the permanent woirkng group of the Highest State Health Authorities (Arbeitsgemeinschaft der Obersten Landesgesundheitsbehörden, AOLG). Ad-Hoc Working Group for Indoor Air Guide Values. Umwelt Bundesamt. 1993. Available online: http://www.umweltbundesamt.de/en/topics/health/commissions-working-groups/ad-hoc-working-group-for-indoor-air-guide-values (accessed on 27 January 2020).

19.	Public Health England. Indoor Air Quality Guidelines for Selected Volatile Organic Compounds (VOCs) in the UK. 2019. Available online: https://www.gov.uk/government/publications/air-quality-uk-guidelines-for-volatile-organic-compounds-in-indoor-spaces (accessed on 28 January 2020).

20. Royal College of Paedriatics and Child Health (RCPCH). Health Effects of Indoor Quality on Children and Young People. 2020. Available online: https://www.rcpch.ac.uk/sites/default/files/2020-01/the-inside-story-report_january-2020.pdf (accessed on 18 February 2020).

21. Agence Nationale de Securitè Sanitaire l'alimentation, de l'environnement et du travail. Valeurs Guides de qualité d'Air Intérieur (VGAI). Le Directeur Général Maisons-Alfort, ANSES. 2011. Available online: http://www.anses.fr/fr/content/valeurs-guides-de-qualit%C3%A9-d%E2%80%99air-int%C3%A9rieur-vgai (accessed on 28 January 2020).

22. France. Décret n° 2011–1727 du 2 décembre 2011 relatif aux valeurs-guides pour l'air intérieur pour le formaldéhyde et le benzene. *J. Officiel République Française* 2011. Available online: https://www.legifrance.gouv.fr/affichTexte.do?cidTexte=JORFTEXT000024909119&categorieLien=id (accessed on 2 December 2011).

23. France. Décret n° 2011–1728 du 2 décembre 2011 relatif à la surveillance de la qualité de l'air intérieur dans certains établissements recevant du public. J. Officiel République Française. 2011. Available online: https://www.legifrance.gouv.fr/affichTexte.do?cidTexte=JORFTEXT000024909128 (accessed on 2 December 2011).

24. Lafon, D. Plan national santé environnement: 2004–2008 National plan for health and environnement. *Arch. Mal. Prof. Environ.* **2005**, *66*, 360–368.

25. Haut Conseil de la Santé publique (HCSP). *Valeurs Reperes D'aide a la gestion dans l'air des espaces clos. Le formaldehyde*; Ministère de la Santè et des Sports: Paris, France, 2009.

26. RIVM-National Institute for Public Health and the Environment. *Health-Based Guideline Values for the Indoor Environment (Report 609021044/2007)*; RIVM: Bilthoven, The Netherlands, 2007.

27. Ahola, M.; Säteri, I.; Sariola, L. Revised Finnish classification of indoor climate 2018. *E3S Web Conf.* **2019**, *111*, 02017. [CrossRef]

28. Hoge Gezondheidsraad. Indoor air quality in Belgium. HGR: Brussel. 2017; Advies nr. 8794. Available online: https://www.health.belgium.be/sites/default/files/uploads/fields/fpshealth_theme_file/hgr_8794_advice_iaq.pdf (accessed on 28 January 2020).

29. Bundesministerium für Land- und Forstwirtschaft, Umwelt und Wasserwirtschaft & Österreichischen Akademie Der Wissenschaften—BMLFUW (2006): Richtlinie zur Bewertung der Innenraumluft, erarbeitet vom Arbeitskreis Innenraumluft am Bundesministerium für Land-und Forstwirtschaft, Umwelt und Wasserwirtschaft und der Österreichischen Akademie der Wissenschaften, Blau-Weiße Reihe (Loseblattsammlung); Österreichische Akademie der Wissenschaften: Wien, Austria. Available online: https://www.bmlrt.gv.at/umwelt/luft-laerm-verkehr/luft/innenraumluft/richtlinie_innenraum.html (accessed on 16 February 2020).

30. Ministério Das Obras Públicas, Transportes e Comunicações. *Decreto-Lei n. 79/2006 de 4 de Abril. Diário da República, I Série-A n. 67*; Ministério Das Obras Públicas, Transportes e Comunicações: Lisboa, Portugal, 2006.

31. Ministérios das Finanças e da Economia e do Emprego. *Decreto n. 60/2013 de 5 de fevereiro de 2013. Diário da República, 2a Série n. 25*; Ministérios das Finanças e da Economia e do Emprego: Lisboa, Portugal, 2013.

32. World Health Organization. *Air quality guidelines for Europe*, 2nd ed.; WHO Regional Publications: Copenhagen, Denmark, 2000.

33. World Health Organization. *Air quality guidelines. Global Update 2005*; WHO Regional Publications: Copenhagen, Denmark, 2006.

34. World Health Organization. Guidelines for indoor air quality: Selected pollutants. Copenhagen: WHO Regional Office for Europe. 2010. Available online: http://www.euro.who.int/__data/assets/pdf_file/0009/128169/e94535.pdf (accessed on 18 January 2020).

35. Norway Ministry of Labour and Social Affairs. Regulations concerning the design and layout of workplaces and work premises (the Workplace Regulations). FOR-2017-04-18-473. January 2013. Available online: https://lovdata.no/dokument/SFE/forskrift/2011-12-06-1356 (accessed on 5 February 2020).

36. Lietuvos Respublikos Sveikatos Apsaugos Ministras. *Įsakymas dėl Lietuvos Higienos Normos Hn 35:2007 "Didžiausia Leidžiama Cheminių Medžiagų (Teršalų) Koncentracija Gyvenamosios Aplinkos Ore" Patvirtinimo. 10 May 2007, nr. V-362*; Lietuvos Respublikos Sveikatos Apsaugos Ministras: Vilnius, Lithuania, 2007.

37. Ordinance of the Polish Minister of Labour and Social Policy of September 26, 1997 on General Safety and Health. *J. Laws* **2003**, *169*, 1650.

38. Fuselli, S.; Pilozzi, A.; Santarsiero, A.; Settimo, G.; Brini, S.; Lepore, A.; de Gennaro, G.; Loiotile, A.D.; Marzocca, A.; de Martino, A.; et al. Monitoring strategies for volatile organic compounds (VOCs) in indoor environments. *Rapp. ISTISAN* **2013**, *13/04*, 31. Available online: http://old.iss.it/binary/publ/cont/13_4_web.pdf (accessed on 1 March 2020).

39. Bonadonna, L.; Briancesco, R.; Brunetto, B.; Coccia, A.M.; De Gironimo, V.; Della Libera, S.; Fuselli, S.; Gucci, P.M.B.; Iacovacci, P.; Lacchetti, I.; et al. Monitoring strategies of biological air pollution in indoor environment. *Rapp. ISTISAN* **2013**, *13/37*, 72. Available online: http://old.iss.it/binary/publ/cont/13_37_web.pdf (accessed on 1 March 2020).

40. Fuselli, S.; Musmeci, L.; Pilozzi, A.; Santarsiero, A.; Settimo, G. Proceedings on Indoor air pollution: Current situation in Italy. Rome, 25 June 2012. *Rapp. ISTISAN* **2013**, *13/39*, 85. Available online: http://old.iss.it/binary/publ/cont/13_39_web.pdf (accessed on 1 March 2020).

41. Santarsiero, A.; Musmeci, L.; Fuselli, S. Proceedings on Indoor air quality: Current national and European situation. The expertise of the National Working Group on indoor air. Rome, 28 May 2014. *Rapp. ISTISAN* **2015**, *15/04*, 134. Available online: http://old.iss.it/binary/publ/cont/15_4_web.pdf (accessed on 1 March 2020).

42. Musmeci, L.; Fuselli, S.; Bruni, B.M.; Sala, O.; Bacci, T.; Somigliana, A.B.; Campopiano, A.; Prandi, S.; Garofani, P.; Martinelli, C.; et al. Monitoring strategies to assess the concentration of airborne asbestos and man-made vitreous fibres in the indoor environment. *Rapp. ISTISAN* **2015**, *15/05*, 37. Available online: http://old.iss.it/binary/publ/cont/15_5_web.pdf (accessed on 1 March 2020).

43. Santarsiero, A.; Musmeci, L.; Ricci, A.; Corasaniti, S.; Coppa, P.; Bovesecchi, G.; Merluzzi, R.; Fuselli, S. Microclimate parameters and indoor air pollution. *Rapp. ISTISAN* **2015**, *15/25*, 62. Available online: http://old.iss.it/binary/publ/cont/15_25_web.pdf (accessed on 1 March 2020).

44. Settimo, G.; Baldassarri, L.T.; Brini, S.; Lepore, A.; Moricci, F.; de Martino, A.; Casto, L.; Musmeci, L.; Nania, M.A.; Costamagna, F.; et al. Presence of CO_2 and H_2S in indoor environments: Current knowledge and scientific field literature. *Rapp. ISTISAN* **2016**, *16/15*, 30. Available online: http://old.iss.it/binary/publ/cont/16_15_web.pdf (accessed on 1 March 2020).

45. Settimo, G.; Musmeci, L.; Marzocca, A.; Cecinato, A.; Cattani, G.; Fuselli, S. Monitoring strategies to PM_{10} and $PM_{2.5}$ in indoor environments: Characterization of inorganic and organic micropollutants. *Rapp. ISTISAN* **2016**, *16/16*, 34. Available online: http://old.iss.it/binary/publ/cont/16_16_web.pdf (accessed on 1 March 2020).

46. Nuccetelli, C.; Risica, S.; Onisei, S.; Leonardi, F.; Trevisi, R. Natural radioactivity in building materials in the European Union: A database of activity concentrations, radon emanations and radon exhalation rates. *Rapp. ISTISAN* **2017**, *17/36*, 70. Available online: http://old.iss.it/binary/publ/cont/17_36_web.pdf (accessed on 1 March 2020).

47. Settimo, G.; Bonadonna, L.; Gherardi, M.; di Gregorio, F.; Cecinato, A. Indoor air quality in healthcare environments: Strategies for monitoring chemical and biological pollutants. *Rapp. ISTISAN* **2019**, *19/17*, 55. Available online: http://old.iss.it/binary/publ/cont/19_17_web.pdf (accessed on 1 March 2020).

48. Settimo, G.; Bonadonna, L.; Gucci, P.M.B.; Gherardi, M.; Cecinato, A.; Brini, S.; De Maio, F.; Lepore, A.; Giardi, G. Qualità dell'aria indoor negli ambienti scolastici: strategie di monitoraggio degli inquinanti chimici e biologici. *Rapp. ISTISAN* **2020**, *20/3*, 67.

49. Pierpaoli, M.; Ruello, M.L. Indoor Air Quality: A bibliometric study. *Sustainability* **2018**, *10*, 3830. [CrossRef]

50. Manigrasso, M.; Vitali, M.; Protano, C.; Avino, P. Ultrafine particles in domestic environments: Regional doses deposited in the human respiratory system. *Environ. Int.* **2018**, *118*, 134–145. [CrossRef]

51. Romagnoli, P.; Balducci, C.; Perilli, M.; Vichi, F.; Imperiali, A.; Cecinato, A. Indoor air quality at life and work environments in Rome, Italy. *Environ. Sci. Pollut. Res.* **2016**, *23*, 3503–3516. [CrossRef]

52. Di Gilio, A.; Farella, G.; Marzocca, A.; Giua, R.; Assennato, G.; Tutino, M.; De Gennaro, G. Indoor/outdoor air quality assessment at school near the steel plant in Taranto (Italy). *Adv. Meteorol.* **2017**, *2017*, 1526209. [CrossRef]

53. Stabile, L.; Massimo, A.; Canale, L.; Russi, A.; Andrade, A.; Dell'Isola, M. The effect of ventilation strategies on indoor air quality and energy consumptions in classrooms. *Buildings* **2019**, *9*, 110. [CrossRef]

54. Stabile, L.; Buonanno, G.; Frattolillo, A.; Dell'Isola, M. The effect of the ventilation retrofit in a school on CO_2, airborne particles, and energy consumptions. *Build. Environ.* **2019**, *156*, 1–11. [CrossRef]

55. Aversa, P.; Settimo, G.; Gorgoglione, M.; Bucci, E.; Padula, G.; de Marco, A. A case study of indoor air quality in a classroom by comparing passive and continuous monitoring. *Environ. Eng. Manag. J.* **2019**, *18*, 2107–2115.

56. Schibuola, L.; Tambani, C. Indoor environmental quality classification of school environments by monitoring PM and CO_2 concentration levels. *Atmos. Pollut. Res.* **2020**, *11*, 332–342. [CrossRef]

57. Azara, A.; Dettori, M.; Castiglia, P.; Piana, A.; Durando, P.; Parodi, V.; Salis, G.; Saderi, L.; Sotgiu, G. Indoor radon exposure in Italian schools. *Int. J. Environ. Res. Pub. Health* **2018**, *15*, 749. [CrossRef] [PubMed]

58. Di Carlo, C.; Lepore, L.; Gugliermetti, L.; Remetti, R. An inexpensive and continuous radon progeny detector for indoor air-quality monitoring. In *WIT Transactions on Ecology and the Environment*; Passerini, G., Borrego, C., Longhurst, J., Lopes, M., Barnes, J., Eds.; WIT Press: Ashurst, UK, 2019; Volume 236, pp. 325–333.

59. De Gennaro, G.; Dambruoso, P.R.; Di Gilio, A.; di Palma, V.; Marzocca, A.; Tutino, M. Discontinuous and continuous indoor air quality monitoring in homes with fireplaces or wood stoves as heating system. *Int. J. Environ. Res. Pub. Health* **2015**, *13*, 78. [CrossRef]

60. Stabile, L.; Buonanno, G.; Avino, P.; Frattolillo, A.; Guerriero, E. Indoor exposure to particles emitted by biomass-burning heating systems and evaluation of dose and lung cancer risk received by population. *Environ. Pollut.* **2018**, *235*, 65–73. [CrossRef]

61. Marchetti, S.; Longhin, E.; Bengalli, R.; Avino, P.; Stabile, L.; Buonanno, G.; Colombo, A.; Camatini, M.; Mantecca, P. In vitro lung toxicity of indoor PM_{10} from a stove fueled with different biomasses. *Sci. Total Environ.* **2019**, *649*, 1422–1433. [CrossRef]

62. Gola, M.; Settimo, G.; Capolongo, S. Chemical pollution in healing spaces: The decalogue of the best practices for adequate indoor air quality in inpatient rooms. *Int. J. Environ. Res. Pub. Health* **2019**, *16*, 4388. [CrossRef]

63. Gola, M.; Settimo, G.; Capolongo, S. Indoor air in healing environments: Monitoring chemical pollution in inpatient rooms. *Facilities* **2019**, *37*, 600–623. [CrossRef]

64. Manigrasso, M.; Guerriero, E.; Avino, P. Ultrafine particles in residential indoors and doses deposited in the human respiratory system. *Atmosphere* **2015**, *6*, 1444–1461. [CrossRef]

65. Manigrasso, M.; Vitali, M.; Protano, C.; Avino, P. Temporal evolution of ultrafine particles and of alveolar deposited surface area from main indoor combustion and non-combustion sources in a model room. *Sci. Total Environ.* **2017**, *598*, 1015–1026. [CrossRef]

66. Avino, P.; Scungio, M.; Stabile, L.; Cortellessa, G.; Buonanno, G.; Manigrasso, M. Second-hand aerosol from tobacco and electronic cigarettes: Evaluation of the smoker emission rates and doses and lung cancer risk of passive smokers and vapers. *Sci. Total Environ.* **2018**, *642*, 137–147. [CrossRef] [PubMed]

67. Zanni, S.; Lalli, F.; Foschi, E.; Bonoli, A.; Mantecchini, L. Indoor air quality real-time monitoring in airport terminal areas: An opportunity for sustainable management of micro-climatic parameters. *Sensors* **2018**, *18*, 3798. [CrossRef] [PubMed]

68. Siani, A.M.; Frasca, F.; Di Michele, M.; Bonacquisti, V.; Fazio, E. Cluster analysis of microclimate data to optimize the number of sensors for the assessment of indoor environment within museums. *Environ. Sci. Pollut. Res.* **2018**, *25*, 28787–28797. [CrossRef] [PubMed]

69. Cincinelli, A.; Martellini, T.; Amore, A.; Dei, L.; Marrazza, G.; Carretti, E.; Belosi, F.; Ravegnani, F.; Leva, P. Measurement of volatile organic compounds (VOCs) in libraries and archives in Florence (Italy). *Sci. Total Environ.* **2016**, *572*, 333–339. [CrossRef]

70. Tirler, W.; Settimo, G. Incense, sparklers and cigarettes are significant contributors to indoor benzene and particle levels. *Ann. I. Super. Sanita* **2015**, *51*, 28–33.

71. SWD. Commission Staff Working Document, 143, Accompanying the Document, Report form the Commission to the European Parliament and the Council, Financial Support for Energy Efficiency in Buildings. 2013. Available online: https://ec.europa.eu/energy/sites/ener/files/documents/swd_2013_143_accomp_report_fin ancing_ee_buildings.pdf (accessed on 18 April 2013).

12

A Contactless Measuring Method of Skin Temperature based on the Skin Sensitivity Index and Deep Learning

Xiaogang Cheng [1,2], Bin Yang [3,4,*], Kaige Tan [5], Erik Isaksson [5], Liren Li [6], Anders Hedman [5], Thomas Olofsson [4] and Haibo Li [1,5]

[1]   College of Telecommunications and Information Engineering, Nanjing University of Posts and Telecommunications, Nanjing 210003, China; chengxg@njupt.edu.cn (X.C.); haiboli@kth.se (H.L.)

[2]   Computer Vision Laboratory (CVL), ETH Zürich, 8092 Zürich, Switzerland

[3]   School of Building Services Science and Engineering, Xi'an University of Architecture and Technology, Xi'an 710055, China

[4]   Department of Applied Physics and Electronics, Umeå University, 90187 Umeå, Sweden; thomas.olofsson@umu.se

[5]   KTH Royal Institute of Technology, 10044 Stockholm, Sweden; kaiget@kth.se (K.T.); erikis@kth.se (E.I.); ahedman@kth.se (A.H.)

[6]   School of computer science and technology, Nanjing Tech University, Nanjing 211816, China; dtspsxzhj@njtech.edu.cn

*   Correspondence: yangbin@xauat.edu.cn

Featured Application: The NISDL method proposed in this paper can be used for real time contactless measuring of human skin temperature, which reflects human body thermal comfort status and can be used for control HVAC devices.

Abstract: In human-centered intelligent building, real-time measurements of human thermal comfort play critical roles and supply feedback control signals for building heating, ventilation, and air conditioning (HVAC) systems. Due to the challenges of intra- and inter-individual differences and skin subtleness variations, there has not been any satisfactory solution for thermal comfort measurements until now. In this paper, a contactless measuring method based on a skin sensitivity index and deep learning (NISDL) was proposed to measure real-time skin temperature. A new evaluating index, named the skin sensitivity index (SSI), was defined to overcome individual differences and skin subtleness variations. To illustrate the effectiveness of SSI proposed, a two multi-layers deep learning framework (NISDL method I and II) was designed and the DenseNet201 was used for extracting features from skin images. The partly personal saturation temperature (NIPST) algorithm was use for algorithm comparisons. Another deep learning algorithm without SSI (DL) was also generated for algorithm comparisons. Finally, a total of 1.44 million image data was used for algorithm validation. The results show that 55.62% and 52.25% error values (NISDL method I, II) are scattered at (0 °C, 0.25 °C), and the same error intervals distribution of NIPST is 35.39%.

Keywords: contactless measurements; skin sensitivity index; thermal comfort; subtleness magnification; deep learning; piecewise stationary time series

---

1. Introduction

Higher economic growth drives increasing energy consumption, and 50% of housing consumption is generated by heating, ventilation and air conditioning (HVAC) systems [1,2]. Furthermore, one of the most important reasons for energy waste is that the actual thermal requirements of indoor occupants are ignored, with the result that overheating and overcooling occur often. Fortunately, real-time thermal comfort perception can provide useful signals to HVAC systems for achieving energy saving and human-centered intelligent control. Therefore, many researchers have been studying thermal comfort measurements for indoor environments in recent decades. Many methods were generated, including the questionnaire survey method [3–5], environmental measurement method [6,7] and contact measuring method of human body physiological parameters [8–18]. In recent years, the semi-contact measuring method [19,20] and contactless measuring method [21–23] for human body physiological parameters were also generated. For example, in references [19,20], an infrared sensor was fixed on the frame of eyeglasses in order to measure skin temperature. In reference [21], a normal vision sensor was also used for measuring skin temperature and two non-linear models were trained. In references [22,23], Kinect was used for recognizing human poses or indoor locations, and then human thermal comfort and dynamic metabolism were estimated, respectively. All these methods are meaningful attempts. However, due to the challenges of measuring thermal comfort which are (1) skin subtleness variation [24], (2) inter-individual differences [14,25] and (3) temporal intra-individual differences [14,26], there is still no satisfactory method for perceiving human thermal comfort.

To overcome the aforementioned challenges, the skin sensitivity index (SSI) was defined in this paper. The SSI is strongly related to skin temperature. A contactless measuring method of skin temperature based on SSI and Deep Learning was proposed, hereinafter referred to as NISDL. Two different deep learning methods of NISDL have been designed and trained, respectively, which are NISDL methods I and II. The main difference between them is that the location of SSI participation in the neural network training is not the same. A total of 1.44 million images were collected for 16 Asian female subjects, and this 'big data' was used for algorithm validation.

The main contributions of this paper are:

(1)   The skin sensitivity index (SSI) was proposed for describing individual sensitivity of thermal comfort, and the index was combined with skin images for deep learning network training.
(2)   A novel contactless measuring algorithm (NISDL) based on SSI was proposed, with two different frameworks of NISDL having been designed for real-time thermal comfort measurement.
(3)   A deep learning algorithm without SSI was also generated and trained. Two comparisons were made: (1) comparison between data-driven methods (deep learning) and model-driven methods (linear models); (2) comparison of measuring effects in the case of SSI participation in training and non-participation in training.

The rest of this paper is organized as follows. Section 2 introduces the related work about thermal comfort. In Section 3, the research methods, including SSI computation, subjective experiments and NISDL methods, are introduced. The results and discussion are shown in Sections 4 and 5. Finally, Section 6 gives the conclusion.

2. Related Work

Since the 1970s, Fanger has explored human thermal comfort and conducted many kinds of subjective experiments. Based on this, he eventually established what is known as Fanger's theory [27]. From then on, many studies about thermal comfort were carried out.

Questionnaire surveys are good as a method to understand the inner feeling of an occupant. With the development of the internet, online questionnaire surveys can also be generated [3,4]. However, it is inconvenient and also difficult to guarantee that occupants will continue to give feedback based on their personal thermal feelings [5]. Therefore, the environment measurement method was also adopted in the building industry [6]. With this kind of method, some objective parameters, such as indoor temperature, airflow and humidity, are often measured. Unfortunately, the goal of the environment measurement method is to meet the thermal comfort needs of a majority of indoor occupants. Therefore, the thermal feelings of a minority were ignored. To overcome this drawback, a kind of nonlinear autoregressive network, still belonging to the environment measurement method, was generated to predict indoor temperature [7]. In fact, human thermal comfort is complicated, and with constant indoor parameters it is difficult to meet each individual's requirements for thermal comfort. As such, some researchers study physiological measurement methods, including the contact measuring method, semi-contact measuring method and contactless measuring method.

For the contact measuring method, skin temperature and heart rate are usually the measured parameters. Wang and Nakayama [8,9] made early attempts at measuring the skin temperature around the human body. Liu [10] conducted subjective experiments and a total of 22 subjects were invited. The data, being local skin temperatures and electrocardiograms, were collected. Based on these collected data, 26 measuring methods of mean skin temperature were assessed. Takada [11] presented a multiple regression equation to predict skin temperature in non-steady state. The multiple regression equation was considered as a function of mean skin temperature. Wrist skin temperatures and upper extremity skin temperatures were also adopted to estimate human thermal sensation, respectively [12,13]. Chaudhuri [14] presented a predicted thermal state (PTS) model, and the capture of peripheral skin temperature. Furthermore, body surface area and clothing insulation were used for analyzing inter- and intra-individual differences. As for the thermal comfort study using heart rate, based on physiological experimentation, Yao [15] investigated the relationship between heart rate variation (HRV) and electroencephalograph (EGG). The results show that HRV and EEG are useful for thermal comfort studies, but further data validation is needed. Moreover, Dai, Chaudhuri and Kim [16–18] combined machine learning with contact measurement, and they are all meaningful attempts. However, as the linear kernel of SVM was used for predicting human thermal sensation in different experiments, sometimes overfitting can happen.

For the semi-contact measuring method, Ghahramani [19] used an infrared sensor to estimate skin temperature of different face points. The infrared sensor was mounted on the frame of eyeglasses. Based on this, a hidden Markov model was constructed to capture personal thermal sensation [20].

In practical application, contact and semi-contact measurement are both difficult to apply widely. The reason is that an occupant needs to wear a sensor, which is uncomfortable and is also not in line with the goal of human-oriented intelligent buildings. For this reason, a kind of contactless measuring method was studied in reference [21]. Based on vision sensors, Cheng [21] extracted the saturation (S) channel from skin images and constructed two saturation-temperature models to estimate skin temperature. The two models are the contactless measuring method of thermal comfort based on saturation-temperature (NIST) and the contactless measuring method of thermal comfort based on partly saturation-temperature (NIPST). Alan [22] proposed a contactless measuring method based on human poses. A total of 12 poses of thermal comfort were defined and Kinect was adopted to estimate human skeleton and poses. Further, Dziedzic [23] also used Kinect to predict human thermal sensation and dynamic metabolic rates.

With the development of machine learning (ML) and computer vision (CV), some thermal comfort perception methods based on ML and CV were proposed. Support Vector Machine (SVM) are often used for analyzing existing databases (RP-884) and captured environmental parameters [24,28,29]. Further, Peng [26] use unsupervised and supervised learning to predict occupants' behavior, applied to three types of offices which are single person offices, multi-person offices, and meeting rooms. The results show that the average energy savings for the entire space is 21% in experimental condition. Li [30] proposed a fuzzy model to predict thermal sensation, skin temperature and heart rate considered as objective parameters. For avoiding overheating, Cosma [31] extracted data from multiple local body parts and analyzed them with four kinds of machine learning algorithms, including SVM, Gaussian process classifier (GPC), k-neighbors classifier (KNC) and random forest classifier (RFC).

The kinds of machine learning adopted in references [24,26,28–31] are traditional algorithms. In recent years, the use of deep neural networks is on the rise [32,33]. In addition, a kind of subtleness magnification technology was presented [34,35]. These provide new directions and opportunities for the measurement of human thermal comfort. Based on this technology, a novel contactless measuring method was generated which will be introduced as follows.

3. Research Methods

3.1. Subjective Physiological Experiments

3.1.1. Subjects Data and Chamber Environments

16 human subjects were invited for experiments and the resulting data volume is 1.44 million images. The experiments were conducted in a chamber with controllable indoor air temperature and relative humidity. The corresponding dry-bulb air temperature is $22.2 \pm 0.2\,°C$ and the relative humidity is $36.9 \pm 2.5\%$. The resolution of vision sensor used for capturing video is 1280×720. The iButton, model DS192H with uncertainty $\pm 0.125\,°C$, was used for measuring skin temperature from the back of subject's hand. All the subjects are Asian females with an average age of 23.9 ± 3.9 years, average weight of 52.2 ± 6.5 kg, and body mass index (BMI) 19.9 ± 2.2 kg/m^2.

3.1.2. Experimental Procedures

The experiment was conducted during winter in Sweden. There are generally three steps in subjective physiological experiments. (1) Preparation stage: The indoor environment parameters were measured and controlled to a suitable level. When the subjects came into the chamber, they should rest for 10 min for adaptation. At the same time, warm water with constant temperature ($45\,°C$) was prepared. (2) Thermal stimulus: After 10 min adaptation, subjects were asked to immerse hands into the water with $45\,°C$. The whole thermal stimulus process lasted for 10 min. (3) Big data collection: After 10 min of stimulus, subjects were asked to sit next to the data collection desk and put her pairs of hands under the vision sensor. The back of the hand is faced up and the data is collected for 50 min. At the same time, skin temperature sensor (iButton) was attached to the back of one hand. The corresponding sampling interval is 1 min. It should be noted that, based on piecewise stationary time series analysis [36], linear interpolation was adopted in this paper, and 11 points were interpolated into 1 min for real skin temperature captured by iButton.

3.2. Skin Sensitivity Index

3.2.1. SSI Definition

When human body encounters thermal stimuli, blood circulation will change which will also be reflected in skin's color and texture. In reference [21], based on the HSV (hue, saturation, value) color space, the S channel was extracted and a linear ST (saturation-temperature) model was established.

$$T = k_i \times S + b_i \tag{1}$$

where i denotes subject number, the k reacts to the change rate of skin temperature, and b denotes the intercept. S and T are skin saturation and temperature respectively. In this paper, k is defined as the skin sensitivity index (SSI). SSI is a high weight coefficient in skin temperature changes and SSI reflects the skin sensitivity level to external thermal stimuli.

3.2.2. SSI Computing

Based on subjective physiological experiments, real skin temperature can be obtained by iButton. The images were also collected from subjects' hands. Therefore, the SSI can be calculated. The steps are as follows: (1) Extracting each frame from captured video; (2) Segment region of interest (ROI); (3) Extracting S channel from ROI images and computing mean values of S for each ROI image; (4) Search SSI value based on real skin temperature and S for each subject.

3.3. NISDL Algorithm

In this paper, considering that SSI is a high weight coefficient for contactless thermal comfort measurement, it will improve the prediction accuracy of skin temperature. Based on SSI, the NISDL algorithm was introduced in this paper. Furthermore, to validate the effectiveness of SSI, two kinds of deep learning frameworks (NISDL method I, II) have been constructed. The main difference between NISDL methods I and II is that the location where the network invokes SSI to participate in the model training is different. The NISDL algorithm constructed is introduced as follows.

3.3.1. Video Pre-Processing

In fact, skin texture variation is subtle and is difficult to perceive. In this paper, for magnifying this kind of subtleness variation, an image subtleness magnification technology known as Euler Video Magnification (EVM) is adopted [34,35]. Based on EVM, let c (x, t) denote skin images which are subtly varied with time t. Suppose that the variation function is formula (2) [34,35].

$$C\ (x, t) = F\ (x + h\ (t))\ \ \ \ \ \ \ \ \ \ \ \ \ \ \ \ (2)$$

where, $h\ (t)$ is variation degree, F is a function which constructs the relationship between C (x, t) and h (t). If the skin image C (x, t) is magnified, and the first-order Taylor series expansion can be handed to $F\ (x + h\ (t))$.

$$C\ (x, t) = F(x + (1 + \xi)\ *\ h(t))\ \ \ \ \ \ \ \ \ \ \ \ (3)$$

where ξ is the magnification coefficient which can be set based on practical application. According to Formula (3), only the variation part was magnified to a magnitude of $1 + \xi$, while the other part of skin texture is not magnified. Therefore, the invisible texture variation is made to be visible.

It should be noted that de-noise processing should be handled before EMV processing. Further, after the video is magnified, ROI is selected and cropped from each frame. The ROI images are imported into NISDL method I and II for model training.

3.3.2. NISDL Method I

As shown in Figure 1, SSI values are used as input data and imported into the deep learning network at the very beginning. The ROI images were also combined with SSI values in the first step. According to the size of ROI images, the SSI value of each ROI image was expanded into a matrix. The matrix is considered as a channel and combined with the 3 channels of ROI images.

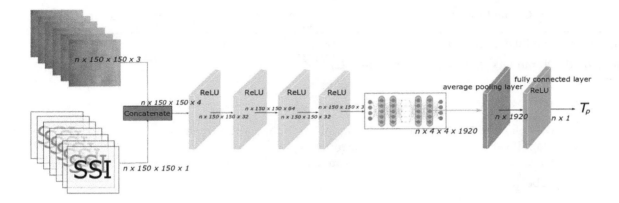

Figure 1. NISDL method I. (SSI values were combined with skin images firstly.).

The merged data between ROI images and SSI values above is inputted into four convolution layers, which are used for dimensionality reduction. In NISDL method I, the DenseNet201 [33] is adopted for features extraction. The last two layers of DenseNet201 are not suitable for skin temperature measurement, hence the two layers are removed. The reason for this is that the activation function of the last layers is softmax. Instead of these two layers, an average pooling layer and a fully connected layer are added behind denseNet201. Based on the deep learning networks designed above, n ROI images and n SSI values were inputted into NISDL method I. Therefore, n skin temperatures can be obtained.

3.3.3. NISDL Method II

Figure 2 is the deep learning framework of NISDL method II. In this method, the ROI images and SSI values were processed for features extraction. Subsequently, the two kinds of features were combined in the second half of the whole framework. An average pooling layer and the DenseNet201 (excluding the last two layers) were also used for features extraction of ROI images. For SSI values, a convolution layer and an average pooling layer were adopted for feature extraction and dimensionality reduction. After features combination, three fully connected layers were constructed in NISDL method II. Therefore, the skin temperature can be obtained. The algorithm in detail, including NISDL method I and II, is shown in Table 1.

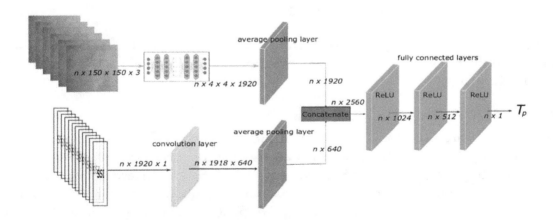

Figure 2. NISDL method II. (SSI values and skin images were processed by deep learning networks. Two kinds of features were extracted and the features were combined.).

Table 1. Contactless measuring method of skin temperature based on skin sensitivity index.

Algorithm: the NISDL algorithm

Output: NISDL model (*.*h5*), skin temperature (°C)
Step:

1. Video pre-processing

    (1) De-noise and handle subtleness magnification for captured video.
    (2) Magnification coefficient ξ is 10 (formula (3)).
    (3) Extracting region of interest (ROI) from each frame of video, the size is 150×150.

2. Making label

    (1) Making numerical interpolation for skin temperatures captured by iButton.
    (2) Uniform interpolation is adopted, plus 11 points/min.
    (3) Establish a correspondence table between ROI images and skin temperatures (after interpolation).

3. Algorithm training

    (1) Commonality between NISDL method I and II

        1) Training set and test set ratio: 12: 4.
        2) Validation set: 500 images.
        3) During network training, 32 images/batch, epoch is 8.
        4) Training ~30000 images, validate once.
        5) Activation function: ReLU
    (2) NISDL method I

        1) SSI values were concatenated with ROI images in the first steps.
        2) Convolutional kernel: 1×1.
    (3) NISDL method II

        1) Features are extracted from SSI values and ROI images, respectively.
        2) Concatenating the two kind of features in the second half of network.
        3) Convolutional kernel: 3×1.

4. Optimizing model parameters

3.3.4. Evaluation Metric

For assessing NISDL algorithm constructed in this paper, the absolute error is adopted.

$$Error = |T_p(i) - T_r(i)| \quad i = 1, 2, 3, \ldots \tag{4}$$

where $T_p(i)$ is the prediction values of skin temperature and obtained from the proposed NISDL algorithm. $T_r(i)$ is the real value of skin temperature and captured by iButton. The parameter i denotes the particular ROI image.

3.3.5. Algorithms for Comparison

Two algorithms are used for comparison in this paper: (1) DL algorithm. The commonality between NISDL method I and II is that they all use SSI for model training. For validating the effectiveness of NISDL algorithm (with SSI), we remove the SSI and corresponding hidden layers for SSI features extraction from NISDL method II, so that it will be another deep learning network (without SSI) and is named the DL algorithm hereinafter. (2) NIPST algorithm. DL algorithm, NISDL method I and II are all nonlinear methods and data driven methods with deep learning networks. For further validating NISDL method I and II, the NIPST algorithm is also used for algorithm comparison, which is a linear and model driven method.

4. Results

16 subjects were invited for subjective physiological experiments and a total of 1.44 million images were captured. Based on this, the NISDL algorithm was validated and compared with the NIPST algorithm and the DL algorithm.

4.1. Hardware Parameters

For this paper, a computer with a GPU was used for images processing and algorithm validation. The GPU is GeForce GTX TITAN X, the CPU is Intel core i5-4460 CPU@3.2Ghz X 4, the RAM is 16G and the word size is 64bit.

4.2. Training of NISDL method I

The size of ROI images are $n \times 150 \times 150 \times 3$, and the size of expanded SSI vlues are $n \times 150 \times 150 \times 1$. The SSI matrix was considered as a channel and concatenated with ROI images, so that the result is $n \times 150 \times 150 \times 4$. The activation function of four convolution layers, shown in Figure 1, are Rectified Linear Units (ReLU) and the size of convolution kernel is 1×1. DenseNet201 was used for feature extraction and its output is a matrix with size of $n \times 4 \times 4 \times 1920$. Based on this, two hidden layers are constructed and the size of the last layer, being a fully connected layer, is $1920 \times n$.

4.3. Training of NISDL Method II

The size of the expanded SSI matrix is $n \times 1920 \times 1$, which differs from that of NISDL method I. The corresponding convolution kernel is 3×1. The features of SSI were extracted by two hidden layers, and the features of ROI images were extracted by DenseNet201. As shown in Figure 2, in the second half of framework, the two kinds of features are concatenated. In order to ensure that the SSI features have a suitable influence on network training (moderate, not too big or too small), the size of ROI image features is set as $n \times 1920$, and size of SSI features is set as $n \times 640$ (triple relationship). Finally, the size of last three hidden layers is 2560×1024, 1024×512 and 512×1, respectively.

4.4. Commonality between NISDL Method I and II

During network training, the same parameters of NISDL method I and II are shown as follows. Based on data of 1.44 million images, the ratio of training set and test set is 12:4, the number of validation set is 500. The epoch is 8, which means that the training set was trained 8 times. The input data batch is 32. When the error of validation set is less than 0.46 °C, the corresponding model (*.h5) will be saved. Further, when 30,0000 images of training set were trained, the corresponding model (*.h5) will also be saved. After the generation of the model, the test set images were inputted into generated model, so that the prediction values of skin temperatures could be obtained.

4.5. Quantitative Comparison

The prediction values of skin temperature are shown in Figure 3. The set of values obtained from iButton is ground truth. The corresponding error statistics, including mean, median, are shown in Figure 4, which is a box-whisker plot. The mean values of NISPT, DL, NISDL method I and II in °C are 0.579, 0.359, 0.335 and 0.265, respectively. In addition, the median values of them in °C are 0.343, 0.309, 0.238 and 0.228, respectively. It was shown that deep learning methods (DL, NISDL method I and II) are all better than the nonlinear model (NIPST) and further that the method with SSI (NISDL method I and II) is better than the method without SSI (DL).

In this paper, the error distributions are given in Figure 5 and Table 2. The errors of DL, NISDL method I and II are mainly concentrated in the range of 0 °C and 0.75 °C. NISDL is better than DL, because two error percentages of NISDL corresponding to [0, 0.25) are 52.25% and 55.62%. In addition, the error percentages of NISDL corresponding to [0.25, 0.5) and [0.5, 0.75) are less than that of DL.

The error percentage of NIPST is increased from the interval of [0.75, 1), meaning that the performance of NIPST is worse than DL and NISDL methods I and II.

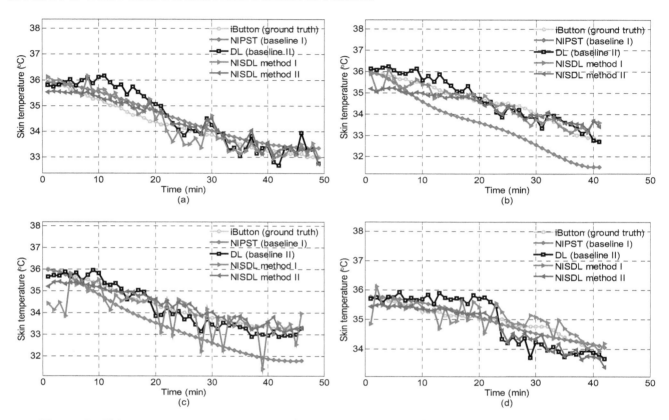

Figure 3. Skin temperatures comparison between ground truth, baseline and NISDL (Images of 12 subjects (0.96 million images) were training set; images of 4 subjects (0.32 million images) were test set. Sub-figures (**a**), (**b**), (**c**) and (**d**) are the comparison results of 4 test set (data of 4 subjects)).

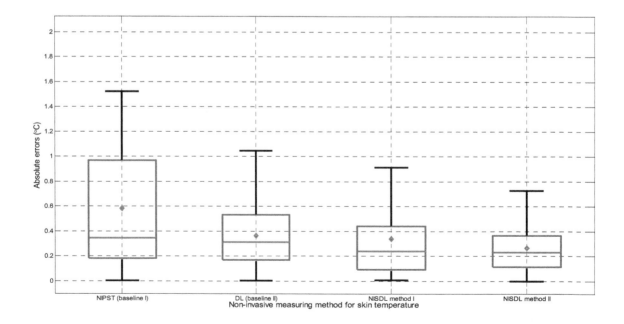

Figure 4. Error statistics (box-whisker plot) comparison between baseline and NISDL (**a**). NIPST was published in references [21]. (**b**). Skin images were trained directly by DenseNet201 to obtain a model and predict skin temperature, and SSI was not involved. (**c**). NISDL method I and NISDL method II all belong to NISDL. The main difference is that SSI values and skin images are combined at different times.

Table 2. Absolute error distribution.

Absolute Error (°C)	NIPST (Baseline, %)	DL (Baseline, %)	NISDL Method I (Figure 1, %)	NISDL Method II (Figure 2, %)
[0, 0.25)	35.39	37.64	52.25	55.62
[0.25, 0.5)	24.16	35.96	28.09	30.34
[0.5, 0.75)	5.06	19.10	10.11	11.80
[0.75, 1.0)	11.24	5.62	3.37	2.250
[>1.0)	24.16	1.69	6.18	0

5. Discussion

5.1. Situation of Overcoming Challenges

NISDL has overcome the three challenges mentioned in Section 1 to some extent. A kind of subtleness magnification technology, which is Euler Video Magnification (EVM), was used for magnifying the skin texture variation, so that the challenges '(1)' given in Section 1 can be overcome. For overcoming challenges '(2)' which is inter-individual difference, the skin sensitivity index (SSI) was proposed and SSI is related with skin saturation. Figures 4 and 5 shows that the performance of NISDL algorithm with SSI is better than that of DL without SSI. In practical application, skin images will be captured in real-time (30 frames/s), the skin temperature variation can always be obtained. Therefore, challenges '(3)' proposed in Section 1 can be overcome. Furthermore, piecewise stationary time series analysis was adopted in this paper for overcoming challenges '(3)'. Considering operability, the breakpoint interval of piecewise stationary signal is set to 5 s, supposing, e.g., that the skin temperature has a constant value during 5 s.

5.2. The Deep Learning Framework

In this paper, NISDL method I and II all belong to the deep learning method. In addition, DL generated for algorithm comparison is also a deep learning method. The error distribution comparison is shown in Figures 4 and 5 and the performance of NISDL II is encouraged. From the perspective of deep learning, the main reason for this is that big data is adopted.

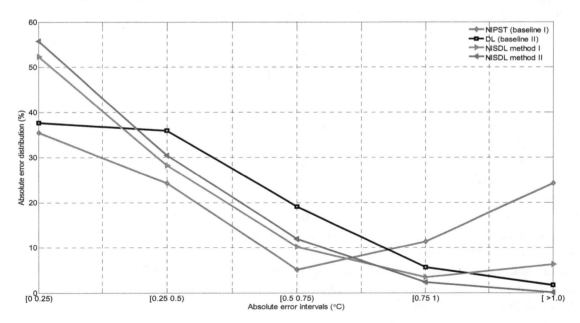

Figure 5. Error distribution comparison between baseline and NISDL (**a**). NIPST was published in [21]. (**b**). Skin images were used in training directly by DenseNet201 to obtain a model and predict skin temperature, and SSI was not involved. (**c**). NISDL method I and NISDL method II all belong to NISDL. The main difference is that SSI values and skin images are combined at different times.).

5.3. The Proposed SSI

Although NISDL and DL are all better than NIPST, there are still a big gap between NISDL and DL. Figures 4 and 5 and Table 2 also show that NISDL is better than DL. The main reason is that SSI is used in NISDL. The NISDL method II is also shown to be better than NISDL method I. Further, when SSI features are extracted and concatenated with ROI images features in the second half of network, the performance will be better.

5.4. Reasons of Designing Two Frameworks for NISDL

Some researchers may ask, why do we design NISDL methods I and II together? The main reason is that we want to extensively confirm the effectiveness of SSI. In this paper, SSI participates in network training from different locations, and the results are all good. When the SSI values are removed from network (DL), the corresponding performance decreased significantly. Based on this, we can know that SSI is helpful for predicting skin temperature through deep learning networks.

5.5. Practical Application

Some researchers may argue that the method proposed in this paper still cannot be applied in practice right now. In fact, a method is always being gradually improved. For example, the NISDL proposed in this paper is better than NIPST which was proposed in 2017. In addition, when more diverse data is captured and used for model training, the performance of NISDL will be better.

Some researchers may argue that the infrared sensor can also be used for measuring skin temperature, so why do we use a vision-based method? In fact, the study [19,20] focused on thermal comfort measurement with an infrared sensor. However, the measurement accuracy is limited. Beyond this, there are other drawbacks in the infrared based measuring method: (1) Distance. The infrared sensor should be placed close to occupant. (2) Cost. The infrared sensor with high accuracy is expensive and the accuracy of an infrared sensor with low cost is also low. (3) Information is limited. From the perspective of the human sensory system, the infrared sensor is 'touch' and vision-based method is 'sight'. The data captured by vision-based methods is much more than that collected by infrared sensor. e.g., human poses can be captured by vision sensor rather than infrared sensor for analyzing human thermal comfort. Based on these three drawbacks, the infrared-based method is difficult to widely apply in practice.

Furthermore, some other researchers may say that vison-based contactless measuring method may have concerns related to personal privacy. In fact, there are at least two options to protect personal privacy issues: (1) Switch button. Based on this switch button, any customer can choose to accept or reject the implementation of real-time personal service of thermal comfort. 92) Information selection. In future practical applications, only the information about human thermal comfort be processed and saved, while other information will be discarded in real-time. (3) Data protection. In order to avoid data protection issues, processed data related to thermal comfort can be directly transferred to the HVAC system instead of being saved. Therefore, from the perspective of human-centeredness, the NISDL algorithm proposed in this paper is helpful.

In future practical applications, the processing capacity of HVAC system should be considered. The HVAC system is usually equipped with computer server. The framework proposed in this paper can be embedded on the computer server directly and GPU is required in computer server. Based on occupant number in the building, it is necessary to prepare one or more GPUs. Further, the best ratio between GPU number and occupant number should be validated and tuned in practical application and will not be considered in this paper.

5.6. Exceptions

While human physiology and human thermal comfort remains a complex issue, the possibility to measure temperature distributions on the body's surface can provide valuable indicators. Therefore, in this paper, we just focus on the prediction of skin temperature. However, some exceptions still should be mentioned. (1) There are some exceptions to the close relationship between skin temperature and thermal comfort. e.g., while sweating occurs, as the reason of sweat evaporates and heat absorption, the skin surface temperature will drop. However, the human perception could be hot. (2) There are some special cases between thermal sensation and thermal comfort. e.g., Occupant sometimes has a warm sensation, but he or she is very comfortable.

5.7. Others

Some potential limitations should be noted. (1) In this paper, all subjects are Asian females. Therefore, maybe we only can say that the NISDL is applicable to Asian women at the present time. Further, more data validation is required. (2) The subject acclimatization time is 10 min in this paper. The result of a longer time, e.g., more than 30 min, will be better than that of 10 min. When the acclimatization time is 30 min, the subjects are more likely to reach a stable starting state. (3) Relative constant parameters were set in an experiment chamber. This means that only one thermal condition was tested. If we handle the physiological experiment in different indoor parameters (e.g., indoor temperature), more valuable data and conclusions can be obtained.

6. Conclusions

In this paper, a kind of contactless measuring method based on skin sensitivity index for thermal comfort (NISDL) is proposed. For validating the effectiveness of SSI, two different deep learning frameworks with SSI were designed. A total of 1.44 million images were used for algorithm validation. The conclusions can be summarized as follows.

(1) SSI is a good and high weight parameter in contactless measurement of skin temperature based on a deep learning network.
(2) The location of SSI participation in NISDL network training has little impact on measuring the performance of skin temperature. Of course, if the SSI features are extracted firstly, and then merged with the features of ROI images, the corresponding effect is slightly better.
(3) The NISDL method proposed in this paper can be used for measuring thermal comfort and more diverse data can help it to improve the measuring accuracy.

In practical application, the inter-difference is very large. How to define and calculate suitable SSI will affect the measuring results. Further, more diverse data comparison is required to improve the algorithm robustness. These areas will be our research directions in the near future.

Author Contributions: Conceptualization, X.C. and B.Y.; Formal analysis, X.C. and B.Y.; Methodology, X.C., B.Y., T.O. and H.L.; Software, X.C. and L.L.; Validation, B.Y., K.T. and E.I.; Writing—original draft, X.C.; Writing—review & editing, B.Y. and A.H.

Acknowledgments: The authors thank William T. Freeman (MIT) for providing his MATLAB code of Euler Video Magnification (EVM).

References

1. U.S. Energy Information Administration. *International Energy Outlook 2017*; IEO2017 Report; U.S. Energy Information Administration (EIA): Washington, DC, USA, 2017.
2. U.S. Energy Information Administration. *Energy Implications of China's Transition toward Consumption-Led Growth*; IEO2018 Report; U.S. Energy Information Administration (EIA): Washington, DC, USA, 2018.

3.   Zagreus, L.; Huizenga, C.; Arens, E.; Lehrer, D. Listening to the occupants: A web-based indoor environmental quality survey. *Indoor Air* **2004**, *14*, 65–74. [CrossRef]

4.   Zhao, Q.; Zhao, Y.; Wang, F.; Wang, J.; Jiang, Y.; Zhang, F. A data-driven method to describe the personalized dynamic thermal comfort in ordinary office environment: From model to application. *Build. Environ.* **2014**, *72*, 309–318. [CrossRef]

5.   Ghahramani, A.; Tang, C.; Becerik-Gerber, B. An online learning approach for quantifying personalized thermal comfort via adaptive stochastic modeling. *Build. Environ.* **2015**, *92*, 86–96. [CrossRef]

6.   Liu, W.; Lian, Z.; Zhao, B. A neural network evaluation model for individual thermal comfort. *Energ. Build.* **2007**, *39*, 1115–1122. [CrossRef]

7.   Afroz, Z.; Urmee, T.; Shafiullah, G.M.; Higgins, G. Real-time prediction model for indoor temperature in a commercial building. *Appl. Energ.* **2018**, *231*, 29–53. [CrossRef]

8.   Wang, D.; Zhang, H.; Arens, E.; Huizenga, C. Observations of upper-extremity skin temperature and corresponding overall-body thermal sensations and comfort. *Build. Environ.* **2007**, *42*, 3933–3943. [CrossRef]

9.   Nakayama, K.; Suzuki, T.; Kameyama, K. Estimation of thermal sensation using human peripheral skin temperature. In Proceedings of the IEEE International Conference on Systems, Man and Cybernetics (SMC2009), San Antonio, TX, USA, 11–14 October 2009; pp. 2872–2877.

10.   Liu, W.; Lian, Z.; Deng, Q.; Liu, Y. Evaluation of calculation methods of mean skin temperature for use in thermal comfort study. *Build. Environ.* **2011**, *46*, 478–488. [CrossRef]

11.   Takada, S.; Matsumoto, S.; Matsushita, T. Prediction of whole-body thermal sensation in the non-steady state based on skin temperature. *Build. Environ.* **2013**, *68*, 123–133. [CrossRef]

12.   Sim, S.Y.; Koh, M.J.; Joo, K.M.; Noh, S.; Park, S.; Kim, Y.H.; Park, K.S. Estimation of thermal sensation based on wrist skin temperatures. *Sensors* **2016**, *16*, 420. [CrossRef] [PubMed]

13.   Wu, Z.; Li, N.; Cui, H.; Peng, J.; Chen, H.; Liu, P. Using upper extremity skin temperatures to assess thermal comfort in office buildings in Changsha, China. *Int. J. Environ. Res. Pu* **2017**, *14*, 1092–1109.

14.   Chaudhuri, T.; Zhai, D.; Soh, Y.C.; Li, H.; Xie, L. Thermal comfort prediction using normalized skin temperature in a uniform built environment. *Energ. Build.* **2018**, *159*, 426–440. [CrossRef]

15.   Yao, Y.; Lian, Z.; Liu, W.; Jiang, C.; Liu, Y.; Lu, H. Heart rate variation and electroencephalograph - the potential physiological factors for thermal comfort study. *Indoor Air* **2009**, *19*, 93–101. [CrossRef] [PubMed]

16.   Dai, C.; Zhang, H.; Arens, E.; Lian, Z. Machine learning approaches to predict thermal demands using skin temperatures: Steady-state conditions. *Build. Environ.* **2017**, *114*, 1–10. [CrossRef]

17.   Chaudhuri, T.; Soh, Y.C.; Li, H.; Xie, L. Machine learning based prediction of thermal comfort in buildings of equatorial Singapore. In Proceedings of the IEEE International Conference on Smart Grid and Smart Cities, Singapore, 23–26 July 2017; pp. 72–77.

18.   Kim, J.; Zhou, Y.; Schiavon, S.; Raftery, P.; Brager, G. Personal comfort models: Predicting individuals' thermal preference using occupant heating and cooling behavior and machine learning. *Build. Environ.* **2018**, *129*, 96–106. [CrossRef]

19.   Ghahramani, A.; Castro, G.; Becerik-Gerber, B.; Yu, X. Infrared thermography of human face for monitoring thermoregulation performance and estimating personal thermal comfort. *Build. Environ.* **2016**, *109*, 1–11. [CrossRef]

20.   Ghahramani, A.; Castro, G.; Karvigh, S.A.; Becerik-Gerber, B. Towards unsupervised learning of thermal comfort using infrared thermography. *Appl. Energ.* **2018**, *211*, 41–49. [CrossRef]

21.   Cheng, X.; Yang, B.; Olofsson, T.; Li, H.; Liu, G. A pilot study of online non-invasive measuring technology based on video magnification to determine skin temperature. *Build. Environ.* **2017**, *121*, 1–10. [CrossRef]

22.   Meier, A.; Dyer, W.; Graham, C. Using human gestures to control a building's heating and cooling System. In Proceedings of the International Conference on Energy-Efficient Domestic Appliances and Lighting (EEDAL 2017), Irvine, CA, USA, 13–15 September 2017.

23.   Dziedzic, J.W.; Yan, D.; Novakovic, V. Real time measurement of dynamic metabolic factor (D-MET). In Proceedings of the 9th International Cold Climate Conference Sustainable new and renovated buildings in cold climates (Cold Climate HVAC 2018), Kiruna, Sweden, 12–15 March 2018.

24.   Peng, B.; Hsieh, S. Data-driven thermal comfort prediction with support vector machine. In Proceedings of the ASME 2017 12th International Manufacturing Science and Engineering Conference (MSEC), Los Angeles, CA, USA, 4–8 June 2017; pp. 1–8.

25. Wang, Z.; de Dear, R.; Luo, M.; Lin, B.; He, Y.; Ghahramani, A.; Zhu, Y. Individual difference in thermal comfort: A literature review. *Build. Environ.* **2018**, *138*, 181–193. [CrossRef]

26. Peng, Y.; Rysanek, A.; Nagy, Z.; Schlüter, A. Using machine learning techniques for occupancy-prediction-based cooling control in office buildings. *Appl. Energ.* **2018**, *211*, 1343–1358. [CrossRef]

27. Fanger, P.O. *Thermal Comfort: Analysis and Application in Environmental Engineering*; Danish Technical Press: Copenhagen, Denmark, 1970.

28. Farhan, A.A.; Pattipati, K.; Wang, B.; Luh, P. Predicting individual thermal comfort using machine learning algorithms. In Proceedings of the IEEE International Conference on Automation Science and Engineering (CASE), Gothenburg, Sweden, 24–28 August 2015; pp. 708–713.

29. Megri, A.; Naqa, I. Prediction of the thermal comfort indices using improved support vector machine classifiers and nonlinear kernel functions. *Indoor Built Environ.* **2016**, *25*, 6–16. [CrossRef]

30. Li, W.; Zhang, J.; Zhao, T.; Wang, J.; Liang, R. Experimental study of human thermal sensation estimation model in built environment based on the Takagi-Sugeno fuzzy model. *Build. Simul.* **2018**. [CrossRef]

31. Cosma, A.C.; Simha, R. Machine learning method for real-time non-invasive prediction of individual thermal preference in transient conditions. *Build. Environ.* **2019**, *148*, 372–383. [CrossRef]

32. LeCun, Y.; Bengio, Y.; Hinton, G. Deep learning. *Nature* **2015**, *521*, 436–444. [CrossRef] [PubMed]

33. Huang, G.; Liu, Z.; Maaten, L.V.D. Densely connected convolutional networks. In Proceedings of the IEEE Conference on Computer Vision and Pattern Recognition (CVPR), Honolulu, HI, USA, 21–26 July 2017.

34. Wu, H.; Rubinstein, M.; Shih, E.; Guttag, J.; Durand, F.; Freeman, W. Eulerian video magnification for revealing subtle changes in the world. *Acm T Graph.* **2012**, *31*, 1–8. [CrossRef]

35. Wadhwa, N.; Wu, H.; Davis, A.; Rubinstein, M.; Shih, E.; Mysore, G.; Chen, J.; Buyukozturk, O.; Guttag, J.; Freeman, W.; et al. Eulerian video magnification and analysis. *Commun. Acm* **2017**, *60*, 87–95. [CrossRef]

36. Cheng, X.; Li, B.; Chen, Q. Online structural breaks estimation for non-stationary time series models. *China Commun.* **2011**, *7*, 95–104.

13

Residents' Perceptions of and Response Behaviors to Particulate Matter

Myung Eun Cho and Mi Jeong Kim *

School of Architecture, Hanyang University, 222 Wangsimni-ro, Seongdong-gu, Seoul 04763, Korea
* Correspondence: mijeongkim@hanyang.ac.kr.

Abstract: This study is interested in understanding the particulate matter perceptions and response behaviors of residents. The purpose of this study was to identify indoor air quality along with the response behaviors of residents in Seoul, to ascertain whether there is a difference in behaviors when particulate matter is present, according to the characteristics of residents and to grasp the nature of this difference. A questionnaire survey of 171 respondents was conducted. The questionnaire measured the indoor air quality perceived by residents, the health symptoms caused by particulate matter, residents' response behaviors to particulate matter and the psychological attributes affecting those response behaviors. Residents of Seoul were divided into college students in their twenties, male workers in their thirties and forties and female housewives in their thirties and forties. The data were calibrated by SPSS 23 using a one-way analysis of variance (ANOVA) and multiple regression analyses. The results show that most people found particulate matter to be an important problem but were unable to do sufficient mitigation action to prevent its presence. Residents showed greater psychological stress resulting in difficulty going out than physical symptoms. The most influential factor on response behaviors was psychological attributes. Participants were aware of the risks of particulate matter but believed it to be generated by external factors; thus, they felt powerless to do anything about it, which proved to be an obstacle to response behaviors.

Keywords: particulate matter; perception; response behavior; psychological attribute

1. Introduction

In recent years, particulate matter (PM) has emerged as a big problem in Korea. According to the Organization for Economic Cooperation and Development (OECD)'s annual report on the concentration of ultra-particulate matter in countries by 2017, the mean population was exposed to $PM_{2.5}$ and with pollution at 25.1 $\mu g/m^3$, Korea was the second worst of the member countries [1], with a level twice as high as the average OECD member countries (12.5 $\mu g/m^3$) and 2.5 times higher than the World Health Organization (WHO)'s annual average recommended concentration (10 $\mu g/m^3$). Based on Korea's PM forecast, the number of "bad" (36–75 $\mu g/m^3$) and "very bad" (more than 76 $\mu g/m^3$) days in metropolitan areas increased from 62 in 2015 to 77 in 2018 [2]. In early March 2019, Korea experienced the most severe PM situation. In Seoul, an 8-day ultra-particulate matter warning ($PM_{2.5}$ with a time-averaged concentration of more than 75 $\mu g/m^3$ for 2 h) and 2 days (March 5 and 6) with an alert level (an average $PM_{2.5}$ of more than 150 $\mu g/m^3$ for 2 h) [3]. As a result, the PM levels became hazardous to health.

PM is a WHO Level 1 carcinogen that has negative effects on health, contributing to cardiovascular and respiratory diseases [4,5]. Choe and Lee [6] investigated the effect of particulate emissions on specific diseases in Seoul and found that the number of hospitalizations for various respiratory diseases increased as the amount of ultra-PM increased. Korea's increase in PM is related to rapid economic growth. Large cities, such as Seoul, have high levels of energy use resulting from the concentration of population and economic activity and their direct emission of air pollutants is high. Further, since its geographical location is on the mid-latitude westerly wind area, seasonal influx of PM from neighboring China also affects the increase of PM in Korea [7].

According to the survey data on the perception of environmental problems among Koreans aged 13 and over, conducted by the Korea National Statistical Office (KNSO), 82.5% of respondents experience anxiety about PM [8]. Kim et al. [9] observed that Koreans regard PM as the most serious social risk factor. Because national concern about PM has been increasing, the government introduced a comprehensive plan for PM in 2017. In 2018, it attempted to reduce PM emissions by enforcing a Special Act on Particulate Matter in major cities across the country, including the capital region [10]. The Ministry of Environment, in consideration of atmospheric environmental standards and health effects, produced a PM forecasting system, which presents the levels of PM as well as countermeasures [11]. However, despite the various risk indicators for PM and notwithstanding the government measures, the residents of Seoul are notably passive in protecting individuals and society from PM despite viewing it as a threat [12]. To prevent and reduce the PM generated by anthropogenic rather than natural factors, public efforts must be accompanied by measures at the national level. Without ensuring that residents understand PM, it is predicted that reduction measures will be ineffective. Therefore, this study aims to identify levels of awareness of PM, recognition of indoor air quality, symptoms of PM exposure experienced by residents and coping behavior in relation to PM in Seoul. Specific research questions are as follows:

First, how do residents perceive the indoor air quality, how do they feel PM symptoms and how do they behave in response to PM?

Second, do the different characteristics of residents produce any variations in behavior in response to PM?

Third, what are the impediments to the proper responses to PM by residents?

In this study, we have sought to understand the perceptions of PM and the subsequent response behaviors of residents, as well as to identify the causes of these behaviors. It is critical to understand and solve the barriers to public engagement to avoid the worst consequences of PM. The results of this study are expected to be used as basic data for effective governmental measures to reduce PM.

2. Status and Risks of Particulate Matter in Korea

2.1. Characteristics of Particulate Matter Generation in Seoul

To identify the characteristics of PM generation in Seoul, we investigated the annual average concentration of PM along with its highest levels of concentration. As shown in Figure 1, the average concentrations of PM_{10} and $PM_{2.5}$ in Seoul were calculated using Seoul's atmospheric environment information from the past 10 years. From 2009 to 2018, the average annual concentration of PM_{10} in Seoul was 53.8 $\mu g/m^3$ and the average annual concentration of $PM_{2.5}$ was 27.2 $\mu g/m^3$ [3]. The average annual concentration of PM_{10} decreased from 76 $\mu g/m^3$ in 2002 to 41 $\mu g/m^3$ in 2012 but this decline has since slowed. Since the government enacted the Special Act on the Improvement of the Air Quality in the Seoul Metropolitan Area and established and performed its Basic Plan for the Management of Air Quality in the Seoul Metropolitan Area in 2003, air pollution as well as PM concentrations have been reduced [7]. However, no further improvement has occurred since 2012.

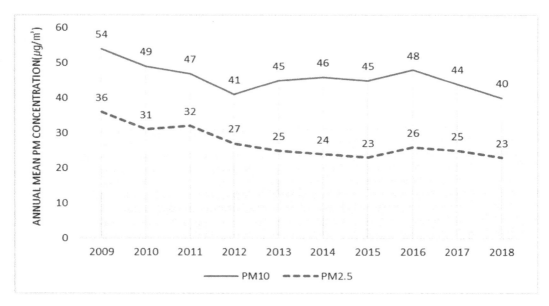

Figure 1. Annual mean particulate matter (PM) concentration in Seoul (2009–2018).

The results of a survey of "bad" and "very bad" days exceeding a $PM_{2.5}$ of 35 $\mu g/m^3$ for an average of 24 h showed that 44 days in 2015, 73 days in 2016, 64 days in 2017 and 61 days annually on average were recorded as "bad" and "very bad". This means that the number of days when the concentration of PM was significant is very large. This can be seen by comparing the number of PM warnings and days of alarm in Seoul. Figure 2 shows the number of PM and ultra-PM warning days using Seoul's atmospheric environment information. If there is a PM_{10} of 150 $\mu g/m^3$ or a $PM_{2.5}$ of 75 $\mu g/m^3$ for more than 2 h, a warning is issued. If there is a PM_{10} of 300 $\mu g/m^3$ or a $PM_{2.5}$ of 150 $\mu g/m^3$ for more than 2 h, an alarm is issued. In 2018, there were 17 PM warning days, 1 PM alarm day and 7 ultra-PM warning days. Seoul residents were exposed to extremely high dust concentrations on these days.

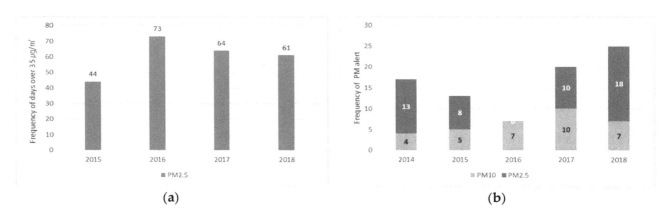

Figure 2. High PM concentration days: (**a**) days exceeding a $PM_{2.5}$ of 35 $\mu g/m^3$ on average for 24 h; (**b**) PM and ultra-PM warning and alarm days (Source: Seoul Atmospheric Environment Information).

This result is closely related to the monsoon season experienced in the geographical location of Korea. Korea's winter atmospheric circulation is affected by the northwestern winds associated with the winter monsoon in East Asia and the location of the barometer over the Korean peninsula. Inflow from China and Mongolia in winter greatly affects the concentration of PM in Korea [13]. Figure 3 shows the changes in the concentrations of PM and ultra-PM from March 2018 to February 2019 [3]. It shows that the levels of PM are high in winter and spring and low in summer. Specifically, they are lowest in September and highest in January. The atmospheric environmental standards for domestic PM are below 50 $\mu g/m^3$ on average for 24 h and below 25 $\mu g/m^3$ on average per year. Korea's levels exceed these standard values in all seasons except summer.

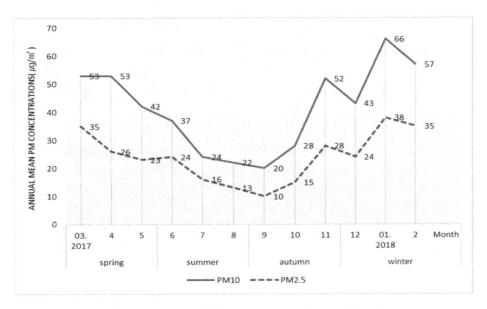

Figure 3. Seasonal dust concentrations in Seoul.

The air quality in Korea deteriorates considerably as the number of days with seasonally high concentrations of PM increase [14].

2.2. Domestic Particulate Matter Standards and Health Protection

The Ministry of Environment [2] proposed enhanced environmental standards for PM in 2018, taking into consideration national health effects, international standards, pollution status and achievability. Table 1 shows the WHO recommendations along with the standards in other regions. The concentration of PM in the domestic environment of higher than 50 $\mu g/m^3$ far exceeds the WHO's recommended level of an annual average PM_{10} of 20 $\mu g/m^3$ and is much greater than that of other regions. In addition, the rate of achieving the annual PM environmental standard was around 60% in 2015 and the rate of achieving the 24 h environmental standard for $PM_{2.5}$ and PM_{10} was very low at 4% and 10.7% respectively.

Table 1. Comparison of air quality standards for PM (Source: Ministry of Environment).

Category	Standard Time	PM Standards				Standard Achievement Rate (2015)
		Korea	WHO	USA	EU	
$PM_{2.5}$ ($\mu g/m^3$)	yearly	15	10	12, 15	25	65.0%
	24 h	35	25	35	-	4.0%
PM_{10} ($\mu g/m^3$)	yearly	50	20	-	40	65.6%
	24 h	100	50	150	50	10.7%

Considering public health, the Ministry of Environment has been conducting PM forecasting from February 2014 and ultra-PM forecasting with a warning system from January 2015 [2]. The PM forecasting system provides forecasts of the PM concentration four times a day for the present day, the following day and the day after that. The PM forecast is graded in four stages: "good", "normal", "bad" and "very bad" (Table 2). The PM warning system works to promptly inform the public when a high concentration of PM occurs and to reduce the damage. It is issued when the air quality is harmful to health. Air pollution alarms are classified into two stages—warning and alarm—as shown in Table 3. At the time of a PM alarm, there are eight countermeasures for people to take: remaining indoors; wearing a health mask; reducing external activities; washing the body after returning home; drinking water and eating fruits and vegetables; undertaking ventilation and indoor water cleaning; managing indoor air quality; and restricting air pollution inducing activities.

Table 2. Particulate matter forecast grade (2018).

PM Concentration (μg/m³, Average for 24 h)	Good	Normal	Bad	Very Bad
PM_{10}	0–30	31–80	81–150	over 151
$PM_{2.5}$	0–15	16–35	36–75	over 76

Table 3. Particulate matter warning and alarm issuing grade (2018).

Category	Warning Issuing	Alarm Issuing
PM_{10}	PM_{10} hourly average concentration over 150 μg/m³ for 2 h	PM_{10} hourly average concentration over 300 μg/m³ for 2 h
$PM_{2.5}$	$PM_{2.5}$ hourly average concentration over 75 μg/m³ for 2 h	$PM_{2.5}$ hourly average concentration over 150 μg/m³ for 2 h

3. Materials and Methods

3.1. Participants and Questionnaire Design

This study surveyed the residents of Seoul. To identify differences in the response to PM associated with gender and age, respondents were divided into three groups. A total of 171 respondents were used for the analysis. The groups were 20-year-old college students (N = 70, gender = 32 male and 38 female, mean age = 21.88, SD = 2.33), 30- to 40-year-old male workers (N = 51, mean age = 41.11, SD = 5.45) and 30- to 40-year-old housewives (N = 50, mean age = 37.06, SD = 4.33). The questionnaire comprised five main parts. The first part included questions about the quality of indoor air perceived by residents in the home. The second part addressed the health of the residents in terms of objective symptoms, subjective symptoms and health behaviors concerning the symptoms. The third section elicited the residents' responses to PM, dividing them into mitigating behavior, adaptive behavior and behavior intentions. The fourth part sought to ascertain the psychological causes of interference with the response behaviors of the residents to PM and the final content consisted of questions investigating the residents' overall knowledge of PM.

3.2. Measures

3.2.1. Measuring Response Behaviors to Particulate Matter

Mitigating behavior and adaptive behavior are countermeasures to deal with risk. Swart and Raes [15] define "mitigation" as an anthropogenic intervention to reduce the sources of air pollution, whereas "adaptation" is an adjustment in natural or human systems in response to climatic stimuli. Mitigation is a way to reduce the cause of hazards associated with PM. Usually, the benefits of mitigating behavior are not seen in the short term, so it is considered a long-term countermeasure [16]. Mitigating behavior is a personal effort to reduce the generation of PM, which includes "using public transportation to reduce atmospheric gas generation", "not using electricity and heating to restrain unnecessary energy use", "using kitchen utensils that generate less harmful gas" and other similar measures. Adaptation refers to controlling the damage caused by PM.

Adaptive behavior can reduce the risks associated with PM by prophylactic and post-exposure measures that minimize the negative effects of PM [17]. In contrast to mitigating behavior, from which long-term effects arise, the effects of adaptive behavior are immediate and are characterized by the matching of subject and beneficiary [18]. Adaptive behavior is action to prevent the damage caused to individuals by high concentrations of dust. Behavioral intentions signify an individual's specific willingness to act in response to PM and include two positive intentions. To achieve sustainable development policies, it is necessary to integrate mitigating actions that directly reduce the concentration of PM and adaptive prevention actions that reduce the impact of the existing dust risk. In this study, mitigating behavior, adaptive behavior and the behavioral intentions of Seoul residents are explored, as shown in Table 4.

Table 4. Responsive behaviors to particulate matter.

Question: What do you do when Particulate Matter Occurs?	
Mitigating behavior	MB1 Using public transportation to reduce atmospheric gas generation MB2 Refraining from using electricity and heating for the reduction of unnecessary energy use MB3 Using kitchen appliances that generate less harmful gas
Adaptive behavior	AB1 Checking the levels of particulate matter concentration every day AB2 Wearing a particulate matter mask on high particulate matter density days AB3 Refraining from going out when the concentration of particulate matter is high AB4 Turning on an air purifiers to reduce particulate matter concentration AB5 Washing the whole body thoroughly in running water after returning home AB6 Wiping dust from the floor with a damp cloth and undertaking indoor water cleaning
Behavioral intentions	BI1 Being ready to suffer immediate damage or inconvenience to reduce particulate matter concentrations. BI2 Being willing to participate in actions to reduce particulate matter BI3 Not feeling any urgency to change my behavior to prevent particulate matter pollution BI4 Thinking that changes in my behavior do not affect particulate matter reduction BI5 Not knowing what concrete action may be taken to reduce particulate matter

3.2.2. Measuring Psychological Attributes that Hinder Reactions to Particulate Matter

Gilfford [19] observed that most people find environmental sustainability to be an important issue but that psychological barriers prevent them from engaging in sufficient actions to address it. These psychological barriers impede the behavioral choices that would facilitate mitigation, adaptation and environmental sustainability [20]. Risk perception is a factor that affects human attitudes and behavioral intentions and that is very important in making individual decisions about a behavior [21]. Brewer et al. [22] explained three dimensions of risk perception. The first is the likelihood that one will be harmed by the hazard; the second is susceptibility, referring to an individual's constitutional vulnerability to a hazard; and the third is severity, indicating the extent of harm a hazard may cause. Psychological distance describes how individuals participate in future events [23]. The perceived distance of events indicates how they are mentally construed. As the perceived distance increases, events are interpreted as more abstract, decontextualized and conceived in generalized terms. When events become closer, they use more specific, contextualized and detailed features [24]. The psychological distance for PM was evaluated using the four distance domains of geography, temporality, socialization and awareness. In addition, this research investigated various barriers increasing the perceived concern of the public, such as information distrust, externalizing responsibility and uncertainty about the causes of PM pollution. In this study, to understand the effects of response behaviors to PM, the psychological attributes of the residents were classified into nine concepts within three domains: "risk perception", "psychological distance", and "perceived concerns" (Table 5).

Table 5. Psychological attributes.

Question: What Are your Thoughts in Response to Particulate Matter?		
Risk perception	Likelihood	1. Contact with particulate matter is a health hazard 2. If I continue to be exposed to particulate matter, I will be damaged in a few years
	Susceptibility	3. Particulate matter is an important problem for me 4. I am more affected by the risk of particulate matter than other risks
	Severity	5. The risks of particulate matter for health are very serious 6. Even short-term contact with particulate matter can increase the likelihood of cancer and early death

Table 5. *Cont.*

		Question: What Are your Thoughts in Response to Particulate Matter?
Psychological distance	Geographic and Temporal	7. The particulate matter condition is more serious in Korea than in other countries 8. The risk of particulate matter is present very often or constantly
	Social	9. Particulate matter will have a significant impact on me and my family 10. Particulate matter is sure to be a serious social problem
	Awareness	11. I am very concerned about particulate matter 12. I am interested in issues related to particulate matter
Perceived concerns	Information distrust	13. I cannot trust the information about particulate matter 14. The risk of particulate matter is greatly exaggerated
	Externalizing responsibility	15. The problem of particulate matter is beyond my ability to solve 16. The solution to the excess of particulate matter should be provided by the government rather than individuals
	Uncertainty	17. The damage caused by particulate matter is unclear 18. It is difficult to measure health damage arising from particulate matter 19. Particulate matter does not affect me right now

3.3. Methodology

This research used a questionnaire to collect its data; the survey was conducted online in March 2019 and the data were analyzed using IBM's SPSS Statistics Program 23. The question concerning residents' perceptions of indoor air quality comprised positive and negative language (the semantic differential method). We distinguished between positive vocabulary and negative vocabulary over five levels, asking residents to assign one level to each question according to their perceptions. The vocabulary used to measure the indoor air quality of residents included "bad", "good", "stuffy", "refreshed", "unpleasant", and "comfortable". A five-point Likert scale was used to assess the symptoms, behaviors and psychological responses of residents, ranging from "very unlikely" (1 point), to "unlikely" (2 points), to "average" (3 points), to likely (4 points), to "very likely" (5 points). Questions measuring residents' knowledge of PM could be answered as "right" (○) or "wrong" (X) and the "right" answers were added to produce a score.

An ANOVA was used for the statistical analysis to determine whether there were differences in behavior as a response to PM, according to the characteristics of residents. Participants were divided into three groups based on their genders and ages: college students in their twenties, male workers in their thirties and forties and female housewives in their thirties and forties. In addition to the response behavior, we compared differences between the groups in terms of environmental exposures to PM, perceived symptoms and psychological attributes. Further, post-hoc tests were conducted to analyze variations between groups. Multiple regression analysis was conducted to identify the factors influencing response behavior.

4. Results

4.1. Perceived Indoor Air Quality and Overall Satisfaction

The perceived indoor air quality and the satisfaction of residents in Seoul were measured using a semantic differential approach. Residents assessed their perceptions of indoor air quality by comparing two opposing pairs of vocabulary in five steps. For example, the evaluation score for the first question on indoor air quality was divided into five levels, spanning two opposing experiences: "a lot of particulate matter" and "no particulate matter". The more PM respondents perceived, the closer to 1 point they scored and the less PM they observed, the closer they scored to 5 points.

A total of 171 residents were analyzed and the results are shown in Table 6. Respondents' answers to the four questions about the quality of indoor air produced an average score of approximately three (2.93–3.25) points. This score comprises the median value of 1 and 5, indicating that the residents of

Seoul understand the indoor air quality to be normal. The evaluations of overall indoor air quality also averaged 3.21 points. However, the average score for the question, "Do you think seriously about or are you interested in the quality of indoor air where you live?" was 4.17 points (where 1 point indicated "not at all" and 5 points denoted "very much"), implying that respondents were interested in the air quality of their living spaces.

Table 6. Perceptions of and satisfaction with air quality.

	Perceptions of		Mean	Std. Deviation
	1 Point	**5 Points**		
Indoor air quality	very bad	very good	3.26	0.91
	a lot of particulate matter	little particulate matter	2.94	0.98
	stuffy	refreshed	2.93	1.02
	unpleasant	comfortable	3.25	0.92
Satisfaction	dissatisfaction	satisfaction	3.21	1.10
	no thought or interest	much thought and interest	4.17	1.10

Two phrases corresponded to 1 point and 5 points respectively and between these two phrases were expressions valued at 2 points (signifying the perception of slightly more PM), 4 points (signifying the perception of slightly less PM) and 3 points (signifying the perception of an average amount of PM).

4.2. Particulate Matter Environmental Exposures and Perceived Symptoms

When measuring the outdoor activity times of respondents, it was found that they stayed outdoors for an average of 4.54 h (SD = 4.31) per day and the analysis of variance showed a statistically significant difference ($p < 0.01$) depending on the subject group. The results are shown in Table 7. The group of 20 college students and the group of 30- to 40-year-old male workers remained outside for an average of 5 h a day—specifically 5.12 h and 5.07 h respectively—while the group of 30- to 40-year-old housewives stayed outdoors for 3 h a day on average.

Table 7. Average daily external activity time.

Group 1		Group 2		Group 3		F (p-Value)
Mean	**Std Deviation**	**Mean**	**Std Deviation**	**Mean**	**Std Deviation**	
5.12 h	3.43	5.07 h	5.49	3.20 h	3.81	3.54 (0.031)

Group 1: 20-year-old college students; Group 2: 30- to 40-year-old male workers; Group 3: 30- to 40-year-old housewives.

The physical symptoms and health problems experienced by residents as a result of PM were verified through objectively judged symptoms, such as cough, allergy and headache, as well as subjectively judged psychological symptoms, such as anxiety and stress. In addition, the extent of any actual treatment at the hospital because of these symptoms was investigated. The respondents' answers ranged from "not at all" (1 point) to "very strongly" (5 points) and the results are shown in Table 8.

The average score for the response, "Thinking that particulate matter has a negative effect on my health" was 4.39, the highest and the average score for the response "Thinking that particulate matter causes problems in my life" was 4.00, the second highest. Symptoms such as respiratory problems and complications with nose, skin and eyes rated 3.69 and 3.66 respectively, revealing that the average scores for the objective symptoms of residents with health problems arising from PM were normal. These findings indicate that the psychological symptoms (Mean = 4.39) were more significant than the objective symptoms (Mean = 3.75); further, the incidence of direct treatment in hospitals (M = 3.15) for such symptoms was not significant (Figure 4). However, ANOVA analysis showed statistically significant differences according to subject groups. A Tukey post-hoc test was conducted to analyze the differences between the groups and the results showed that the perceived symptoms were greater among the group of housewives in their thirties and forties than in the group of college students in their twenties or in the group of male workers in their thirties and forties. Notably, the differences in psychological symptoms were pronounced among these groups (Table 8).

Table 8. Health-related symptoms experienced by residents as a consequence of particulate matter.

	Perceived	Mean				F (p-Value)
		Group 1	Group 2	Group 3	Total	
Objective symptom	Suffering from cough, hoarseness, bronchitis, respiratory distress and difficulty breathing	3.54	3.50	4.08	3.69	3.33 (0.038 *)
	Experiencing clogged nose, rhinitis, sneezing, allergy and atopic symptoms, such as itchy or swollen skin or dry eyes	3.51	3.43	4.10	3.66	3.76 (0.025 *)
	Suffering from headache, arthritis and other pain	2.61	2.92	3.38	2.92	4.38 (0.014 *)
	Experiencing the deterioration of an already present illness	2.42	2.88	3.38	2.84	7.57 (0.001 **)
Subjective symptom	Suffering increased stress and fatigue	3.34	3.50	4.14	3.62	7.23 (0.001 **)
	Feeling worried and anxious	3.18	3.60	4.30	3.63	12.84 (0.000 ***)
	Thinking that particulate matter has a negative effect on my health	4.21	4.27	4.78	**4.39**	8.55 (0.000 ***)
	Thinking that particulate matter causes problems in my life	3.71	3.68	4.72	**4.00**	17.55 (0.000 ***)
	Thinking that particulate matter reduces my performance and concentration	3.20	3.31	4.02	3.47	6.80 (0.001 **)
Remedy	I have recently been to the hospital for any of the above symptoms	1.94	2.15	3.00	2.31	8.96 (0.000 ***)
	I have recently been taking medication for some of the above symptoms	2.12	2.31	3.04	2.45	6.19 (0.003 **)
	I have recently taken a break at home because of some of the above symptoms	2.51	2.39	3.42	2.74	7.80 (0.001 **)

* $p < 0.05$, ** $p < 0.01$, *** $p < 0.001$. Group 1: 20-year-old college students; Group 2: 30- to 40-year-old male workers; Group 3: 30- to 40-year-old housewives.

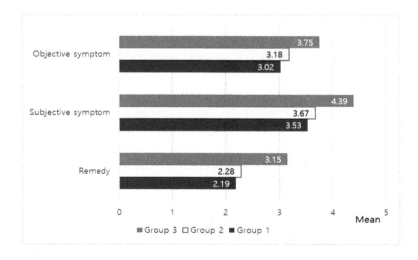

Figure 4. Physical symptoms of particulate matter experienced by residents.

4.3. Behavioral Changes in Response to Particulate Matter

Residents' response behaviors to the occurrence of PM were analyzed and the results indicate that adaptive behaviors (M = 3.66, SD = 0.95) controlling the damage caused by PM were more prevalent than mitigating behaviors (M = 2.98, SD = 1.03) to reduce the fundamental causes of PM, as shown in Table 9. The most common PM adaptive response behaviors were "AB1: Checking the PM concentration every day" (4.00 points) and "AB5: Washing your whole body thoroughly in running water after returning home" (3.78 points). Regarding behavioral intentions, the most common PM response behaviors were "BI2: Being willing to participate in actions to reduce particulate matter" (4.00 points) and "BI1: Being ready to experience immediate damage or inconvenience to reduce the particulate matter concentration" (3.69 points). For other responses, scores ranging from 2.50 to 3.50 were obtained, indicating that residents' responses to PM were not significant. However, there were statistically significant differences among the groups (Table 9). In particular, the group of 30- to 40-year-old housewives showed higher scores for adaptive behavior and behavioral intentions than the other groups (Figure 5).

Table 9. Residents' response behaviors to particulate matter.

Behavior Type		Mean				F (*p*-Value)
		Group 1	Group 2	Group 3	Total	
Mitigating behavior	MB1	4.04	2.82	2.60	3.25	19.63 (0.000 ***)
	MB2	2.80	3.00	3.00	2.91	0.49 (0.613)
	MB3	2.55	2.94	3.12	2.83	2.91 (0.057)
Adaptive behavior	AB1	3.55	3.80	4.84	**4.00**	19.63 (0.000 ***)
	AB2	2.77	3.35	4.28	3.38	20.36 (0.000 ***)
	AB3	2.82	3.41	4.32	3.43	19.53 (0.000 ***)
	AB4	2.77	3.92	4.32	3.56	19.03 (0.000 ***)
	AB5	3.48	3.76	4.24	3.78	5.55 (0.005 **)
	AB6	2.91	3.35	3.98	3.35	13.37 (0.000 ***)
Behavioral intentions	BI1	3.48	3.45	4.24	3.69	12.11 (0.000 ***)
	BI2	3.91	3.86	4.28	**4.00**	3.72 (0.026 *)
	BI3 [1]	−2.72	−2.82	−1.82	−2.49	12.44 (0.000 ***)
	BI4 [1]	−2.54	−2.82	−2.12	−2.50	4.88 (0.009 **)
	BI5 [1]	−3.05	−2.94	−2.86	−2.96	0.42 (0.657)

* $p < 0.05$, ** $p < 0.01$, *** $p < 0.001$, [1] Reverse scored for negative perceptions. Group 1: 20-year-old college students; Group 2: 30- to 40-year-old male workers; Group 3: 30- to 40-year-old housewives.

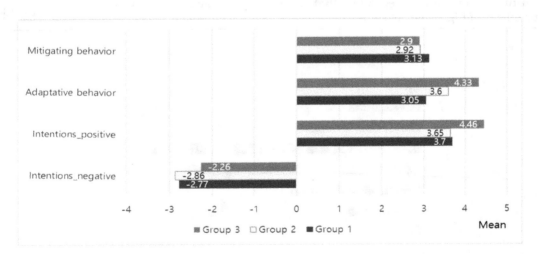

Figure 5. Residents' response behaviors to particulate matter.

4.4. Barriers to Mild Behavioral Changes as a Response to Particulate Matter

We investigated the residents' understandings of PM and their psychological causes as obstacles to behavioral changes to reduce and prevent PM.

4.4.1. Lack of Knowledge

The items indicating the residents' knowledge of PM consisted of 10 authentic scales (true/false) based on data from experts and the Ministry of Environment. The respondents selected "correct" or "incorrect" in answer to each item and answers of "correct" were treated as single points, producing a total score between 0 and 10 points. Table 10 shows the results for the 171 respondents. The average knowledge of PM was 6.56 (SD = 1.44) and the results of the analysis are shown in Figure 6. Knowledge of PM did not vary among the groups.

Table 10. Residents' knowledge of particulate matter.

Knowledge	Number of Correct Answers (%)
1. The influx of particulate matter is mainly caused by artificial actors, such as boilers, automobiles and power generation facilities	138 (80.7)
2. More than 80% of particulate matter is from abroad, as a result of the yellow dust and smog from China	39 (22.8)
3. Preliminary reduction measures against emissions of particulate matter are implemented nationwide and are targeted based on particulate matter concentration	43 (25.1)
4. Ultra-particulate matter comprises very thin, small particles but causes deterioration of visibility in places where the flow of air is stagnant, creating obstacles to traffic and navigation	118 (69.0)
5. When there is a particulate matter alarm, schools can prohibit outdoor classes, adjust the times of travel to and from school and temporarily shut down	151 (88.3)
6. Domestic particulate matter concentrations are similar to those in major cities of other OECD countries, such as New York and London	141 (82.5)
7. Even when exposure occurs for only a short time, particulate matter can penetrate directly into the alveoli, resulting in asthma, lung disease and even death	119 (69.6)
8. The mask is a quasi-drug certified by the Food and Drug Administration that can only prevent exposure to particulate matter if the product is marked "KF94" or "KF80"	133 (77.8)
9. Generally, high concentrations of particulate matter occur in spring and summer	89 (52.0)
10. It is helpful to eat fruits and vegetables that are rich in water and vitamin C to combat the effects of high dust concentrations	152 (88.9)

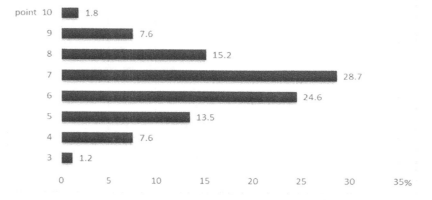

Figure 6. Residents' total knowledge of particulate matter.

4.4.2. Psychological Attributes of Particulate Matter

The results of the survey of the psychological influences on behavioral responses to PM show that risk perception (Mean = 4.01, SD = 0.76) and psychological distance (Mean = 4.32, SD = 0.73) achieved a higher than average score of 4.0 points (where 1 point = "not at all" and 5 points = "very much"). Table 11 shows the results of the analysis. Question 2 related to likelihood when considering residents' risk perception—"If you continue to be exposed to particulate matter, you will be affected in a few years" (Item 4.5)—as did Question 1, "Contact with particulate matter is harmful to health" (4.35 points)'; both of these questions received the highest scores. The score for the residents' perception of the psychological distance with which they regarded PM was higher than that for their risk perception. All questions concerning psychological distance were rated at greater than 4.0 and among them, the scores for Question 10, relating to the social implications of PM—"particulate matter is sure to be a serious social problem" (4.57 points)—and for Question 8, pertaining to the geographic and temporal significance of PM—"the risk of particulate matter is present very often or consistently" (4.47 points)—were the highest. Questions about perceived concerns, such as information distrust, the externalizing of responsibility and uncertainty, were reverse scored. Among them, Question 16, relating to the externalizing of responsibility—"The solution to the excess of particulate matter should be provided by the government rather than individuals" (−4.35 points)—and Question 15—"The problem of particulate matter is beyond my ability to solve" (−4.02 points)—received the highest scores. There was a statistically significant difference between the groups (Table 11). The group of housewives aged in their thirties and forties showed higher scores for risk perception and psychological distance than the other groups (Figure 7).

Table 11. Psychological influences on residents' response behaviors.

Psychological Attributes			Mean				F (p-Value)
			Group 1	Group 2	Group 3	Total	
Risk perception	Likelihood	1	4.08	4.33	4.74	4.35	8.58 (0.000 ***)
		2	4.44	4.29	4.82	**4.50**	6.22 (0.002 **)
	Susceptibility	3	4.10	4.19	4.62	4.28	5.11 (0.007 **)
		4	2.84	3.07	3.68	3.15	6.70 (0.002 **)
	Severity	5	4.10	4.15	4.58	4.27	4.69 (0.010 *)
		6	3.34	3.41	3.86	3.51	3.10 (0.048 *)
Psychological distance	Geographic and temporal	7	4.18	4.19	4.60	4.30	2.99 (0.053)
		8	4.41	4.25	4.80	4.47	8.13 (0.000 ***)
	Social	9	3.92	4.19	4.70	4.23	11.01 (0.000 ***)
		10	4.44	4.43	4.92	**4.57**	10.55 (0.000 ***)
	Awareness	11	4.02	4.13	4.76	4.27	9.96 (0.000 ***)
		12	3.70	4.00	4.60	4.05	11.86 (0.000 ***)
Perceived concerns [1]	Information distrust	13	−2.97	−3.62	−3.80	−3.40	10.84 (0.000 ***)
		14	−2.40	−2.43	−1.52	−2.15	13.65 (0.000 ***)
	Externalizing responsibility	15	−4.04	−3.96	−4.08	**−4.02**	0.30 (0.741)
		16	−3.98	−4.52	−4.70	**−4.35**	11.91 (0.000 ***)
	Uncertainty	17	−2.68	−2.88	−2.78	−2.77	0.36 (0.692)
		18	−2.97	−3.01	−2.88	−2.95	0.15 (0.857)
		19	−2.40	−2.50	−1.88	−2.28	4.08 (0.019 *)

* $p < 0.05$, ** $p < 0.01$, *** $p < 0.001$, [1] Reverse scored for negative perceptions. Group 1: 20-year-old college students; Group 2: 30- to 40-year-old male workers; Group 3: 30- to 40-year-old housewives.

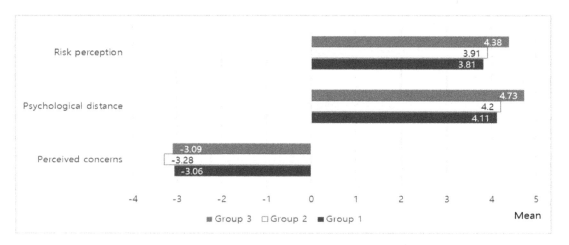

Figure 7. Residents' psychological influences.

4.5. Factors Affecting Response Behaviors to Paticulate Matter

A multiple regression analysis was conducted to determine the factors with the greatest effect on residents' response behaviors. Knowledge of PM, psychological attributes and PM-related symptoms were selected as factors influencing response behaviors and the analytical result was statistically significant ($p < 0.05$; F = 25.306, p-value = 0.000). Table 12 shows the evaluation of the contribution and statistical significance of the individual independent variables. Psychological attributes were the most influential factor on residents' responses to PM, followed by PM-related symptoms.

Table 12. Multiple regression analysis of the response behaviors to particulate matter.

Variable	B	Std Error	Beta	t Value	p-Value
Constant	8.103	3.577		2.265	0.025
Knowledge of PM	0.630	0.443	0.095	1.424	0.156
Psychological attributes	0.373	0.082	0.375	4.555	0.000
PM-related symptoms	0.191	0.070	0.219	2.737	0.007

Note: R = 0.55, R^2 = 0.31, Adj. R^2 = 0.30.

5. Discussion and Conclusions

In Seoul, it is inadvisable to go out on any day, except in summer, as a result of the PM concentration and its effect on human health. When the WHO standard is applied, the concentration of PM in Seoul is revealed to be remarkably high and cases of excessive concentration exceeding the daily average atmospheric environment standard are also frequent. The government has been pursuing various policies to reduce PM concentration but the concentration of PM has not improved to the extent that people recognize any change. Research has shown that most people believe PM to be an important problem but do not undertake enough mitigating behaviors to prevent its occurrence. Reliable research should be preceded by an analysis of cause and effect before the preparation of PM countermeasures. We sought to understand residents' perceptions of PM, their response behaviors and the psychological causes of these behaviors; the main findings of this study are as follows.

First, to identify the perceived indoor air quality of the residents, the questionnaire was distributed at a time when the concentration of PM had been significant throughout the year; thus, residents' interest in indoor air quality was great. However, despite high outdoor concentrations of PM, the indoor air quality was not considered poor or dissatisfactory. The symptoms that residents attributed to PM were not physical, such as allergies or bronchitis. They more commonly experienced psychological stress or anxiety, feeling that PM had a negative effect on health and was interfering more frequently with daily life, in that residents were going out less often to avoid it.

Second, the problem with the PM-related behaviors of residents in Seoul was that many people did not take actual action despite their willingness to do so. Mitigating behavior to try to minimize PM generation was less common than adaptive behavior to protect individuals from the risks of PM. This may have been because of the nature of mitigating behavior, which often entails loss or inconvenience. Even if the risks associated with PM were largely recognized, it would be necessary to compensate for the loss when the habitual behavior of individuals was abandoned in favor of following the recommended behavior. Unlike preventive actions to protect individuals from PM, compensation for mitigating behavior is only likely to be achieved in the distant future and even then, such compensation would provide a social rather than an individual benefit. This suggests that more persuasive strategies should be employed to stimulate changes in individual behaviors for collective interests.

Third, the psychological characteristics of individuals can either hinder or promote action for environmental sustainability. The greater the perceived risk and the more psychological that risk, the more likely it is that an active response will occur. Analysis of the psychological factors affecting residents in Seoul revealed that they perceived the risk of PM to be very close (in terms of psychological distance). However, despite revealing significant levels of risk perception and psychological distance, the results of this study did not show active response behaviors. Attention must be paid to the externalizing of responsibility among the psychological causes of this lack of action. This study found distrust of and uncertainty about PM information to be largely unrecognized as causes of apathy; on the contrary, understanding PM saturation to be externally produced was the main reason for inaction. If external factors are emphasized as the cause of the problem, it is possible that individuals perceive the risk as uncontrollable, which may nullify their will to respond. The results of this study also suggest that residents' psychological barriers have a negative effect on their response behaviors for the reduction of PM.

Fourth, the results of the correlation analysis show that the most influential factor in PM-related behavior was psychological attributes but that physical symptoms also affect response behaviors to PM. This result is meaningful in that it indicates desirable directions for PM policy. Based on the analysis of psychological factors, it was found that, because PM is also produced by individuals, active prevention is likely to occur if the importance of personal responses is conveyed. While knowledge of PM was not found to significantly affect response behaviors, residents in Seoul showed an average score of 6.56 out of 10 for PM knowledge. Lack of knowledge and understanding of the PM problem hinders appropriate response behavior. It is necessary to provide accurate and correct PM education and continuously available reliable information to enhance the knowledge of PM. In addition,

There was significant variation in residents' perceptions, response behaviors and psychological attributes concerning PM based on gender and age. Conversely, the time during which residents undertook external activities exposed to PM did not significantly affect PM-related symptoms or behavior. In contrast to college students in their twenties and male workers in their thirties and forties who remained outdoors for 5 h on average, women in their thirties and forties who stayed outdoors for only 3 h on average responded more sensitively to PM-related symptoms and their commitment to PM reduction and mitigating actions was much greater. Similarly, the results of the psychological factor analysis for PM showed that the group of women in their thirties and forties were more aware of the risks than the other groups and felt those risks to be closer psychologically. However, it is the younger generations who will become the subjects of society in future and will need to face these problems and solve them. There is an urgent need for plans to increase awareness of the risks and the will of the public to act.

Government efforts to reduce PM have focused primarily on expensive communication campaigns. However, the results of this study suggest that the mere encouragement of attitudinal change is not effective; public engagement in terms of accurate PM knowledge, health education and the study of various personal characteristics and psychological causes is required.

This study identified residents' thoughts, response behaviors and psychological factors relating to PM and showed that personal characteristics, cognition and emotions affect behavioral intentions. Unlike previous studies, the significance of this research is that it seeks to understand residents' behaviors in a multidimensional way. Psychological factors are expected to be used effectively to motivate individuals to protect themselves from the risks of PM exposure. The residents of Seoul involved in this study were well aware of the risks of PM but felt that its generation and any possible solutions were external and therefore beyond their control. This situation is likely to prevent public action to protect individuals and the society. It is recommended that we shift our focus from risk and attention to participation and action. It is expected that a proper direction for efficient PM reduction can be established using these results.

Author Contributions: Conceptualization, M.E.C. and M.J.K.; methodology, M.E.C. and M.J.K.; formal analysis, M.E.C.; investigation, M.E.C.; data curation, M.J.K.; writing—original draft preparation, M.E.C.; writing—review and editing, M.J.K.; supervision, M.J.K.; funding acquisition, M.J.K.

References

1.   OECD. OECD Statistics. 2019. Available online: https://stats.oecd.org/index.aspx?queryid=72722 (accessed on 24 March 2019).

2.   Ministry of Environment. Particulate Matter Status. 2019. Available online: http://www.me.go.kr/cleanair/sub02.do#2 (accessed on 12 March 2019).

3.   Seoul Metropolitan Government. Seoul Atmospheric Environment Information. 2019. Available online: http://cleanair.seoul.go.kr/air_pollution.htm?method=average (accessed on 29 April 2019).

4.   Sacks, J.D.; Stanek, L.W.; Luben, T.J.; Johns, D.O.; Buckley, B.J.; Brown, J.S.; Ross, M. Particulate matter—Induced health effects: Who is susceptible? *Environ. Health Perspect.* **2011**, *119*, 446–454. [CrossRef] [PubMed]

5.   Pope, C.A., III; Burnett, R.T.; Thun, M.J.; Calle, E.E.; Krewski, D.; Ito, K.; Thurston, G.D. Lung cancer, cardiopulmonary mortality, and long-term exposure to fine particulate air pollution. *JAMA* **2002**, *287*, 1132–1141. [CrossRef] [PubMed]

6.   Choe, J.I.; Lee, Y.S. A study on the impact of PM2.5 emissions on respiratory diseases. *Korea Environ. Policy Adm. Soc.* **2015**, *23*, 155–172. [CrossRef]

7.   Lee, H.J.; Jeong, Y.M.; Kim, S.T.; Lee, W.S. Atmospheric circulation patterns associated with particulate matter over south Korea and their future projection. *J. Clim. Chang. Res.* **2018**, *9*, 423–433. [CrossRef]

8.   Korean Statistical Information Service. Recognition of Environmental Problems (PM Inflow). 2019. Available online: http://kosis.kr/statisticsList/statisticsListIndex.do? (accessed on 1 April 2019).

9.   Kim, Y.; Lee, H.; Kim, H.; Moon, H. Exploring message strategies for encouraging coping behaviors against particulate matter. *Korea J. Commun. Inf.* **2018**, *92*, 7–44.

10.   Hwang, I.C. Particulate matter management policy of Seoul: Achievements and limitations. *Korea Assoc. Policy Stud.* **2018**, *27*, 27–51.

11.   Han, H.; Jung, C.H.; Kim, H.S.; Kim, Y.P. The revisit on the PM10 reduction policy in Korea. *J. Korea Environ. Policy Adm.* **2017**, *25*, 49–79.

12.   Yang, W.H. Changes in air pollutant concentrations due to climate change and the health effect of exposure to particulate matter. *Health Welf. Policy Forum* **2019**, *269*, 20–31.

13.   Park, S.; Shin, H. Analysis of the factors influencing PM2.5 in Korea. *Korea Environ. Policy Adm. Soc.* **2017**, *25*, 227–248.

14.   Kim, Y.P. Air pollution in Seoul caused by aerosols. *Korea Soc. Atmos. Environ.* **2006**, *22*, 535–553.

15.   Swart, R.O.B.; Raes, F. Making integration of adaptation and mitigation work: Mainstreaming into sustainable development policies? *Clim. Policy* **2007**, *7*, 288–303. [CrossRef]

16.   Klein, R.J.T.; Schipper, E.L.F.; Dessai, S. Integrating mitigation and adaptation into climate and development policy: Three research questions. *Environ. Sci. Policy* **2005**, *8*, 579–588. [CrossRef]

17.   Brouwer, R.; Schaafsma, M. Modelling risk adaptation and mitigation behaviour under different climate change scenarios. *Clim. Chang.* **2013**, *117*, 11–29. [CrossRef]

18. Wilbanks, T.J.; Sathaye, J. Integrating mitigation and adaptation as responses to climate change: A synthesis. *Mitig. Adapt. Strateg. Glob. Chang.* **2007**, *12*, 957–962. [CrossRef]
19. Gifford, R. The dragons of inaction: Psychological barriers that limit climate change mitigation and adaptation. *Am. Psychol.* **2011**, *66*, 290–302. [CrossRef] [PubMed]
20. Finell, E.; Tolvanen, A.; Pekkanen, J.; Minkkinen, J.; Ståhl, T.; Rimpelä, A. Psychosocial problems, indoor air-related symptoms, and perceived indoor air quality among students in schools without indoor air problems: A longitudinal study. *Int. J. Environ. Res. Public Health* **2018**, *15*, 1497. [CrossRef] [PubMed]
21. Ferrer, R.; Klein, W.M. Risk perceptions and health behavior. *Curr. Opin. Psychol.* **2015**, *5*, 85–89. [CrossRef] [PubMed]
22. Brewer, N.T.; Chapman, G.B.; Gibbons, F.X.; Gerrard, M.; McCaul, K.D.; Weinstein, N.D. Meta-analysis of the relationship between risk perception and health behavior: The example of vaccination. *Health Psychol.* **2007**, *26*, 136–145. [CrossRef] [PubMed]
23. Menon, G.; Chandran, S. When a day means more than a year: Effects of temporal framing on judgments of health risk. *J. Consum. Res.* **2004**, *31*, 375–389.
24. Jones, C.; Hine, D.W.; Marks, A.D.G. The future is now: Reducing psychological distance to increase public engagement with climate change. *Risk Anal.* **2017**, *37*, 331–341. [CrossRef]

PERMISSIONS

LIST OF CONTRIBUTORS

Gonçalo Marques and Rui Pitarma
Unit for Inland Development, Polytechnic Institute of Guarda, Avenida Doutor Francisco Sá Carneiro Nº 50, 6300-559 Guarda, Portugal

Marlene Pacharra
MSH Medical School Hamburg, University of Applied Sciences and Medical University, Am Kaiserkai 1, D-20457 Hamburg, Germany
Leibniz Research Centre for Working Environment and Human Factors at TU Dortmund University, Ardeystr. 67, D-44139 Dortmund, Germany

Stefan Kleinbeck, Michael Schäper, Christine I. Hucke and Christoph van Thriel
Leibniz Research Centre for Working Environment and Human Factors at TU Dortmund University, Ardeystr. 67, D-44139 Dortmund, Germany

Mehmet Tastan and Hayrettin Gökozan
Department of Electronic, Turgutlu Vocational School, Manisa Celal Bayar University, 45400 Manisa, Turkey

Jolanda Palmisani, Alessia Di Gilio, Laura Palmieri and Gianluigi de Gennaro
Department of Biology, University of Bari Aldo Moro, via Orabona 4, 70125 Bari, Italy

Carmelo Abenavoli, Marco Famele and Rosa Draisci
National Institute of Health, National Centre for Chemicals, Cosmetic Products and Consumer Health Protection, Viale Regina Elena 299, 00161 Rome, Italy

Junjie Li , Shuo Tian and Yichun Jin
School of Architecture and Design, Beijing Jiaotong University, Beijing 100044, China

Shuai Lu
School of Architecture and Urban Planning, Shenzhen University, Shenzhen 518060, China

Qingguo Wang
China Design and Research Group, Beijing 100042, China

Modeste Kameni Nematchoua
Beneficiary of an AXA Research Fund Postdoctoral Grant, Research Leaders Fellowships, AXA SA 25 avenue Matignon, 75008 Paris, France
LEMA, UEE, ArGEnCo Department, University of Liège, 4000 Liège, Belgium

Department of Architectural Engineering, 104 Engineering Unit A, Pennsylvania State University, State College, PA 16802-1416, USA
The University of Sydney, Indoor Environmental Quality Lab, School of Architecture, Design and Planning, Sydney, NSW 2006, Australia

Matthieu Sevin and Sigrid Reiter
LEMA, UEE, ArGEnCo Department, University of Liège, 4000 Liège, Belgium

Yu-Chuan Yen, Chun-Yuh Yang and Yu-Ting Cheng
Department of Public Health, College of Health Science, Kaohsiung Medical University, Kaohsiung City 807, Taiwan

Kristina Dawn Mena
Epidemiology, Human Genetics, and Environmental Sciences, School of Public Health, University of Texas Health Science Center at Houston, Houston, TX 77046, USA

Pei-Shih Chen
Department of Public Health, College of Health Science, Kaohsiung Medical University, Kaohsiung City 807, Taiwan
Institute of Environmental Engineering, College of Engineering, National Sun Yat-Sen University, Kaohsiung City 807, Taiwan
Department of Medical Research, Kaohsiung Medical University Hospital, Kaohsiung City 807, Taiwan
Research Center for Environmental Medicine, Kaohsiung Medical University, Kaohsiung City 807, Taiwan

Tareq Hussein
Department of Physics, The University of Jordan, Amman 11942, Jordan
Institute for Atmospheric and Earth System Research (INAR), University of Helsinki, PL 64, FI-00014 UHEL, Helsinki, Finland

Ali Alamee, Omar Jaghbeir and Kolthoum Albeitshaweesh
Department of Physics, The University of Jordan, Amman 11942, Jordan

Mazen Malkawi
Regional Office for the Eastern Mediterranean (EMRO), Centre for Environmental Health Action (CEHA), World Health Organization (WHO), Amman 11181, Jordan

Brandon E. Boor
Lyles School of Civil Engineering, Purdue University, West Lafayette, IN 47907, USA
Ray W. Herrick Laboratories, Center for High Performance Buildings, Purdue University, West Lafayette, IN 47907, USA

Antti Joonas Koivisto
Institute for Atmospheric and Earth System Research (INAR), University of Helsinki, PL 64, FI-00014 UHEL, Helsinki, Finland

Jakob Löndahl
Department of Design Sciences, Lund University, SE-221 00 Lund, Sweden

Osama Alrifai
Validation and Calibration Department, Savypharma, Amman 11140, Jordan

Afnan Al-Hunaiti
Department of Chemistry, The University of Jordan, Amman 11942, Jordan

Thomas Maggos and Panagiotis Panagopoulos
Environmental Research Laboratory, NCSR "Demokritos", 15310 Ag. Paraskevi, Athens, Greece

Vassilios Binas and George Kiriakidis
Institute of Electronic Structure and Laser, Foundation for Research and Technology, 70013 Heraclion Crete, Greece

Vasileios Siaperas and Antypas Terzopoulos
691 Industrial Base Factory, 19011 Avlonas, Greece

Jongseong Gwak
Institute of Industrial Science, The University of Tokyo, Tokyo 153-8505, Japan

Motoki Shino and Minoru Kamata
Department of Human and Engineered Environment Studies, Graduate School of Frontier Sciences, The University of Tokyo, Chiba 277-8563, Japan

Kazutaka Ueda
Department of Mechanical Engineering, The University of Tokyo, Tokyo 113-8656, Japan

Gaetano Settimo
Italian Institute of Health, viale Regina Elena 299, I-00185 Rome, Italy

Maurizio Manigrasso
Department of Technological Innovations, INAIL, via Roberto Ferruzzi 38, I-00143 Rome, Italy

Pasquale Avino
Department of Agriculture, Environmental and Food Sciences, University of Molise, via F. De Sanctis, I-86100 Campobasso, Italy

Xiaogang Cheng
College of Telecommunications and Information Engineering, Nanjing University of Posts and Telecommunications, Nanjing 210003, China
Computer Vision Laboratory (CVL), ETH Zürich, 8092 Zürich, Switzerland

Bin Yang
School of Building Services Science and Engineering, Xi'an University of Architecture and Technology, Xi'an 710055, China
Department of Applied Physics and Electronics, Umeå University, 90187 Umeå, Sweden

Kaige Ta, Erik Isaksson and Anders Hedman
KTH Royal Institute of Technology, 10044 Stockholm, Sweden

Liren Li
School of computer science and technology, Nanjing Tech University, Nanjing 211816, China

Thomas Olofsson
Department of Applied Physics and Electronics, Umeå University, 90187 Umeå, Sweden

Haibo Li
College of Telecommunications and Information Engineering, Nanjing University of Posts and Telecommunications, Nanjing 210003, China
KTH Royal Institute of Technology, 10044 Stockholm, Sweden

Myung Eun Cho and Mi Jeong Kim
School of Architecture, Hanyang University, 222 Wangsimni-ro, Seongdong-gu, Seoul 04763, Korea

Index